WILDLIFE OF PENNSYLVANIA

AND THE NORTHEAST

Charles Fergus

illustrations by Amelia Hansen

**STACKPOLE
BOOKS**

Published by
STACKPOLE BOOKS
5067 Ritter Road
Mechanicsburg, PA 17055
www.stackpolebooks.com

First edition

10 9 8 7 6 5 4 3 2

Printed in the United States of America

Cover photograph of red fox by McDonald Wildlife Photography, golden-crowned kinglet by Tom Vezo, and wood frog by Larry West.

Cover design by Wendy Reynolds

Some of the material in this book is based on Wildlife Notes published by the Pennsylvania Game Commission, which have been expanded and updated by the author for publication here.

Library of Congress Cataloging-in-Publication Data

Fergus, Charles.
    Wildlife of Pennsylvania / Charles Fergus ; illustrations by Amelia Hansen.— 1st ed.
        p.   cm.
    Includes bibliographical references (p.).
    ISBN 0-8117-2899-4 (pb)
    1. Vertebrates—Pennsylvania.   I. Title.

QL606.52.U6 F47   2000
596'.09748—dc21   99-052440

CONTENTS

BIRDS

AMPHIBIANS AND REPTILES

AMPHIBIANS

REPTILES

ACKNOWLEDGMENTS

I would like to thank the specialists who reviewed the manuscript of this book, pointed out inaccuracies, and made many helpful suggestions. Dr. Carolyn Mahan of Penn State University's Altoona Campus reviewed the section on mammals. Daniel Brauning of the Pennsylvania Game Commission checked over the entries on birds. Andrew Shiels, Nongame and Endangered Species Unit Leader for the Pennsylvania Fish and Boat Commission, and Chris Urban, herpetologist, reviewed the amphibians and reptiles. Dr. Howard Reinert of Trenton College of New Jersey read the chapters on the snakes. Any errors that remain in the book are mine.

Thanks are due to my wife, Nancy Marie Brown, who suggested that I write *Wildlife of Pennsylvania*. I would also like to thank Scott Weidensaul, Dr. Robert Brooks, Jerry Hassinger, Bob Bell, Carl Graybill, Arnie Hayden, Dr. Tom Serfass, Dr. Gary Alt, Clark Shiffer, Dr. Margaret Brittingham, Bob Boyd, Nick Bolgiano, and the librarians in Penn State's Biological Sciences library.

I have not personally discovered any of the facts related in this book concerning the life histories and habitats of our wildlife. I have relied on accounts and studies published in many journals and books. In partial repayment for borrowing this knowledge, I would like to dedicate *Wildlife of Pennsylvania* to the biologists, zoologists, mammalogists, ornithologists, and herpetologists who are out there in nature monitoring the animals, and who, with their passion, patience, powers of observation and deduction, and admirable commitment, represent the best hope for our environment and the many wonderful wild creatures that enliven it.

INTRODUCTION

One of my earliest memories centers on wildlife. I could not have been much older than four or five; I was walking with my father on a trail through the woods. We were almost certainly at Alan Seeger Natural Area in the Rothrock State Forest in Huntingdon County, Pennsylvania. My father was a professor of plant pathology at Penn State, and he often collected fungi at Alan Seeger for the university's mycological herbarium. Many times he took me with him.

I'm poking along, eyes on the ground, searching for mushrooms: red-capped, like the ones that elves sit on in fairy-tale books; black tongues jutting up rudely from the earth; coral, colored ivory, pink, or yellow; or the pure-white *Amanita* that looks as fragile as cake frosting but is sufficiently poisonous that a piece the size of my thumb (so my dad has warned) can kill me . . . make me dead, whatever that means.

Suddenly the quiet forest erupts. Black creatures flog into the air all around us, huge dark shapes rising, and my heart hammers wildly and every nerve in my body shrills as I clutch for Dad's pantleg—

They were the first wild turkeys I had ever seen. There must have been a dozen, probably a hen and her nearly mature brood. Today, almost forty-five years later, I remember the way those birds scared me, and how relieved I felt when they flew away and the forest simplified itself back into rocks, mushrooms, leaves, trees. Today I still feel a thrill when, on a hike, I spot a flock of turkeys legging it off through the woods or soaring down the mountain in heavy, ungainly flight, twitching their heads to and fro while looking for a place to land.

Over the years, I've been fortunate to have spent many days out-of-doors. Beyond his fungi, my father was only mildly interested in things natural, and my mother was thoroughly rooted in town. I wonder what they thought of a son who ventured out into the countryside with binoculars slung around his neck, field guides and a canteen in a small rucksack. I wasn't a systematic, studiously attentive naturalist (I'm still not); I simply liked watching wildlife. It thrilled me to see animals in their natural environment, to notice the way they moved, what they ate, the signs they left, and how they reacted to others of their own kind and to creatures of different species. I learned to creep along quietly. Stop and listen. Stand or sit still for long spells as I connected complex, lilting calls to shy thrushes in the underbrush, spied on pileated woodpeckers hammering apart rotten trees, exchanged stares with woodchucks peering from their dens, watched box turtles trundling through the leaf litter. Few of my sightings were as dramatic as those turkeys—although the rattlesnake I nearly stepped on came close, as did the bear that stood manlike to scratch its back on a tree, and the coyote that ghosted up out of its day bed and disappeared into the mountain laurel not ten paces in front of me.

From 1973 to 1976, I worked for the Pennsylvania Game Commission. During that period I started writing a series of Wildlife Notes, informational leaflets describing the birds and mammals in the state. I was fascinated by what I learned. I continued to work on the series into the mid-1980s, by which time I had left the Game Commission and taken a job in my native Centre County, back among streams and marshes and mountains and valleys. Today the Game Commission continues to hand out Wildlife Notes to people with questions about animals—questions arising from school projects, or out of a lifetime interest in wildlife, or perhaps because of a chance encounter like the one my father and I had with the turkeys.

This book, *Wildlife of Pennsylvania,* expands and updates the material originally presented in the Wildlife Notes. It covers mammals and birds and also branches out to include the reptiles and amphibians. The descriptions are intended for a general audience. I hope the text answers most of the common questions about Pennsylvania's animals—and I hope it inspires people to get out into nature to see the creatures the book details. *Wildlife of Pennsylvania* is written to be a companion volume to the many field guides that help an observer identify various species of wildlife. For amateur naturalists who want to delve more deeply into the lives and habits of different species, I have cited key references.

Pennsylvania is a rich and varied state whose diversity is reflected in the relatively large number of creatures found here. We have in residence predominantly northern species as well as southern ones, whose ranges overlap in the commonwealth. We have wildlife of the lowlands and uplands, animals that inhabit farming areas and grasslands, and others that prosper in boggy places, brushy lands, or deep woods. The foresight of conservationists has preserved for us a large network of public lands—over four million acres in state game lands, forest lands, parks, and natural areas, and one national forest. In addition, there are many thousands of acres of private land where people can observe wildlife. The full title of this book is *Wildlife of Pennsylvania and the Northeast,* and most of the species I describe also inhabit nearby states such as New York, New Jersey, Delaware, Maryland, Virginia, West Virginia, and Ohio, and much of New England.

As the years have passed, I've found myself preoccupied with other concerns—building a house, beginning a family, earning a living—so that at times I have drifted away from the passionate observing of wildlife. And yet the animals are still there, to thrill me, reassure me, even heal me. I love to walk in the woods, canoe on streams and lakes, go camping, and simply steal a few quiet moments on our forested acres. I have a son of my own— William, age twelve—and if, in this age of computers and instant news, of human overpopulation and often calamitous environmental change, I can plant in him and in others the notion that nature and wildlife are intriguing and worth preserving, I will be pleased.

Charles Fergus
Port Matilda, Pennsylvania

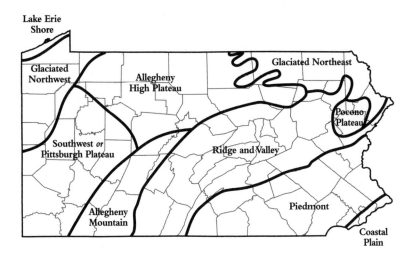

Pennsylvania can be considered to have ten physiographic provinces based on geology and landform.

The **Lake Erie Shore** includes Presque Isle, a sandy spit jutting out into Lake Erie and a key stopover point for migrating birds.

The **Glaciated Northwest** has many natural wetlands in depressions left by glaciers that retreated from the area thousands of years ago.

The **Southwest** or **Pittsburgh Plateau** is lower in elevation than the mountains to the north and east; it includes forests, farms, and urbanized areas.

The **Allegheny Mountain** province is an undulating plateau where surface mining for coal has greatly changed the landscape and habitats for wildlife.

Extensive forests cloak the mountainous **Allegheny High Plateau.** Many species of woodland wildlife thrive in this area, which has a large amount of public land.

In the **Ridge and Valley** province, rocky, wooded ridges parallel low, fertile farming valleys, providing a wide range of habitats and many edge zones.

The **Glaciated Northeast** includes many lakes and natural wetlands.

The high **Pocono Plateau** has oak forests, northern hardwoods, and lakes and swamps.

The **Piedmont** takes in much of southeastern Pennsylvania; it has few forests and many intensively farmed areas, cities, and suburbs.

A narrow strip of **Coastal Plain** coincides with the Delaware River floodplain; much of the area has been urbanized, but important pockets of freshwater tidal marsh remain.

MAMMALS

The vertebrates include fishes, amphibians, reptiles, birds, and mammals, and of these, the mammals are the most highly evolved. The name *mammal* refers to the mammary glands, or breasts, which females in this group use to feed milk to their young. Mammals are "warm-blooded," able to maintain a near-constant body temperature, which allows them to stay active despite changes in the temperature of their surroundings. Mammals have hair; in many species, the hair serves as insulation and provides protective coloration. The different species possess teeth of various shapes to efficiently perform tasks such as shearing, tearing, gnawing, and grinding. Mammals have relatively large skulls to house their well-developed brains.

Great variation exists in mammals' size and modes of living. Tiny shrews eat even tinier insects. Huge bears feed on prey ranging from mice to deer and supplement their diets with nuts and berries. Mammals have adapted to life on land, in trees, and in the water. Some species have evolved the ability to fly (the bats) and to glide (the flying squirrels).

During the Pleistocene Epoch, which began around two million years ago and lasted until about ten thousand years before the present, a staggering variety of mammals occupied North America: all of the species that exist today, plus hundreds of others, including lions, tigers, sloths, wild horses, asses, llamas, tapirs, and camels. A wave of extinctions—perhaps caused by climate change, by a catastrophe such as a meteor impact, or by the rise of prehistoric human hunters—cut down on the variety and number of species, particularly of the large mammals.

Today, about 350 mammal species live in North America north of Mexico. More than 60 species are found in Pennsylvania. Most mammals are sedentary, spending their lives in a limited home range; others, such as certain bats, breed in our state and winter many hundreds of miles away.

In historic times, largely through trapping, hunting, and logging, humans have extinguished the following species in Pennsylvania: the marten *(Martes americana)*; fisher *(Martes pennanti)*, now being restored to the state; mountain lion *(Felis concolor)*; Canada lynx *(Lynx canadensis)*; gray wolf *(Canis lupus)*; wolverine *(Gulo gulo)*; moose *(Alces alces)*; bison *(Bison bison)*; and elk *(Cervus elaphus)*, a small herd of which has been reestablished in Pennsylvania.

Doutt, J. K., C. A. Heppenstall, and J. E. Guilday. *Mammals of Pennsylvania.* 4th ed. Harrisburg, PA: Pennsylvania Game Commission, 1997.
Gardner, A. L. *Wild Mammals of North America: Biology, Management, and Economics.* Baltimore: Johns Hopkins University Press, 1982.
Godin, A. J. *Wild Mammals of New England.* Baltimore: Johns Hopkins University Press, 1977.
Merritt, J. F. *Guide to Mammals of Pennsylvania.* Pittsburgh: University of Pittsburgh Press, 1987.
Whitaker, J. O., and W. J. Hamilton. *Mammals of Eastern United States.* Ithaca, NY: Cornell University Press, 1998.

OPOSSUM

The Virginia opossum *(Didelphis virginiana)* is one of the world's most primitive living mammals and the only marsupial inhabiting America north of Mexico. Marsupials give birth to tiny, almost embryonic young, which their mother carries in a pouch on her abdomen. The fossil record suggests that marsupials may have originated in the New World. Today, many marsupials are found in South America and Australia. Physically, marsupials have changed little in millions of years. The opossum's relatives date back ninety million years to the Cretaceous Period; however, the opossum did not appear in North America until the Pleistocene, less than a million years ago, when it moved north out of the tropics.

Opossum is an Algonquian Indian word meaning "white animal." A creature without a specialized body structure or limiting food requirements, the opossum thrives in many settings. It is found throughout Pennsylvania.

Biology. Adults are 24 to 32 inches long, including a 9- to 13-inch tail, and weigh 4 to 12 pounds. Males are larger and heavier than females. An opossum has a cone-shaped head ending in a long, pointed snout with fifty teeth (more than any other North American mammal). The eyes are small and dark, the ears rounded, leathery, and unfurred. The tapering tail is naked and scaly like a rat's. The feet have five toes; the great toe on each hind foot has no claw and is long and opposable.

The coarse body fur is pale gray, with the outer hairs tipped yellow-brown. An opossum's legs and feet are black or dark brown. Males, females, and immatures are colored alike. You would not call an opossum a particularly handsome animal, nor a physically imposing one. Opossums walk with an ungainly shuffle, averaging 0.7 mile per hour; their running speed is 4 miles per hour. Capable climbers, they ascend hand over hand, using their prehensile tails and the opposable toes on their hind feet for gripping and balancing. An opossum can hang on to a branch using its tail alone.

Although an opossum's brain is small and primitive in structure, *Didelphis virginiana* is not considered to be stupid: studies have put its intelligence on a par with that of the pig and dog. An opossum's senses of smell and touch are well developed, but its hearing is not especially keen, and its eyesight is weak. When moving about, an opossum sniffs the air and periodically rises on its hind legs to look around. Normally silent, it will growl, hiss, screech, or click its teeth when afraid, excited, or annoyed.

If, when threatened, an opossum cannot climb a tree or hide, it may crouch and defend itself—or, more likely, feign death. When feigning death (also called "playing 'possum"), an opossum collapses into a limp, motion-less state. Its eyes and mouth hang open, its tongue lolls out, and its breath-ing becomes shallow; it drools and may defecate or exude a musky, repellent fluid from its anal glands. This deathlike trance may last from a few minutes to more than an hour and may help an opossum survive an attack, since some predators reject dead prey. Feigning is probably involuntary, caused by a nervous paralysis brought on by shock.

Opossums are omnivorous feeders and scavengers. They eat many ani-mals, including earthworms, insects, fish, salamanders, snakes, toads, shrews, mice, voles, and young birds and mammals, as well as birds' eggs. They con-sume apples, berries (grapes, pokeberries, blackberries), fruits, mushrooms, acorns, grasses, grains, corn, and other cultivated plants. They rely more on animal than on plant food, and they eat garbage and carrion, including crea-tures killed on highways. Opossums seem to be immune to pitviper venom and can safely prey on rattlesnakes, copperheads, and water moccasins.

Opossums are shy and secretive. Sometimes they forage by day, but they are much more active at night. They shelter beneath porches and building foundations and in old woodchuck and skunk burrows, hollow logs and trees, rock crevices, and the abandoned leaf nests of squirrels. Opossums do not dig burrows, although they line existing cavities with leaves, which are carried grasped in the coils of the tail. An opossum generally uses several dens in its home range.

For most of the year, opossums are solitary. The sexes come in contact during the breeding season, late February and March in Pennsylvania. Males fight and hiss at one another, and they make clicking sounds when pursu-ing females. After mating, the male plays no role in raising the young. Ges-tation is extremely short: twelve or thirteen days. The newborn young are hairless, pink-skinned, blind, about 0.5 inch long (smaller than a honeybee), and weigh just 0.005 ounce. It has been reported that an entire litter of baby opossums can fit into a teaspoon. A newborn's hind limbs are rudi-mentary, but its front limbs and feet are well developed and equipped with claws. The baby opossum crawls upward, with overhand strokes as if swim-ming, through its mother's fur to a pouch in the skin on her belly.

Litters vary from five to thirteen young, with an average of six to nine, although as many as twenty-five have been reported. The mother's pouch is lined with fur, and inside it are thirteen mammaries. When a baby opos-sum begins to nurse, the nipple expands, forming a bulb on the end that swells inside the baby's mouth and helps it stay attached. Should a newborn fail to find a teat, it will die.

The Virginia opossum uses a prehensile tail and opposable toes on its hind feet to brace itself when climbing.

The young increase their weight tenfold and double their length in seven to ten days. After a month, they are the size of house mice. At eight to nine weeks, their eyes open and they let go of the mammaries for the first time. They begin leaving the pouch for short periods, riding on their mother's back. When three to four months old, they start looking for food on their own. The young stop nursing but may remain with their mother for several weeks longer. Females generally breed again, from May to mid-June; at least two weeks pass between the weaning of the first litter and the birth of the second. Females first breed when a year old.

In fall and winter, opossums devote almost twice as much time to feeding and improving their nests as they do during the rest of the year. Opossums do not hibernate, but they may den up through cold or snowy periods. An opossum adds a thick layer of body fat, but it does not grow a winter coat, and its thin fur provides little insulation. Pennsylvania is near the species' northern limit, and many opossums here lose the tips of their ears and tails to frostbite.

Foxes, coyotes, bobcats, dogs, hawks, owls, and snakes prey on opossums. I heard a shriek outside my house one morning at dawn; on the lane leading up to the road, I found the foreleg of an opossum and a small pile of entrails. I'm certain a great horned owl had made a kill there a moment earlier. Many opossums are hit by cars, often when feeding on other road-killed animals. An opossum's life expectancy in the wild is less than two years, with a few individuals reaching age four or five.

Habitat. Opossums live in farmland, brushy areas, and open woods, in dry and wet terrain, and at varying elevations. They invade suburbs and even towns and cities where food and cover are available. An ideal habitat is rich bottomland woods veined with streams.

Where food is plentiful, an opossum may range no farther than a few hundred yards; in intensely cultivated areas, individuals must travel much farther—up to 2 miles. Opossums do not defend individual territories. In Maryland, biologists found that opossums had elongated rather than circular ranges (circular being the pattern of most other land mammals), following the edges of rivers and streams. In a sample of twenty-five individuals, the average home range was 0.6 mile in length.

Population. The opossum is one of Pennsylvania's most abundant mammals, even though it apparently didn't live here until after about 1500. *Didelphis virginiana* has been expanding to the north and west during the past century, and today the species ranges from southern Ontario and New England south to Florida, from Minnesota south through Nebraska and Texas to Middle America, and along the West Coast.

Hartman, C. G. *Possums.* Austin, TX: University of Texas Press, 1962.

SHREWS AND MOLES

Shrews and moles belong to order Insectivora, a diverse group considered to be the most primitive of the true placental mammals. As their name suggests, insectivores feed mainly on insects. Shrews hunt for insects aboveground and in tunnels in the leaf mold and the debris at the earth's surface. Moles live underground, from a few inches to 2 feet down, although they sometimes venture out of their tunnels. Eight species of shrews and three species of moles inhabit Pennsylvania.

Shrews. Shrews have long, pointed noses, beady eyes, and narrow skulls. Their fur is soft and velvety, and their small ears are concealed by fur. Shrews differ from mice (order Rodentia) in the following ways: A shrew has five toes on each foot. (A mouse has four toes on its front feet.) A shrew's teeth are sharp and pointed and are stained a reddish or chestnut color. (A mouse's incisors are chisel-shaped and lack the reddish staining.) And a shrew's eyes are beadier and its nose more pointed than those of a mouse. Most Pennsylvania shrews look fairly similar, and it can take an expert to tell some of the species apart.

Active year-round, shrews have terrific metabolic rates and must eat almost constantly. When a shrew is in motion, its heart may beat 1,200 times per minute, and it may take 750 breaths. Shrews are most active at dawn and dusk and during the night, but they also move about by day. Several times while sitting in the woods in broad daylight, I've watched shrews hustle past in search of prey—once a shrew startled me by racing across my boot. Shrews are quick and aggressive and attack animals larger than themselves. Their eyesight is poor, mainly limited to detecting light, but their senses of smell, hearing, and touch are keen. One species, the northern short-tailed shrew, has poisonous saliva, a rare example of toxicity in mammals; delivered with a bite, the toxin slows down prey and kills small animals outright.

Shrews are short-lived, dying from floods, starvation, rapid temperature changes, accidents, fights with other shrews, and even the shock of a sudden fright. Some predators catch and kill them, perhaps mistaking them for mice, but because shrews give off an objectionable musky odor, the predators do not always eat them. In fact, that's the way people usually find shrews: lying dead, discarded by a predator that has long since gone on its way.

Shrews raise several litters each year. Gestation lasts for around three weeks. The young are born helpless and unfurred, but they grow rapidly and reach adult size in four to six weeks. Few shrews live longer than a year, and most probably die much sooner.

Masked Shrew *(Sorex cinereus)*. The masked shrew is named for a scarcely distinguishable band of dark fur across its eyes. The most widely distributed shrew in North America, *Sorex cinereus* ranges over almost all of the continent's northern half. It is found throughout Pennsylvania and south in the Appalachian Mountains to Georgia. Overall length is 3 to 4.3 inches; tail, 1.4 to 1.8 inches; weight, 0.12 to 0.2 ounce. The summer pelt is grayish brown above, paler below.

Masked shrews live in wooded areas, under rocks, logs, and in the leaf litter, often in swamps and along stream banks and spring runs. They also inhabit grassy fields, hedgerows, and stone walls, in places where the ground is moist. Masked shrews spend most of their lives in underground runways and in the tunnels of mice and other small mammals. They utter high chipping notes, probably a form of echolocation used to find their way in darkness. Their prey includes insects, worms, centipedes, slugs, snails, and spiders; they eat vegetable matter, such as moss and seeds, and carrion. One observer reported that a captive masked shrew ate more than three times its body weight daily.

Masked shrews breed from March to September. Females give birth in fist-size nests of leaves, grass, and fine rootlets under logs, stumps, and rocks. Masked shrews usually have three litters each year, with four to ten young per litter. Owls, hawks, herons, shrikes, weasels, foxes, cats, and the larger shrews kill masked shrews, few of which reach the maximum life span of eighteen months.

Long-Tailed Shrew *(Sorex dispar)*. The long-tailed shrew inhabits the Appalachians from Maine to North Carolina and Tennessee. It occurs in a diagonal band across upland Pennsylvania from the northeast to the south-central, living in cool, damp forests of deciduous or mixed tree species and along mountain streams. *Sorex dispar* is also called the rock shrew because it forages in rock slides, threading its long, slender form through crevices and natural tunnels among the boulders.

Long-tailed shrews are dark gray with paler underparts in summer and an overall slate gray in winter. Total length, 3.9 to 5.3 inches; tail, 2 to 2.3 inches; weight, 0.14 to 0.2 ounce. Long-tailed shrews eat flies, beetles, cave crickets, centipedes, and spiders. We know little about the reproduction and life history of this shy, relatively uncommon species, but they may be similar to those of the masked shrew and smoky shrew.

Maryland Shrew *(Sorex fontinalis)*. In appearance and habits, the Maryland shrew is similar to the masked shrew, *Sorex cinereus*. Originally described as a separate species, it was then lumped in with *Sorex cinereus* until the 1970s, when biologists found the two types overlapping in southeastern Pennsylvania but apparently not breeding with each other. Taxonomists are still unsure whether *cinereus* and *fontinalis* should be considered distinct species or simply subspecies (races) of the same species. The Maryland shrew inhabits southeastern and south-central Pennsylvania and adjacent parts of Delaware, Maryland, and West Virginia. Biologists have found *Sorex fontinalis* in moist meadows, oak forests, clear-cut forests, hedgerows, lowlands,

and at midslope elevations and on the tops of ridges. Its life history probably parallels that of the masked shrew.

Smoky Shrew *(Sorex fumeus)*. The smoky shrew occurs throughout the
East from Nova Scotia to North Carolina. It inhabits most of Pennsylvania
but is scarce or absent along the state's southeastern border. Its coloration is
a uniform dull smoky brown, except for the bicolored tail (brown above
and yellowish below) and the pale-buff feet. Overall length is 3.7 to 5
inches; tail, 1.4 to 2 inches; and weight, 0.2 to 0.35 ounce, about one-third
the weight of a house mouse.

Smoky shrews prefer cool, damp woods, including shaded hemlock
ravines, northern hardwood forests, spruce and sphagnum bogs, and stream
borders with moss-covered boulders and logs. They build nests of shredded
leaves or grasses in rock crevices and under rotting logs and stumps. The
breeding season runs from March to September. Females bear two or, less
often, three litters per year of two to eight (usually five or six) young. From
five to fourteen smoky shrews may live on a typical wooded acre.

Active at all hours, smoky shrews burrow through the leaf mold and
travel in other animals' tunnels. Foraging, they utter a continual faint twit-
ter that is probably a form of echolocation. Smoky shrews eat earthworms,
insects, small salamanders, sowbugs, and spiders. They fall prey to short-
tailed shrews, weasels, foxes, bobcats, hawks, and owls. The maximum life
span is fourteen to seventeen months.

Pygmy Shrew *(Sorex hoyi)*. The pygmy shrew ranges across much of
northern North America. Although it has been captured in only three
Pennsylvania counties (Franklin, Centre, and Clearfield), it probably occurs
in most of the state except perhaps the eastern third. *Sorex hoyi* is the smallest mammal in the Keystone State, and it may be the tiniest land mammal
in the world (the bumblebee bat of Thailand is slightly smaller). Overall
length is 3.2 to 3.8 inches; tail, 1.1 to 1.3 inches; and weight, 0.08 to 0.14
ounce (half the weight of a large earthworm).

Pygmy shrews live in wet or closely mingled wet and dry habitats,
under old stumps and rotting logs, among the ground litter in areas grown
with sedges and ferns, in aspen clumps and hardwood forests, and in thick
stands of conifers near water. They eat slugs, earthworms, insects, spiders,
and carrion. Pygmy shrews can swim, and in winter they have been seen
running on the surface of the snow.

Water Shrew *(Sorex palustris)*. The water shrew ranges through northern North America; it is found in New England, across northern Pennsylvania, and in the southern Appalachian Mountains. *Sorex palustris* is our state's second largest shrew: overall length, 5.3 to 6.1 inches; tail, 2.4 to 3.5 inches; and weight, 0.35 to 0.6 ounce. The water shrew inhabits forested areas and is adapted for a semiaquatic life. The banks of cold, clear streams provide optimum habitat. Water shrews use small surface runways under bank overhangs, fallen logs, and brush piles. They live in bogs and around springs and sometimes take shelter in beaver lodges or muskrat houses during winter.

The water shrew uses its large hind feet, which are fringed with short, stiff hairs, to paddle about underwater. It can dive and stay submerged for fifteen seconds; water does not penetrate its dense pelt. A water shrew can run short distances across the surface of a quiet stream or pond, buoyed up by air globules like a water bug.

Water shrews eat insects and other small invertebrates, including stonefly, mayfly, caddisfly, and cranefly larvae; slugs, snails, and earthworms; small fish and fish eggs; and salamanders. They tend to be nocturnal but are also active at dusk, on cloudy days, and in the shade on sunny days. Water shrews are nabbed and eaten by weasels, minks, otters, hawks, owls, snakes, fish (including black bass, trout, and pickerel), and large frogs. *Sorex palustris* breeds from March to September, producing two or three litters of four to eight young each.

Northern Short-Tailed Shrew *(Blarina brevicauda)*. One of the commonest shrews and most abundant small mammals in its range, the short-tailed shrew inhabits eastern America from southern Canada to Georgia and west to Nebraska. In Pennsylvania, it occurs statewide. Short-tailed shrews are a dark slate color above and a paler gray below. *Blarina brevicauda* is the largest and most robust of Pennsylvania's shrews, with an overall length of 4.1 to 5.2 inches, a 0.7- to 1.2-inch tail, and a weight of 0.4 to 0.8 ounce.

The short-tailed shrew possesses poisonous saliva that gets into a prey animal's system through cuts caused by the shrew's sharp teeth. The toxin weakens or kills warm-blooded creatures such as mice, voles, songbirds, and other shrews. Short-tailed shrews also feed on earthworms (the single most important food item), slugs, snails, insects, salamanders, small snakes, mice, carrion, subterranean fungi, plant roots, nuts, fruits, and berries. They cache food, including comatose rodents and paralyzed snails, in their burrows to be eaten later. The short-tailed shrew emits ultrasonic calls that help it echolocate objects, including prey, in dark tunnels in the leaf litter.

The northern short-tailed shrew is an indefatigable hunter
of small rodents, insects, and other prey.

The species lives in various moist habitats, including woods, stream banks, tall grass, and brush. Short-tailed shrews frequent the top few inches of the soil and leaf litter, digging their own tunnels and using those of other animals; they burrow through the snow in winter. An individual's home range varies from 0.5 to 1 acre. *Blarina brevicauda* is well adapted to cold weather; in winter, it restricts its activity to lessen the amount of food it must find, and it draws on a special high-energy tissue called "brown fat" stored between its shoulder blades.

Short-tailed shrews weave dry plant matter and hairs into two types of nests: a resting nest and a larger mating chamber, both about a foot below the ground or under a log, stump, or rock. Short-tailed shrews breed from March to September, with three to nine young (usually four to six) born following twenty-one days' gestation. Three, and perhaps four, litters are raised each year. Biologists estimate population densities of one to fifty short-tailed shrews per acre. Individuals may live up to two years in the wild.

Least Shrew *(Cryptotis parva)*. The least shrew ranges from eastern North America and the Midwest south into Mexico and Central America. It is found throughout Pennsylvania. It is cinnamon to brown above and ashy gray below. The overall length is 2.7 to 3.5 inches, including a 0.5- to 0.8-inch tail; the least shrew weighs 0.1 to 0.2 ounce.

Unlike most other Pennsylvania shrews, the least shrew favors open, dry settings such as old pastures, meadows, and woods edges. Least shrews are scattered in local colonies throughout suitable habitats. More gregarious than other shrews, they gather in groups of a dozen or more in winter nests to conserve body heat.

Least shrews are active mainly at night. They move about in runways about 0.5 inch in diameter. They eat insects, earthworms, centipedes, millipedes, snails, frogs, and carrion; their habit of entering beehives to feed on larvae has earned them the name "bee shrew." Least shrews breed from March through November. They raise several litters per year, each of three to six young. Biologists believe that both parents care for the young, which are weaned at three weeks of age. Remains of least shrews sometimes show up in owl pellets. In Indiana, in the winter of 1949, a biologist found twenty-seven least shrews and one short-tailed shrew in the digestive system of a rough-legged hawk. Little is known about *Cryptotis parva* in Pennsylvania, where it is classified as an endangered species.

Moles. Moles are insectivores that live deeper in the soil than their close relatives, the shrews. Well suited to a burrowing life, moles have strong shoulders and front legs; large, spadelike forefeet; and stout, strong claws. Their hind limbs and pelvic bones are reduced in size, letting them turn around more easily in tunnels.

Moles do not have external ears. Their eyes are tiny, dim-seeing, and hidden away under fur. They use their needle-sharp teeth to eat insects and other invertebrates. Moles stay active year-round. Although gardeners and homeowners may resent their burrowing, moles perform valuable ecological functions by aerating the upper layers of the soil and keeping insect populations in check. The most obvious signs of moles are the ridges pushed up by their tunneling, and the molehills, mounds of dirt where the tunnels have broken through to the ground's surface.

Moles have only one litter each year, born in an underground chamber. Since they rarely venture aboveground, moles are not taken by predators as frequently as shrews are. Shrews, mice, and voles often use mole tunnels.

Hairy-Tailed Mole *(Parascalops breweri)*. The species inhabits New England and the Northeast; it occurs statewide in Pennsylvania except for the extreme southeast. *Parascalops breweri* also ranges south through West Virginia and Ohio to Kentucky, Tennessee, and North Carolina. It is distinguished from our other two moles by its tail, which is covered with fur rather than scaly and naked. The coat is soft and velvety, dark gray or black with slightly paler underparts. Hairy-tailed moles are 5.5 to 6 inches long, including a tail of 0.9 to 1.4 inches; weights vary from 1.4 to 2.2 ounces.

Hairy-tailed moles live in woods, meadows, and brushy lands, in loose, sandy, well-drained soil with ample vegetative ground cover. Their tunnels branch and squiggle across the ground. They eat earthworms, beetle larvae, other insects, snails, spiders, millipedes, centipedes, sowbugs, and roots. Individuals sometimes search for prey aboveground, particularly earthworms.

One to twelve hairy-tailed moles may live on a single acre. An individual's home range is thought to be around 0.2 acre. Hairy-tailed moles build subterranean nests for resting, breeding, and wintering. The breeding and wintering chambers lie from 10 to 20 inches underground and are insulated with grass and leaves. Females breed in March and, after a month of gestation, bear four or five young. Blind and naked at first, the babies grow rapidly; within a month they are weaned and able to fend for themselves.

Eastern Mole *(Scalopus aquaticus)*. The eastern mole ranges from southern New England, Michigan, Wisconsin, and Wyoming south to Florida and Texas. It is absent from the Appalachian Mountains. *Scalopus aquaticus* inhabits the eastern third of Pennsylvania. The eastern mole has a pointed, movable snout that is naked at the tip, and its eyes are covered with a thin membrane of skin. The tail—short, thick, rounded, and nearly naked—may serve as a tactile organ when the animal backs up underground. The fur is black to brownish black in winter, pale on the undersides, and generally paler in summer. An adult eastern mole is 5.6 to 8 inches long, including a 0.6- to 1.2-inch tail; it weighs 1.8 to 4.2 ounces.

The species' Latin name is misleading: *Scalopus aquaticus* is the least aquatic of our moles. It prefers well-drained sandy or light loamy soils where it can dig easily and find earthworms to eat. It lives in forests, fields, rich bottomlands, meadows, golf courses, and lawns. The eastern mole eats earthworms, insects, slugs, centipedes, ants, spiders, and vegetable foods, including corn, tomatoes, potatoes, wheat, and grass seeds. Because of its offensive musky odor, *Scalopus aquaticus* has few natural predators.

Eastern moles do not hibernate, and individuals are active both night and day. They dig two kinds of burrows: permanent ones down to 10 inches below the ground, for nesting; and shallow, temporary foraging tunnels that bulge the surface of the earth. When digging, an eastern mole moves its broad forefeet sideways, passing dirt back along its body to the hind feet, which kick it to the rear. It tunnels at the rate of 10 to 20 feet an hour and can proceed 100 feet per day.

Scalopus aquaticus breeds in March and April. After six weeks of gestation, the female bears two to five young. The young are on their own after a month and are sexually mature by the following spring. One study found that the home ranges of male eastern moles averaged 2.7 acres, and those of females, 0.7 acre. Longevity in the wild may be as long as six years.

Star-Nosed Mole *(Condylura cristata).* The star-nosed mole is named for a cluster of twenty-two pink, fleshy, tentaclelike feelers at the end of its snout. It ranges from southern Labrador and Ontario south to the Carolinas and inhabits Pennsylvania statewide. The coat is a deep blackish brown; immatures are paler and browner. Adults are 6.8 to 8.1 inches long, including a 2.7- to 3.3-inch tail. Weights range from 1.4 to 2.7 ounces.

The star-nosed mole prefers deep, mucky soils in wet meadows, bottomlands, marshes, and swamps. Its tunnels may open directly into the water, for the star-nosed mole is a capable swimmer. It propels itself through the water by alternately stroking with its broad forepaws and sculling with

The star-nosed mole uses its great clawed feet to tunnel through the soil. Twenty-two fleshy, tentaclelike feelers help it find prey, such as earthworms and aquatic invertebrates.

its long, hairless tail. In winter, people have seen star-nosed moles swimming beneath the ice. These moles pick up aquatic invertebrates—caddisfly, midge, and stonefly larvae—from the bottoms of streams and catch small fish and crustaceans. On land, star-nosed moles eat earthworms (often the single most important food item), slugs, and grubs. When a mole is foraging, the tentacles on its snout move and twitch constantly. Naturalists have long supposed that these appendages are touch sensors. But recent experiments indicate that star-nosed moles use their tentacles to sense the electrical fields produced by earthworms.

Condylura cristata is a fairly social animal, living in small colonies and sharing burrow systems. Males and females pair in the fall. In May or June, about forty-five days after mating, the female bears three to seven young in an underground nest lined with dead leaves and grass. At the age of three weeks, when they are two-thirds grown and well furred, the juveniles disperse, and by September they reach adult weight. Star-nosed moles are preyed on by great horned owls, barn owls, screech-owls, red-tailed hawks, skunks, foxes, weasels, and snakes, and, when swimming, by large fish. Individuals may live as long as three or four years.

Gorman, M. L., and R. D. Stone. *The Natural History of Moles.* Ithaca, NY: Cornell University Press, 1990.
Merritt, J. F., G. L. Kirkland, Jr., and R. K. Rose. *Advances in the Biology of Shrews.* Pittsburgh: Carnegie Museum of Natural History, Special Publication 18, 1994.

BATS

Bats are the only mammals that fly. Their wings are thin membranes of unfurred skin stretching from forelegs to hind legs and from hind legs to tail. The name of the taxonomic order to which they belong, Chiroptera, means "hand-winged": bats' long, slender finger bones act as wing struts, stretching the skin taut for flying and, when closed, folding the wings neatly against the body.

Ten species of bats may be seen in Pennsylvania. All of our bats belong to Vespertilionidae, also called the evening bats and mouse-eared bats, and the most common bat family in North America. Bats are insect predators and commonly take their prey on the wing. Often they feed over water, and some bats occasionally land and seize insects on the ground. Voracious eaters, bats may consume up to a quarter of their body weight at a single feeding.

Bats' eyes are small and of limited use, but their ears are large, well developed, and critical to their life habits. In flight, a bat utters a series of high-pitched squeaks (so high-pitched that they're almost always inaudible to humans) that echo off nearby objects—branches, fences, telephone wires, insects—and bounce back to the bat's ears. Split-second reflexes help the bat dodge obstructions and intercept prey. A bat uses its mouth to scoop small insects out of the air. It may disable a larger insect with a quick bite, cradle it in a basket formed by the wings and tail, and carry it to the ground or to a perch for eating. If an insect takes evasive action, the bat may flick out a wing and snare it. Bats have sharp teeth with which they chew their food into tiny, easily digested pieces.

Most bats mate in late summer or early fall, although some breed in winter. The male's sperm is stored in the female's reproductive system until spring, when fertilization takes place. The young are born in summer. Naked, blind, and helpless, they are nursed by their mothers, and by six weeks of age most are self-sufficient and nearly adult size. The reproductive potential of bats is low, with most species bearing a single young per year, although the larger species may have up to four.

None of Pennsylvania's bats are active during bright daylight; they make their feeding flights in late evening, at night, and in early morning. They spend the day roosting—singly, in pairs, in small groups, or in large colonies, depending on sex and species—hanging by their clawed hind feet, upside down in dark, secluded places such as caves, abandoned mines, rock crevices, and hollow trees. Some bats also congregate in vacant buildings, barns, church steeples, and attics; others hide among the leaves of trees. Occasionally I've found bats roosting in exposed places such as huckleberry shrubs and thistle patches.

In fall, winter, and early spring, insects aren't readily available to bats in the Northeast. Several species migrate south, but the majority remain in our area and go into hibernation, usually inside caves. Bats are true hibernators. Throughout the winter, they eat nothing, surviving by slowly burning the fat accumulated during summer and fall. The body temperature of a hibernating bat drops close to the air temperature; its respiration and heartbeat slow; and chemical changes occur in its blood. Bats can be roused fairly easily from hibernation and often are able to fly within a few minutes of wakening. Most bats favor caves for hibernation sites, or hibernacula, particularly those having the lowest stable temperature above freezing. During winter, bats will move about within a cave to find zones of optimum temperature and perhaps to exercise their flight muscles. Often bats of several species hibernate in the same cave.

People hold many misconceptions about bats: bats are prone to rabies; their droppings are a dangerous source of tuberculosis and other diseases; they attack people; they are dirty and lice-ridden. In fact, bats are no more apt to contract rabies than are other warm-blooded animals. (People should not, however, handle bats, especially those found on the ground or in the open during the day.) There is no evidence that bats or their droppings, called guano, transmit tuberculosis to man. A host of scientific studies have shown that healthy bats do not attack people, and even rabid bats are rarely aggressive. To be able to fly, bats must keep themselves clean; they host no more parasites than do other wild animals, and the parasites that afflict bats are specialized ones posing no problems to humans.

To make up for their low reproductive rates, bats are relatively long-lived. Some, banded and released, have been recaptured more than thirty years later, although the average life span is five to ten years. Because they feed in midair and are active at dusk and at night, bats are not often caught by predators. Owls and hawks take some bats, as do house cats, raccoons, and foxes. Rat snakes occasionally eat hibernating bats, and snakes and birds prey on unattended young. Other causes of mortality are cave flooding, accidents, and long periods of rain and high winds that prevent bats from feeding.

In general, bat populations are falling, and the greatest threat by far is from humans. In winter, hibernating bats may be aroused by people exploring caves; repeated disturbances force bats to squander precious calories needed for overwintering. Caves may be destroyed by mining, flooded by dams, or dynamited shut. Loss of feeding habitat also imperils bats, and widespread use of insecticides cuts down on the numbers of their prey.

Bats perform a valuable service by keeping insect pests in check, including leafhoppers, stinkbugs, cucumber beetles, and the adult moths of potato worms, cutworms, corn borers, and tomato hornworms. Several species of bats use humans' dwellings for their nesting colonies, and their noise and droppings can become a nuisance. To rid a building of bats, wait until winter when the creatures are hibernating in caves and close or screen off any cracks or openings that the bats are using to get inside. Replace the habitat that you have now made inaccessible (or attract bats to the area around your home) by putting up a bat house. The *Bat House Builder's Handbook* can be obtained, for a small donation, from Bat Conservation International, P. O.

Box 162603, Austin, TX 78716. A Pennsylvania Game Commission publication, *Woodworking for Wildlife,* also has plans for bat houses.

Little Brown Bat *(Myotis lucifugus).* Pennsylvania's most common bat, the little brown bat, is found statewide in forests, farms, and towns. Length, including the tail, is 3 to 3.7 inches; wingspread, 8.6 to 10.5 inches; and weight, 0.2 to 0.4 ounce, greatest just before hibernation. Females are slightly larger than males. Little brown bats are a rich bronze-brown, usually with a dark spot on the shoulders. The fur is dense, fine, and glossy, the wings black and bare.

The little brown bat leaves its roost at dusk and usually flies to a stream or pond, where it skims the surface for a drink before beginning to hunt. It flies at speeds of 13 to 22 miles per hour and eats a wide variety of insects, including nocturnal moths, bugs, beetles, flies, mosquitoes, and midges. Insects are usually caught with the wing or tail membrane and transferred to the mouth. Sometimes little brown bats feed by simply fluttering back and forth through swarms of small insects, rather than targeting individual prey. Bats examined within twenty minutes of taking flight frequently have insect-filled stomachs. The little brown bat makes several feeding flights each night.

In October and November, little brown bats leave their summer roosts and move to tunnels, mine shafts, and caves, where they hibernate, clinging to ceilings and clustered against one another. They return to the same hibernation sites each year, from which they emerge in April or May. In summer, the males are solitary, roosting in hollow trees, under loose bark, behind loose house siding and shingles, and in rock crevices. Females gather in nursery colonies of ten to ten thousand individuals in hollow trees, attics, barns, and other dark, hot, secluded places. Each female bears a single baby in June or early July, following fifty to sixty days of gestation. Biologists have found no evidence that mother bats ever carry their babies along on hunting trips. After four weeks, a young bat is fully grown and ready to leave the roost and fly nightly with the adults. Females mature sexually at about eight months; males mature in their second summer. Little brown bats have been documented to live for up to thirty-four years.

Northern Bat *(Myotis septentrionalis).* Formerly known as Keen's bat, this species is similar in size and color to the little brown bat but with a longer tail and narrower, longer ears. The northern bat ranges throughout Pennsylvania but is much less common than the little brown bat. Length is 3.1 to 3.5 inches; wingspread, 9 to 10.7 inches; and weight, 0.2 to 0.4 ounce.

We know little about the ecology and feeding behavior of *Myotis septentrionalis,* although they may be similar to those of the little brown bat. Northern bats roost singly or in small groups; in winter, they share caves with little brown bats, big brown bats, eastern pipistrelles, and Indiana bats. Females gather in nursery colonies in attics, barns, and tree cavities. Each female gives birth to a single baby in July.

Indiana Bat *(Myotis sodalis).* The Indiana bat strongly resembles the little brown bat but is colored a dull pinkish brown. Length is 2.9 to 3.7 inches; wingspread, 9.4 to 10.3 inches; and weight, 0.2 to 0.4 ounce. Longevity may be twenty years in the wild.

In summer, Indiana bats roost in hollow trees and beneath loose bark on standing dead trees; they seem not to roost in buildings. They forage in both upland and bottomland woods, taking moths, caddisflies, flies (including many mosquitoes), and beetles. The female bears a single young in late June or early July. In winter, Indiana bats hibernate in tight clusters of up to 250 individuals per square foot on the ceilings and walls of caves and mine shafts. The dense formations are particularly vulnerable to disturbance by humans: when a bat on the edge of the cluster is awakened, it moves about, starting a ripple of activity that spreads through the group. A winter of

Indiana bats hibernate clustered together on the ceilings
and walls of caves and mine shafts.

repeated disturbances can cause bats to burn vital fat stores, and they may run out of energy before spring.

The range of *Myotis sodalis* is centered on Missouri, Kentucky, Indiana, and Illinois. An estimated 95 percent of the total population winters in fifteen large caves in those states. Pennsylvania is in the eastern part of the species' range. In our state in recent years, the number of Indiana bats has fallen to an estimated 150. Populations are dwindling throughout the range, and *Myotis sodalis* is on the federal endangered species list.

Eastern Small-Footed Bat *(Myotis leibii).* Also known as Leib's myotis, this is one of North America's smallest bats: length, 2.8 to 3.3 inches; wingspread, 8.3 to 9.7 inches; weight, 0.1 to 0.3 ounce. The small-footed bat resembles the little brown bat but has a golden tint to its fur.

The flight of the small-footed bat is slow and fluttering, like that of a large moth. *Myotis leibii* inhabits wooded mountainous country. Its feeding and breeding habits probably parallel those of the other small, closely related bats. The small-footed bat may wait until as late as mid-November before entering caves and mine shafts to hibernate. Individuals cling to narrow cracks in the roof, wall, or floor, singly and in groups of fifty or more. They usually stay close to the entrance where the temperature remains just above freezing. The small-footed bat emerges from its hibernation quite early, in March or April. *Myotis leibii* is rare in Pennsylvania, and its population is believed to be decreasing; the state Biological Survey has classified it as a threatened species.

Silver-Haired Bat *(Lasionycteris noctivagans).* This medium-size bat is 3.7 to 4.5 inches long, its wingspread is 10.5 to 12 inches, and it weighs 0.2 to 0.4 ounce. The long, soft fur is blackish brown tipped with white, yielding a frosted appearance.

The silver-haired bat inhabits wooded areas bordering lakes and streams. It roosts in dense foliage, behind loose bark, in woodpecker holes, in hollow trees—rarely in a cave. The silver-haired bat begins feeding earlier in the day than most other bats, often before sunset. It hunts along the edges of streams, lakes, and ponds in a slow, gliding flight full of twists and turns. Females bear twins in June or July. Biologists are not sure whether the species actually breeds in Pennsylvania or simply migrates through the state in spring and fall. Silver-haired bats hibernate in a winter range extending from southern Illinois to coastal New Jersey and south to Mississippi, Alabama, and Georgia. Males may remain in the southern wintering range all year long.

Eastern Pipistrelle *(Pipistrellus subflavus)*. The eastern pipistrelle is also called the pygmy bat because of its small size: length, 2.9 to 3.5 inches; wingspread, 8.1 to 10.1 inches; and weight, 0.2 to 0.4 ounce. The pipistrelle's fur is yellowish brown, darker on the back. The back hairs are tricolored: gray at the base, then a band of yellowish brown, and dark brown at the tip. The species inhabits eastern North America and Central America and is found throughout Pennsylvania.

Pipistrelles take wing in early evening and make short, elliptical flights at treetop level. In summer, they inhabit open woods and forest edges near water, rock and cliff crevices, buildings, and caves. They hibernate from October or November until April or early May, roosting singly in caves where the temperature hovers between 52 and 55 degrees Fahrenheit. They sleep soundly and may dangle in the same spot for weeks on end. Pipistrelles eat flies, small moths, leafhoppers, beetles, and wasps. They are preyed on by owls and by larger bats. *Pipistrellus subflavus* breeds in autumn, and females bear two young in June or July.

Big Brown Bat *(Eptesicus fuscus)*. A large bat, 4.1 to 4.8 inches long, the big brown has a wingspread of over 12 inches and a weight of 0.4 to 0.6 ounce. The fur is sepia brown; the face, ears, and flight membranes are blackish. This common bat lives across the United States in rural and urban areas, with summer roosts in attics, belfries, and barns, behind shutters, and in hollow trees. Unbothered by noise and commotion, big brown bats have been spotted hunting for insects at night above crowded city streets.

Big brown bats take wing at dusk and generally use the same feeding grounds each night. They fly at a leisurely speed along a nearly straight course 20 to 30 feet in the air, often emitting an audible, high-pitched echolocation chatter. Their prey includes wasps, flying ants, houseflies, leafhoppers, and large beetles. Banding studies indicate that most big brown bats stay within a 30-mile radius of their summer and winter roosts. Among the last bats to enter hibernation, big brown bats return to caves, buildings, mines, and storm sewers in November and December. They prefer cool, dry sites near tunnel entrances and do not congregate in large groups. They emerge in March and April. Females usually bear two young in June. Because of their ability to roost and hibernate in man-made structures, big

Big brown bat.

brown bats may be more abundant today than before Europeans settled North America.

Eastern Red Bat *(Lasiurus borealis)*. Eastern red bats are covered with soft, fluffy fur. Females are a dull buffy chestnut and males a bright rusty red. A large bat, the eastern red is 3.7 to 4.8 inches long, with a wingspread of 11.3 to 12.9 inches; weight is 0.2 to 0.5 ounce. Individuals roost singly in trees (except for females with young), often on forest edges, in hedgerows, and along shrubby borders. Rarely do they use caves or buildings. Some eastern red bats live in cities, where they catch flying insects attracted to artificial lights.

Red bats start feeding early in the evening, preying on moths, flies, bugs, beetles, crickets, and cicadas, which they take from the air, foliage, and ground. Strong fliers, they average 8 miles per hour on their long, pointed wings and can reach speeds of 40 miles per hour. Females bear three or four young in treetop roosts. Young red bats are able to fly at three to four weeks of age and are weaned after five or six weeks. Blue jays may prey on the babies, and others die after falling from the roost. Red bats depart from Pennsylvania in September or October, migrating as individuals. The species winters as far south as the Caribbean and Central America; we know little about their winter quarters.

Hoary Bat *(Lasiurus cinereus)*. The largest bat of the Eastern forests, the hoary bat is 5.1 to 6 inches long, has a wingspread of 14.5 to 16.5 inches, and weighs 0.9 to 1.6 ounces. Its thick, dark brown fur is heavily tinged with white. Although the species ranges across the state, it is not common.

Hoary bats roost in trees—in the woods, along forest edges, and in farmland. Solitary when roosting, individuals choose protected sites 10 to 15 feet above the ground. Hoary bats are strong, swift fliers that feed primarily on moths and also take beetles and dragonflies. Sometimes hoary bats catch small bats such as pipistrelles.

Pennsylvania is near the southern edge of the species' summer range and perhaps within the northern boundary of its winter range. The hoary bat's migration is complicated and poorly understood. Some individuals may hibernate in the Northeast, but most go to the southern and western states and to Central America; sometimes they migrate in flocks. In spring, the females return north to bear young. Between mid-May and early July, the mother gives birth to two babies while hanging in a tree.

Evening Bat *(Nycticeius humeralis)*. The evening bat is dark brown with black ears and black flight membranes. It is about 3.6 inches long, including a 1.5-inch tail, and it weighs 0.4 ounce. The evening bat ranges across the South. It is rare in Pennsylvania, having been reported from only Bucks, Cumberland, and Greene counties, with a maternity colony on the campus of Waynesburg College. Females bear two young (rarely one or three) in mid-June. Evening bats roost in hollow trees, woodpecker holes, and buildings. In slow and stately flight, they forage over open areas, often cornfields, taking beetles, moths, and leafhoppers. Biologists estimated that a colony of 300 evening bats in Indiana ate 6.3 million insects per year.

Barbour, R. W., and W. H. Davis. *Bats of America*. Lexington, KY: University Press of Kentucky, 1969.
Leen, N., and A. Novick. *The World of Bats*. New York: Holt, Rinehart, and Winston, 1969.

COTTONTAIL RABBIT

The eastern cottontail *(Sylvilagus floridanus)* is one of the best known wild animals. This common, widespread rabbit lives in brushy areas, farmland, old fields, woods edges, towns, and suburbs. The species ranges across eastern and midwestern North America to Florida, Mexico, and Central America. Some taxonomists assert that the cottontails in Pennsylvania belong to two separate species: the eastern cottontail and the Allegheny cottontail *(Sylvilagus obscurus),* which lives in high heaths densely grown with mountain laurel and blueberry shrubs. Other taxonomists say that the Allegheny cottontail is a subspecies or race of the eastern cottontail. A third subspecies, the New England cottontail *(Sylvilagus transitionalis),* is now believed to exist no farther west than eastern New York, although in the past it was considered to be a resident of Pennsylvania. All of these rabbits resemble one another in appearance and habits.

Cottontails belong to the family Leporidae, along with the hares. Hares are born precocial (well furred, with their eyes open, and capable of running shortly after birth), but rabbits are born altricial: naked and helpless and with their eyes closed. Rabbits and hares have chisel-shaped, ever-growing front incisors, as do their close relatives the rodents, but rabbits and hares also possess a second pair of teeth behind the first, which the rodents lack.

Biology. The cottontail is a long-eared, small- to medium-size mammal. Because its hind legs are longer than its front legs, it hops when running. A cottontail's fur is very soft, brownish above with gray and black hairs intermixed, and white or pale buff on the undersides. The soles of the large hind feet are densely furred. The puffy white tail is about 2 inches in diameter. A white ring encircles each eye, and about half of all individuals have a white blaze on the forehead. Cottontails molt in spring and fall and do not turn white in winter. Eastern cottontails are 15 to 18 inches long and weigh 2 to 3 pounds, with females (does) slightly heavier than males (bucks).

A cottontail has sharp hearing and a keen sense of smell. Its eyes, set well back on the sides of the head, monitor a wide field of view and alert the animal to attack from ground level or above. Cottontails do most of their feeding in the evening, at night, and early in the morning; at times during the day, they like to bask in the sun. Individuals scratch out forms, shallow depressions in the ground hidden by brush, weeds, or tall grass, where they spend most of the daylight hours. They do not dig dens. During stormy or cold weather, they take shelter in woodchuck burrows, brush piles, and unraveling stone walls, and beneath debris in dumps and junkyards.

When pursued, a cottontail tries to escape by using a burst of speed and a zigzag running pattern. It can hit a top speed of 18 miles per hour but cannot run steadily over long distances. Usually a cottontail sprints for cover and then crouches there, immobile. The white tail may act as a focal point for a predator's attack; sometimes a cottontail leaves its tail in a predator's talons or jaws but escapes with its life. Or the rabbit stops suddenly and tucks its tail, causing the predator to lose track of its target. When scent-trailed by a slow-moving creature like a hound, a cottontail will double back to stay in its home range. At times, cottontails duck into woodchuck burrows to hide.

Summer foods include low, broad-leaved weeds, clover, grass, leaves of various plants, fallen fruit, and garden vegetables. Captive wild cottontails eat grass equivalent to more than 40 percent of their weight daily during the summer. In winter, cottontails switch to blackberry and raspberry canes, bark, buds, tender twigs of bushy plants, poison ivy vines, and dried leaves and stems of herbaceous plants. Other important winter foods include the shoots and bark of gray birch, red maple, and smooth sumac. Cottontails leave a clean, diagonal cut when cropping twigs; deer, which lack upper incisors, break the twigs off, leaving ragged ends.

There are two kinds of rabbit droppings: small, hard, dark brown pellets, and larger, softer, green ones. The dark pellets are the end result of normal

Eastern cottontail rabbit.

digestion, in which the body removes most of the nutrients from the food. The green pellets are part of an alternative digestion process called reingestion or coprophagy. Reingestion lets a rabbit eat a lot of food quickly, then return to the safety of a thicket or brush pile; there the rabbit expels the soft pellets, which still contain much nutritive value and large amounts of vitamin B. Hidden from predators, the rabbit eats the soft pellets—often taking them directly from its anus—and digests its food completely.

Sylvilagus floridanus has a high reproductive rate. A female may have up to seven litters per year (four is the average), with the first arriving as early as March and the last coming in September. About half of all litters are born in May and June. During breeding, males chase females, and both males and females leap high into the air in exuberant display. More than one male may breed with a single female, and the sexes do not form a lasting pair bond. Following a four-week gestation, the female bears two to nine young, with five the norm. Females become receptive and breed again soon after giving birth. A study in Missouri found that a typical adult female cottontail bore twenty-five young over the course of a year. Juvenile females born in early spring are sexually mature by late summer and often mate and rear a September litter; males generally do not breed until their second year.

Before giving birth, a doe scratches out a cup-shaped cavity about 5 inches across and 4 to 6 inches deep, lining it with dried grasses and fur plucked from her belly and breast. Pulling out fur exposes her mammary glands, making it easier for her young to nurse on them. The doe visits her nest twice a day, at dawn and at dusk, and feeds her litter. She may protect her offspring by bluffing predators or by attacking them, kicking out with her hind feet; she may move the litter if danger threatens. Young cottontails develop rapidly. Their eyes open after four or five days. After about sixteen days, they are weaned, fully furred, and on their own.

Predators of many sorts catch and eat rabbits: snakes, hawks, owls, foxes, bobcats, coyotes, domestic dogs, house cats, and, in hunting season, humans. Skunks, raccoons, crows, and other predators eat baby rabbits found in the nest. Spring floods, heavy rains, and farm machinery also kill cottontails, and many die when struck by automobiles. Few cottontails live more than a year in the wild, although their life span in captivity may reach ten years.

Habitat. Eastern cottontails live in thickets, brier patches, swamps, recently timbered woods, meadows, weedy fields, overgrown fencerows, forest edges, and woods openings. Seldom are they found in deep woods. Heavily cultivated land may provide food but not enough protective cover. In towns and cities, eastern cottontails live in vacant lots, grassy fields, parks, and cemeteries.

A cottontail's home range may be as small as 0.5 acre or as large as 20 acres, depending on the availability of food and cover; the average is from 2.5 to 7 acres. A cottontail rarely leaves its home, where it intimately knows the best food sources and escape routes.

Population. The rabbit population at the opening of the twenty-first century is not as large as it was in the past. The main reason for this decline is an ongoing loss of habitat. Farmers have cleared and cultivated fencerows and brushy places that once held rabbits. New roads and expanding cities and towns continually reduce the acreage available for cottontails and other wildlife.

In the early 1900s, many thousands of forested acres were logged off in Pennsylvania. Those lands grew back in brush, creating prime habitat that supported tremendous cottontail populations in the 1920s and 1930s. In the last several decades, maturing second-growth forests have shaded out understory vegetation, causing the cottontail population to decline.

A typical population density in an excellent habitat would be three to five rabbits per acre. In summer, when litters are being born and food is plentiful, the population peaks. By fall, summer's surplus of young rabbits has been thinned by predation, accidents, and disease. The population is at an ebb in late winter after hunters, predators, and harsh weather have taken their toll.

SNOWSHOE HARE

The snowshoe hare *(Lepus americanus)* is an uncommon mammal in Pennsylvania. Its several names refer to its haunts and biology. One name is varying hare: "varying" marks the animal's twice-a-year changes in pelt coloration, and "hare" identifies it as a member of genus *Lepus,* related to rabbits but differing from them in several ways. "Snowshoe" describes the animal's large, furry hind feet, which bear it easily over deep snow. The species is sometimes called the swamp jackrabbit, linking it to a preferred habitat and to the jackrabbit, a hare of western North America.

Lepus americanus ranges from Newfoundland to Alaska, and south in the Appalachians to Tennessee and North Carolina, and in the Rocky Mountains to New Mexico. In Pennsylvania, the snowshoe hare is restricted to the wooded northern mountains.

Biology. Although closely related to the more abundant cottontail, the snowshoe is not a rabbit. As a hare, its digestive tract differs structurally from that of a rabbit, and its newborn young are precocial (fairly well developed) in contrast to the rabbit's hairless, blind young. Adult snowshoe hares are about 19 inches long and weigh 3 to 4.5 pounds, with females slightly larger than males. A snowshoe's body conformation is similar to that of a cottontail, although the snowshoe has longer ears, larger feet (the hind foot measures 5.5 inches, compared to 3 to 4 inches for the cottontail), a rangier build, and a shorter, smaller tail.

In summer, a snowshoe's fur is gray-brown, darker on the rump and down the middle of the back, the throat buffy, and the tail dark brown above and white below. In October, the brown hairs begin to fall out, and white hairs replace them. The molt begins on the feet and ears and proceeds upward and toward the rear until the entire pelt is white, except for the tips of the ears, which remain black or dusky year-round. A complete changeover takes about ten weeks. In March, another molt begins, with brown hairs replacing white. Temperature and background color do not trigger the molts: the changes are in response to variation in the amount of light. As days grow shorter in fall, a hare's eyes receive light for shorter and shorter periods each day, triggering changes in the pituitary gland, located at the base of the brain. During the molt, the pituitary shuts off pigment production in the new fur, which therefore grows in white. In spring, lengthening days stimulate the reverse effect.

A snowshoe hare has excellent hearing, and the animal frequently twitches its large ears about to catch sounds. Its eyesight is keen; the eyes, on the sides of the head, provide limited depth perception but take in a wide field of view. A hare will sometimes stand up on its hind legs to see and hear better.

The four toes on a snowshoe's hind feet are large and can be spread far apart. The bottoms of the toes and the soles of the feet are covered with dense, coarse hair that grows long in winter, forming "snowshoes" that support the animal in deep snow and provide traction on an icy crust.

Should a predator threaten, a hare can burst out of a sitting position into a dead run. It can race up to 30 miles per hour over ground and snow, leap a dozen feet at a bound, dodge with agility, and swim if forced into the water. A hare circles like a cottontail, because it does not want to leave its familiar home range. A Wisconsin biologist once tracked a hare in snow for over an hour; despite a continuous, noisy pursuit, the animal stayed within 10 acres. Unlike cottontails, hares rarely duck down holes when chased. Based on studies in western states, biologists estimated that the home range of an adult male is 25 acres, and that of a female, 19 acres, with individuals of both sexes using about 4 acres on a typical day. Hares do not build nests or dig dens, although they may shelter from rain or snow in a hollow log or a rock crevice. In areas where hares are common, especially in swamps, their movements beat down runways in the vegetation.

During the day, a hare keeps to its form, a small depression in the leaf litter or in the ground, either natural or caused by the weight of the animal's body resting there repeatedly. The form may be on a small rise, letting the hare see its surroundings. To protect against predators, a hare may site its form beneath overhanging branches, in a clump of shrubs or tall weeds, or at the base of a stump. A hare may have several forms in its home range. When resting, the creature sits with its head and neck drawn in close to the body and all four feet gathered beneath it.

In summer, snowshoes eat grasses, clover, dandelions, ferns, berries, and the buds and growing tips of low, woody plants, including hazel, willow, aspen, alder, and birch. *Lepus americanus* practices coprophagy, or reingestion, sometimes eating its own droppings to glean all the nutrients from its food (see the account of the eastern cottontail for a more detailed explanation). After frosts kill herbaceous growth, hares feed on the dried remains. In winter, they depend on woody plants, eating twigs and bark as high as they can reach when standing on their hind legs; deep snow makes a plat-

The white winter coat of the snowshoe hare blends in with the snow.

form, putting them closer to their food. Hares feed on many kinds of trees. Sometimes they eat carrion.

Both males and females thump with their hind feet to communicate with one another. Snowshoe hares make grunting and snorting noises, utter a low clicking or chirping sound like a person saying "tsk, tsk," and scream or bleat when frightened or wounded.

Courtship begins in early March, when males (bucks) fight furiously for females (does). The males kick with their powerful hind feet, sometimes injuring or even killing each other. During courtship, a buck chases a doe, with both jumping high in the air and then simultaneously crouching opposite each other, sometimes touching noses. When she is fully receptive, the doe stops running away and lets the buck mount her. Several males may breed with one female; males and females do not form lasting pair bonds.

The female bears one to seven young (the average is two or three) after thirty-six days of gestation. The mother gives birth while sitting in her form. Young hares, called leverets, can walk and hop within a day of their birth. The young spend their days in separate hiding places, gathering together when the doe returns to nurse them, usually during the evening twilight. The leverets start eating green vegetation within ten days of their birth and are weaned and independent after four to six weeks. They stay in an area of less than 2 acres until they are six weeks old, when they expand their range to about the size of the doe's home range. Later they disperse and find home ranges of their own.

Females have one to four litters each year. Snowshoes have potential life spans of eight to nine years, but only an estimated 30 percent live out one year and perhaps 15 percent reach age two. They fall prey to foxes, bobcats, coyotes, and fishers; young are taken by weasels, red-tailed hawks, goshawks, and great horned owls. Human hunters also take some hares.

Habitat. Snowshoe hares inhabit mixed hardwood forests with conifers and a dense understory, such as rhododendron and mountain laurel. They favor brushy areas where woods have been logged off or burned within the last ten years. They live in wetlands among cedar, spruce, and larch, and they inhabit dense stands of aspen or alder interspersed with conifers. In Pennsylvania, the high country of ridgetops, mountain slopes, upland swamps and bogs (particularly in the Pocono region), and extensively wooded plateaus harbor the greatest numbers of hares.

As do cottontail rabbits, snowshoes move into areas where the forest canopy has been opened up by fires, wind and ice storms, and clear-cutting.

Cottontails build up their populations in such areas within a year or two, but snowshoes, with a slower reproductive rate and differing food and cover requirements, need up to seven years to take hold. Hares stocked in fresh clear-cuts do not stay there; perhaps an innate fear of aerial predators prompts them to search for taller, denser cover.

Population. Because Pennsylvania is near the southern edge of the species' range, the snowshoe is not as abundant here as it is in New England and Canada. In the far north, snowshoes show a cycling in the population level, with numbers peaking every nine or ten years.

Snowshoe hares prosper in brushy habitat, which has decreased over the last several decades in Pennsylvania as forests cut around 1900 have matured. The snowshoe shares its habitat with the white-tailed deer, and the Keystone State's large deer population has severely browsed back escape cover and forage in some areas. The Pennsylvania Biological Survey has classified *Lepus americanus* as a "vulnerable" species and recommends that wildlife biologists map existing populations, draw up and implement a habitat improvement plan, and adjust hunting seasons to protect snowshoe hares and allow their numbers to increase.

CHIPMUNK

The eastern chipmunk *(Tamias striatus)* is a small, lively, ground-dwelling rodent found throughout Pennsylvania. A member of the squirrel family, Sciuridae, the chipmunk is closely related to the red, gray, fox, and flying squirrels, and to the woodchuck. The eastern chipmunk ranges from Quebec south to northern Florida and from the East Coast to Saskatchewan and the Midwest.

Biology. Adult chipmunks are 8 to 10 inches long, including a 3- to 4-inch tail, and weigh 2.3 to 4.4 ounces. A chipmunk's head is blunt, with rounded erect ears. The legs are short. Each hind foot has five clawed toes, and each forefoot is equipped with four clawed toes plus a thumblike digit with a soft, rounded nail. The tail is well furred and flattened. The front incisor teeth are broad and chisel-shaped like those of all rodents. A chipmunk has internal cheek pouches, used for transporting food and removing dirt excavated from the burrow.

The short, dense body fur is colored alike for both sexes: reddish brown, sprinkled with black and white hairs, and brightest on the rump and flanks. The cheeks and sides of the body are grayish tan to tawny brown, and the underparts are white. The most prominent markings are five blackish brown stripes on the back and sides; the narrowest stripe centers on the backbone, and on each side from shoulder to rump run two other dark stripes sandwiching a cream-colored band. On each cheek, a black line runs through the eye, and two white or buffy stripes outline it.

Chipmunks are graceful and spry, quick to dart for cover when startled. They run with their bushy tails held straight up. Rarely will they run straight to their burrows; instinctively they protect the locations of their holes by using other refuges scattered throughout their home range. Although they spend most of their time on the ground, chipmunks occasionally climb trees. Their senses of sight and hearing are keen. When eating, they often perch on stumps, rocks, or logs to keep an eye on their surroundings. Chipmunks have three calls: a loud *chip* resembling a robin's note; a soft *cuck-cuck* that may be repeated for several minutes; and a chip following by trilling, *chip-r-r-r.* A chipmunk uses these sounds as alarm calls and to warn other chipmunks out of its home territory.

Omnivorous feeders, chipmunks eat wild nuts (white oak acorns are a particular favorite), seeds of woody and herbaceous plants (including cherry, maple, shadbush, dogwood, viburnum, ragweed, wintergreen, and wild geranium), mushrooms, berries, corn, apples, peaches, pears, and garden vegetables. They also eat insects, snails, earthworms, millipedes, salamanders, small snakes, frogs, birds' eggs, and young mice and birds. Chipmunks eat food on the spot—evidenced by piles of shelled seeds or nut fragments on or below stumps—or carry it off and store it. A chipmunk can transport surprisingly large amounts of food in its cheek pouches: one observer counted thirty-two beechnuts carried by a chipmunk. When full, the cheek pouches bulge out noticeably on each side of the chipmunk's head.

A typical home range is half an acre. Individual home ranges overlap, with each animal defending an area around its burrow entrance. The burrow itself may be simple or complex. The opening, about 2 inches wide, is hidden under a rock, stump, log, or at the base of a fence post or a wall. The tunnel plunges straight down for several inches, then levels off; it may extend for up to 30 feet, sometimes branching into offshoot tunnels with separate entries. The chipmunk excavates its burrow, pushing or carrying dirt away from the entrance. It lines a nest chamber, about a foot in diameter, with crumbled dry leaves and grass. Food may be stashed in the nest chamber or in a storage chamber nearby, which is often quite large and capable of holding up to half a bushel of nuts and seeds. Storage and nest cavities may be as deep as 3 feet below ground.

Eastern chipmunk.

Chipmunks are diurnal (active during the day), although hot summer weather will send them into their cool burrows. In autumn, chipmunks gather food for winter, storing it in their burrows and in aboveground caches in their home ranges. In October or November, they go underground, plug their den entrances, and live on stored food until spring. Chipmunks do not build up a thick layer of body fat. During winter, they become torpid, with their respiration rate falling from sixty to less than twenty breaths per minute and their body temperature dropping from around 100 degrees Fahrenheit to 45 degrees. Periodically they waken and eat. On sunny days in winter, they may leave the den briefly; some individuals emerge to breed in late February, although snow and cold usually drive them back underground.

By late March, breeding is in full swing. Gestation takes around one month; females bear three to five young from mid-April to mid-May, using their nests as brood chambers. Some females bear young in late July or early August; biologists believe that these late litters are produced by females born the preceding year who failed to breed during the spring. In the northern part of its range, *Tamias striatus* has one litter per year, and individuals do not breed during the year of their birth.

Newborn chipmunks are blind and naked, about 2.5 inches long and weighing 0.1 ounce. After around a week, as their fur grows in, the characteristic body stripes become visible. Their eyes open after a month; over the next thirty days the half-grown juveniles begin leaving the nest and foraging with their mother. When they disperse, juveniles may travel over half a mile before setting up their own home ranges. The male plays no role in raising the young, and adults without young are solitary.

Weasels are major predators of chipmunks; the weasels' sinuous bodies let them pursue chipmunks inside their burrows. Hawks, foxes, bobcats, house cats, raccoons, and snakes also prey on chipmunks. Many are killed by cars on rural roads and suburban streets. Longevity in the wild is two to three years, and there are records of chipmunks living up to thirteen years.

Habitat. An optimum habitat is open deciduous woods with plenty of stumps and logs. Chipmunks are found along woods edges with thick understory and briers; among rocky ledges covered with vines and brush; in and around stone walls; in fencerows; in farm woodlots; on brush land; in rubbish heaps and dumps; under barns and outbuildings; and in city parks, cemeteries, and patches of cover in the suburbs. Chipmunks are not common in coniferous woods and swamps.

Habitats, for chipmunks and most other forms of wildlife, are constantly changing. If a beaver dam floods an area of forest, chipmunks will move out—and may be picked off by predators as they try to find territories in land already colonized by other chipmunks. If a farmer lets a pasture grow into woods, chipmunks—usually young animals dispersing from their mothers' home ranges—will gradually move in from bordering woods and fencerows.

Population. The eastern chipmunk is common throughout its range, with six to around fifteen individuals per acre of suitable habitat. The number of individuals in a given area may fluctuate from year to year, usually in relation to the food supply. Predation does not have a great effect on local populations, since chipmunks are prolific breeders and tend to keep the available habitats filled.

Wishner, L. *Eastern Chipmunks: Secrets of Their Solitary Lives.* Washington, DC: Smithsonian Institution Press, 1982.

WOODCHUCK

The woodchuck, *Marmota monax,* is a rodent and the largest member of the squirrel family in Pennsylvania. The woodchuck ranges from Labrador and Nova Scotia south to Georgia, through southern Canada and the Midwest, and northwest to Alaska. It is found across Pennsylvania and is one of the state's most common mammals. The woodchuck—also called groundhog, marmot, or whistle pig—is a prolific digger whose burrows provide escape cover, temporary shelter, and homes for many other creatures.

Biology. Adults are 20 to 26 inches long, including a bristly 6-inch tail. Weights range from 5 to 10 pounds, with extremely large individuals weighing up to 15 pounds. Males are slightly larger than females. A woodchuck's weight fluctuates through the year, with the animal at its heaviest by summer's end.

A woodchuck's coarse fur is yellowish brown to blackish brown; light-colored hairs in the coat give many woodchucks a grizzled appearance. The belly fur is sparser and paler than the fur on the back. A woodchuck's feet are dark brown or black. Melanism, or black coloration, is more common in *Marmota monax* than in most other mammals. The white or yellowish front incisors are broad and chisel-shaped, like those of beavers, mice, and squirrels. These teeth, an upper pair in opposition with a lower pair, grow constantly and are kept worn down by grinding against one another.

Individuals are usually solitary and aggressive, but occasionally two or three will feed or sun themselves together. A woodchuck's life centers on its burrow, to which it quickly resorts when danger threatens. The size of the animal's home range depends on the local food supply; it may be no broader than 20 yards from the den in rich farmland, or it may spread over 7 or 8 acres. A male's territory usually encompasses the home ranges of one or two females.

Woodchucks eat a variety of vegetation, including grasses, weeds, clover, alfalfa, young corn plants, garden vegetables, and fruits such as apples and pears. Woodchucks do not require standing water to drink, and many live far from streams or ponds. Like rabbits, they get enough moisture from succulent plants, dew, and water left standing after rainfalls.

Woodchucks are active during the day, particularly in the cool hours of morning and evening during summer. They peek out of their holes, cautiously raise their heads to look about, emerge, and stand upright on the burrow's mound; if satisfied that nothing dangerous is nearby, they lower to all fours and move off. When feeding, a woodchuck lifts its head every few seconds to check on its surroundings. Most of the time the animal waddles along slowly, but it can run as fast as 10 miles per hour for short distances. Its senses of smell and hearing are acute, and its eyes pick up motion readily. The animal's sensory organs are all located near the top of the skull, letting a woodchuck sample its surroundings simply by raising its crown out of the burrow.

A solid, muscular body, short and powerful legs, and sturdy paws and claws equip the woodchuck for digging. A woodchuck often sites its burrow in a well-drained, sloping area. The animal uses its forefeet to loosen the soil, then kicks the earth behind it with its hind feet. It pushes loose dirt out of the main burrow entrance, where the soil may mound up considerably. A woodchuck's tunnel descends sharply, then levels off and narrows; it may be 3 feet below the ground and 50 feet long. Side tunnels are dug, as well as two or three back entrances or "drop holes," inconspicuous openings down which the creature can scurry and from which it can surreptitiously check for danger. In the burrow system are nest chambers, a hibernating chamber, and areas where the animal defecates. A woodchuck will also bury its feces in the dirt of the entrance mound, which is renewed several times a week when the animal cleans loose dirt out of the burrow.

Woodchucks climb trees—ascending and descending headfirst—to escape enemies and to reach fruits and succulent leaves. Often they walk

Woodchucks create habitat for many types of wildlife with the tunnels they dig; creatures from salamanders to foxes find shelter in the burrows.

along on wooden fence rails. They use their front paws to clutch stems of plants and to hold fruits while feeding. Woodchucks swim capably. They make several sounds. A sharp whistle (actually produced by the vocal cords) signals other woodchucks to watch out for danger; when feeding, an individual may grunt or make a *chuck-chuck* sound; and when angry or cornered, a woodchuck will chatter its teeth. Woodchucks are formidible fighters. Humans, dogs, coyotes, and foxes are about the only predators that can kill adult woodchucks, although the young are taken by hawks and owls. The red fox is the major predator of *Marmota monax*.

Woodchucks feed heavily throughout summer and early fall to build up body fat equaling about a third of their weight; they do not store food. With the hard frosts of October, they begin denning up, and few remain active past the first of November. Some woodchucks use different summer and winter dens, and some use the same den in both seasons. Before beginning its hibernation, a woodchuck plugs its den entrances with dirt to keep out unwelcome visitors. The woodchuck curls itself into a ball in its nest chamber and falls into a profound sleep. Its body temperature drops from 96 to 47 degrees Fahrenheit, its heart rate slows from a hundred beats a minute to fifteen, and other metabolic processes lag as the animal lives on its body fat. Some woodchucks waken and leave their dens briefly during warm spells in winter. Woodchucks lose from 20 to 37 percent of their body weight over winter.

Starting in February, males emerge from hibernation and begin looking for mates, often leaving muddy tracks in the snow as they go from burrow to burrow. Males fight with each other to establish dominance. Most females leave their dens in March. It's believed that each female mates with only one male. After a thirty-one-day gestation period, the female gives birth to two to eight young (the average is four or five) from mid-April into May. Newborn woodchucks are blind and helpless; they lack fur, and their skin is a dark pink. Their eyes open after about four weeks, and their mother begins bringing green food into the den for them to eat. After five or six weeks, she has weaned the litter.

By mid-June or early July, juveniles are ready to leave the home burrow and establish their own territories. The move is a perilous one, and many young woodchucks are killed by vehicles and predators. The young take up residence in abandoned dens or dig their own burrows. As fall approaches, they feed heavily to build up the fat reserves that must get them through the winter. Woodchucks mature sexually in the year following their birth, but only about one-quarter of them breed; yearlings that breed do so later in the season than other woodchucks. *Marmota monax* has only one litter per year.

In the wild, a woodchuck may live five or six years. In a study conducted at the Philadelphia Zoo, researchers found that captive woodchucks died of many causes, including cancer of the liver, heart attacks, and cerebral strokes caused by hardening of the arteries. Rarely, a woodchuck may be afflicted with malocclusion, a condition in which the front incisors fail to meet and therefore cannot grind themselves down. This misalignment can result in a tooth growing in a complete circle, sometimes penetrating the skull cavity and killing the animal.

Habitat. Woodchucks live in fields, meadows, pastures, fencerows, orchards, woodlots, forests, stream banks, suburban areas, near and beneath farm buildings, and in the deep woods. An ideal habitat would be a thick fencerow bordering cropfields and pastures. A woodchuck may dig its burrow in the center of an open field, but usually the animal chooses a more protected location such as a field edge, a hedgerow, at the base of a tree or stump, or under a stone wall.

Woodchucks provide a great amount of wildlife habitat with the tunnels they dig. Skunks, opossums, raccoons, and foxes occupy and remodel vacant burrows and use them to rear young. Rabbits hide in the burrows. Many smaller animals, including small rodents, reptiles, and amphibians, spend much or a part of their lives in woodchuck burrows. Woodchucks' digging also loosens and aerates the soil.

Population. Woodchuck numbers depend on food availability, soil type, and whether people actively try to suppress them. (Farmers may shoot woodchucks, whose burrows can damage agricultural equipment and injure cows and other livestock.) Population densities vary from one woodchuck per 2 acres to one woodchuck per 11 acres. *Marmota monax* has generally benefited from human activity: when Pennsylvania was a largely forested wilderness, with many fewer acres of open land, the woodchuck population was much smaller than it is today.

SQUIRRELS

Pennsylvania has five species of tree squirrels: the gray squirrel, the most common species, lives in towns and forests; the fox squirrel inhabits woods bordering agricultural land and farm fencerows; the red squirrel usually lives in or near conifers; and two species of flying squirrels live in woodlands.

Agile, acrobatic mammals, squirrels belong to the rodent order and are closely related to the chipmunk, woodchuck, beaver, mice, and rats. Tree squirrels have keen senses of sight, hearing, and smell. They are strong for their size and very quick. They are diurnal—active during the day—except for the flying squirrels, which are nocturnal. Squirrels give birth to blind, hairless young that are nursed by their mother and remain dependent on her for up to two months.

Squirrels are permanent residents within their relatively small home ranges and spend much of their time in trees. In winter, they do not hibernate; they den in tree cavities, where they may remain inactive during storms and bitterly cold weather. Their diet is generally vegetarian, with an emphasis on mast (wild nuts) and seeds, although they also feed on small animals when the opportunity presents itself. In turn, they are preyed on by many aerial and ground predators.

Gray Squirrel *(Sciurus carolinensis)*. The gray squirrel is found statewide in Pennsylvania. The species ranges across the eastern half of North America. Total length is 17 to 21 inches, including a bushy 7- to 10-inch tail. Adults have slender, muscular bodies and weigh from 1 to 2 pounds. Their hind legs are longer than their front legs, an adaptation that helps them jump; their sharp claws grip and cling to tree bark. Gray squirrels can run as fast as 14 miles per hour on the ground; they can leap 6 feet between trees, using their tails to keep their balance.

Most gray squirrels are gray above and off-white below, often with rusty or brownish markings on the sides and tail. There are two molts each year; the winter pelt is denser and more silvery than the thinner, more yellowish summer coat. Albinism is rare in gray squirrels, but melanism—black coloration—is fairly common, especially in northern Pennsylvania. "Black squirrels" may be any shade from dark gray to nearly jet black, often tinged with brown.

Gray squirrels inhabit hardwood and mixed coniferous–hardwood forests where trees are mature enough to produce nuts. They also live in farm woodlots, city parks, and residential areas. The species uses two kinds

Gray squirrels feed on the ground as well as in the trees.

of nests. One is a cavity in a tree—often a woodpecker hole that the squirrel has enlarged, or a rotted area cleared out behind a branch stub—insulated with shredded bark and leaves. The second type is a leaf nest, a rounded, bulky affair 1.5 feet in diameter, secured to tree branches, built out of twigs and leaves and lined with plant material. Tree cavities are used in winter, and leaf nests provide cooler summer quarters.

Sciurus carolinensis has a number of calls, including a scolding chatter and a repetitive note that ends in a descending, squalling exhalation rather like a cat's meow: this last call has earned the gray squirrel the folk name "cat squirrel."

Gray squirrels mainly eat acorns, hickory nuts, walnuts, butternuts, and beechnuts. Other foods include berries, mushrooms, pine seeds, grains (especially field corn), and fruits of dogwood, wild cherry, and black gum trees. In spring, gray squirrels eat buds, which are high-energy foods. They consume the buds and flowers of oaks and red and sugar maples and, later, the winged samaras of the maples. In autumn, the squirrels often cut nuts from trees and then bury them beneath the leaf litter and, weeks or months later, sniff them out and eat them. One study revealed that gray squirrels found 85 percent of the nuts they had buried. Unrecovered nuts sprout into seedlings; in this way, squirrels help promote forest regrowth. Gray squirrels occasionally eat insects and their larvae, birds' eggs, and young birds. In winter, they feed on sunflower seeds gotten from bird feeders.

I often see gray squirrels from my writing desk as they move through the forest on our land. I particularly enjoy their late-winter mating activities: as males chase females, the gymnastics can become spectacular, with pairs racing up trunks, circling as they ascend, running out on limbs, vaulting from one tree to the next, and scurrying down the trees almost as quickly as they climb them. At times, two or more males may chase one female, to the accompaniment of loud chattering and buzzing calls.

The gestation period is forty-four days, with young born from February to April. Litters range from one to six, with two to four the average. At birth, baby squirrels weigh about half an ounce. Their eyes open after a week, and at one month they are fully furred; they are weaned by two months. Young from first breedings disperse when their mother mates again or when she bears her second annual litter in July or August. Young of the second litter often stay with the female through the winter. The first litter of the year usually is raised in a tree cavity, and the second is reared in a leaf nest.

Sedentary creatures, gray squirrels may live out their lives on a single wooded acre, although some individuals range over 6 to 7 acres. A good habitat with a mix of nut-bearing tree species can support as many as twenty gray squirrels per acre, although three per acre is more usual. In the fall, individuals may shift from one area to another where food is more plentiful, generally short-range movements of a few hundred yards but occasionally migrations measured in miles.

Gray squirrels are preyed on by hawks (particularly red-tailed hawks), great horned owls, foxes, bobcats, coyotes, weasels, and snakes, and over a million are bagged by hunters in Pennsylvania each year. Raccoons, black rat snakes, and red squirrels prey on gray squirrel nestlings. Gypsy moth infestations that denude thousands of acres of oak forest, destroying the mast crop and stripping away the protective leaf canopy, can cause gray squirrel populations to plummet. In the wild, a gray squirrel may live for six to twelve years. Each year the mortality rate is about 50 percent, which means that about half of the population perishes.

Fox Squirrel (*Sciurus niger*). A large and handsome species, the fox squirrel is near the northern limit of its range in Pennsylvania, where it is found mainly in the southern and western parts of the state. (It also lives in the Southeast, the Midwest, and the oak-and-hickory groves of the prairie states.) The fox squirrel gets its name from its coloration, which resembles that of the red fox. The back and sides are a rich orangish brown, and the underparts are a pale yellowish brown, buff, or pale gray. The fox squirrel is noticeably larger and more thickset than the gray squirrel, with which it sometimes shares a habitat. Adults are 20 to 23 inches long, including an 8- to 11-inch tail, and weigh around 2 pounds.

Fox squirrels prefer open, parklike woods with sparse ground cover; they are rare in mountains, deep forests, and brushland. Farm woodlots, fencerows, and town parks are favored habitats. Fox squirrels live in hardwood forests with oak, hickory, beech, walnut, and butternut trees. Often they hang by their hind limbs while picking or holding nuts with their forepaws. Like gray squirrels, fox squirrels "scatterhoard" nuts by burying them under the leaf litter. In addition to eating mast (nuts), fox squirrels consume buds, the inner bark of trees, fruits and berries, the flowers and young fruit of tuliptrees, fungi, insects, birds' eggs, and young birds.

The fox squirrel's home range covers 3 to 25 acres. In an Ohio study, fox squirrels traveled 0.7 mile daily between woodlots. The species lives in both leaf nests and tree cavities. Active year-round, fox squirrels do not hibernate in winter but will hole up and sleep soundly through several days of snow or extreme cold. They mate in January, and one to six (usually two to four) young are born in late February or early March; some females produce a second litter in June or July.

The fox squirrel is slower and more sluggish than the gray squirrel, and it spends more time on the ground, moving with a slow, loping gait and stopping frequently to sniff the air and look for food. *Sciurus niger* is preyed on by foxes, coyotes, raccoons, domestic cats, snakes, hawks, and owls. Tagging studies have shown that fox squirrels live up to six years in the wild. In Pennsylvania, the fox squirrel is uncommon and spotty in the eastern part of the state; slightly more common in the central counties; and abundant and widespread in the western part of the state.

Red Squirrel *(Tamiasciurus hudsonicus)*. The red squirrel lives across northern North America; in the East, its range extends south in the Appalachian Mountains to Georgia. The species occupies its favored habitat—mature forests of white pine and hemlock—statewide in Pennsylvania. It also lives in mixed forests and in pure hardwood stands. In winter, red squirrels are colored chestnut above, grading to olive gray on the sides, with a black line separating the fur on the back and sides from that on the belly; the ears have prominent tufts of hair. The ear tufts are absent in the sleeker summer coat, which is a rich rusty brown. The eyes are ringed with white and the undersides are grayish white in all seasons.

The red squirrel is 11 to 14 inches long, including a 5- to 6-inch tail, and weighs 5 to 9 ounces. This squirrel is sometimes called the chickaree, for its long, vibrant, chittering call, *chir-rir-rir-rir,* given to advertise its territory; it is also known as the pine squirrel, reflecting its preference for nesting and feeding in conifers.

Piles of stripped pine cones show where a red squirrel has been feeding. Red squirrels eat immature green cones of white pine; seeds of hemlock, red pine, pitch pine, jack pine, larch, and other conifers; acorns and other nuts; maple sap and seeds; witch hazel seeds; fruits; buds; twigs; and mushrooms. They store green cones, nuts, and fruits in middens—caches located in hollow trees or in underground dens. They prey on insects, snails, the eggs and young of birds, and immature gray squirrels. Red squirrels have been observed killing and eating baby rabbits.

Alert, raucous, and energetic, red squirrels dash nimbly through tree branches. They vigorously defend their home ranges and middens from other red squirrels and also from gray squirrels, which they drive off even though the gray squirrels are nearly twice their size. Red squirrels nest in tree cavities, and they build exterior nests by laying a platform of twigs across the base of one or several branches and weaving an outer shell out of leaves and twigs. Nests are lined with shredded bark, moss, dried leaves, and grass. A leaf nest may be as high as 65 feet above the ground. Red squirrels also live in barns, deserted buildings, and rock piles. In winter, they may hole up for a few days during storms, visiting their middens for food. When the snow is deep, red squirrels use burrows beneath the snow.

Red squirrel.

In Pennsylvania, the breeding season is from February to September, with two litters born each year. Gestation is about thirty-five days. The average litter has three to seven young, each weighing 0.25 ounce. Juveniles are weaned and on their own after nine to eleven weeks. An adult female sometimes bequeathes her home territory to her offspring, with the young squirrels sectioning out their own territories from it. Most individual territories cover 3 to 4 acres. Minks, bobcats, house cats, coyotes, foxes, weasels, hawks, owls, and tree-climbing snakes take red squirrels. The fisher, a large weasel currently being reintroduced into Pennsylvania, often preys on red squirrels. The typical life span for *Tamiasciurus hudsonicus* is two to three years.

Southern Flying Squirrel *(Glaucomys volans)*. The southern flying squirrel is statewide in Pennsylvania; it occurs from southern Maine to Florida and from Minnesota to Texas, with isolated populations in Mexico and Central America. The southern flying squirrel is slightly smaller than the closely related northern flying squirrel, and much more common in the Keystone State.

Adults are 8 to 10 inches long, including a 3- to 5-inch tail. Weights range from 1.5 to 3 ounces. The soft, velvety fur is grayish brown on the back and pearly white on the belly. The large, dark brown eyes are adapted for night vision. The so-called flying membrane is a loose flap of skin between the fore and hind legs on either side of the body; when a flying squirrel extends its legs, they stretch the membrane taut, making an airfoil on which the animal can glide from one tree to another or from a tree to the ground. A flying squirrel can sail up to 40 yards in a downward direction. It uses its broad, flat tail as a rudder.

Flying squirrels are mainly arboreal, although they also forage on the ground. They are rarely seen, since they are nocturnal. They nest in hollow tree limbs and woodpecker cavities and sometimes in large birds' nests, which they cap with shredded bark and leaves. After a gestation period of about forty days, two to seven young (on average, three or four) are born in April, May, or June. The young are weaned after about two months. Adult males do not help the females rear the young. The southern flying squirrel may produce two litters per year, with the second litter arriving in September.

Flying squirrels may be more common than you think. A good way to look for them is to rap a stick against trees or branches that have cavities; the squirrels may stick their heads out or emerge to see what's going on. I was hiking on a trail on our land on a cloudy, drizzly morning when I spotted four young flying squirrels clinging to the bark above a nest box which I'd nailed to a tree for birds; the little squirrels looked at me seemingly with more curiosity than fear. We usually have a resident family in our shed, and I've seen them eating sunflower seeds at our bird feeder at twilight. Then there was the luckless one found drowned in a toilet. (How it got inside the house, I haven't a clue.)

Flying squirrels eat nuts; seeds; winter buds of hemlock, maple, and beech; tree blossoms and sap; fruits; berries; ferns; and fungi, both above-ground and subterranean types. They store surplus nuts in their dens and also bury them in the ground. Although small and apparently docile, flying squirrels are the most predaceous of the tree squirrels, eating moths, beetles, insect larvae, spiders, birds and their eggs, small mice and shrews, and carrion. One naturalist in the 1920s put a yellow-bellied sapsucker into a roomy cage with a flying squirrel, which, even though it had ample food, killed the bird and ate it. Owls and house cats are major predators of flying squirrels; foxes, coyotes, weasels, skunks, raccoons, and black rat snakes also take them. The average life span is estimated at five years.

Active year-round, flying squirrels are quite sociable, and in cold weather several individuals may share a tree cavity, sleeping snuggled together for warmth; up to fifty southern flying squirrels have been found in one nest. The southern flying squirrel may become torpid during the coldest part of the winter. One to three individuals typically live on an acre of suitable wooded habitat.

Northern Flying Squirrel *(Glaucomys sabrinus).* The northern flying squirrel is slightly darker and browner than its southern counterpart. The two species' ecology and habits are similar, although the northern flying squirrel shows a greater affinity for conifers, and the southern flying squirrel favors nut-producing hardwoods. The northern flying squirrel inhabits New England as far south as the mountains of northern Pennsylvania; there are also disjunct populations in parts of the Appalachians to the south. *Glaucomys sabrinus* ranges from the Great Lakes states west to Alaska.

Northern flying squirrels inhabit old-growth and mature forests, particularly northern hardwoods (beech, birch, maple) interspersed with hemlock, spruce, and fir. Home ranges are estimated at 5 to 19 acres and perhaps larger. Adults do most of their foraging between dusk and midnight, and for one to three hours before dawn. They feed heavily on lichens and fungi, including many underground ones, truffles and their relatives, which the squirrels sniff out. They also eat seeds, buds, fruits, and insects. They den in tree cavities or woodpecker holes during the winter, and in the summer, they may build nests out of leaves and shredded bark, in crotches of conifers high above the ground. Females produce one litter per year, usually with two to four young.

The loss and fragmentation of old-growth forest may be causing a decline in the northern flying squirrel population in Pennsylvania. *Glaucomys sabrinus* is considered a threatened species in the state.

MacClintock, D. *Squirrels of North America.* New York: Van Nostrand Reinhold, 1970.
Wells-Gosling, N. *Flying Squirrels: Gliders in the Dark.* Washington, DC: Smithsonian Institution Press, 1985.

BEAVER

The beaver, *Castor canadensis,* is North America's largest rodent. Before European settlement, the beaver was abundant in suitable aquatic habitat across the continent. The creature's fur is thick and beautiful, useful for making coats and hats; in the seventeenth and eighteenth centuries, the demand for beaver pelts sent trapping expeditions throughout the unexplored West, stimulating expansion of the American nation. By the late 1800s, uncontrolled trapping had almost wiped the beaver out.

In Pennsylvania, beavers were considered extinct in 1912. Soon after that, conservationists began reintroducing the species, using beavers from other states, and today, *Castor canadensis* is found statewide. On a continental scale, the beaver has reoccupied much of its former range.

Biology. The beaver has a blunt head with a skull and teeth suitably massive for cutting through wood. It possesses a short neck and legs and a stocky, muscular build. Adults weigh 30 to 70 pounds and are 32 to 48 inches long. The coat is a uniform reddish to blackish brown, and the sexes are colored alike.

Beavers are highly adapted to living in and near the water. The coat has two layers: dense, short, waterproof underfur beneath an outer layer of longer guard hairs. The two inner toes on each hind foot have specialized slit nails employed by the beaver to comb its coat and to waterproof its fur using oil produced by a pair of abdominal glands. The hind feet, broad and webbed between the toes, act as paddles when the beaver swims. The nose, ears, and mouth have valves that shut when the animal dives. Nictitating membranes protect the eyes. The lips can close behind the chisel-shaped incisor teeth, letting the beaver gnaw while underwater without having to open its mouth. Its large lungs and liver let a beaver store enough air and oxygenated blood to stay submerged for fifteen minutes.

The flat, oval-shaped tail is covered with scaly, leathery skin. It is quite large in relation to the animal's body: 12 to 20 inches long by 5 or 6 inches wide. The beaver uses its tail as a rudder and as a sculling oar in the water; as a support when the animal sits erect on dry land; and as a counterbalance when the creature walks upright while carrying twigs and branches in its forepaws. The tail also plays a role in temperature regulation. Fat is stored in the tail during summer, and the beaver draws on this reserve during winter, so that the tail is at its smallest in spring. A sharp slap of the tail on the water's surface (the sound is like a big rock plunked into the water) warns other beavers of danger. Contrary to popular belief, the tail is not employed as a hod or trowel for carrying mud.

The beaver uses its dextrous front feet for digging, manipulating food, and working on dams. A beaver's vision is weak, but its hearing and sense of smell are acute. Most food is located by smell. Beavers waddle along on dry land and can run for short distances at about the speed of a walking human. In the water, beavers swim slowly, around 2 miles per hour, holding their forefeet close to the body and paddling with the hind feet. Beavers are mainly nocturnal but may be active in daylight as well.

A beaver's musk glands, or castors, produce an oily, heavily scented substance, castoreum, which the animals deposit on prominent sticks and mounds of mud to mark their territory. Social animals, beavers live in colonies composed of a mated pair and their offspring. The colony builds a dam across a slow stream in a wooded area. The beavers gnaw down trees and place branches, sticks, leaves, grass, and mud across the stream until the flow is stopped and the water level starts to rise. The beavers continue to lay down sticks in a crisscross pattern until the dam is high and strong enough to back up a sizable pond. From this protected area, the beavers fan out and cut trees, both for food and for use in maintaining the dam.

Most dams are about a yard high and a hundred feet long; they are not watertight and allow most of the stream flow to seep through. The colony may raise their dam to flood more woodland, letting them reach food without leaving the water, or they may build additional dams upstream for the same purpose. In the pond, usually on a slight rise on the bottom, the beavers pile sticks to build up a lodge: dome-shaped, 5 to 6 feet high and 20 to 30 feet across, with a central chamber 6 to 8 feet wide and 2 feet above water level. A small air hole extends to the top of the lodge, and one or more tunnels exit from the living chamber directly into the pond. The floor of the lodge is fairly dry; some naturalists believe it is bare, and some assert that the beavers carpet it with grasses, sedges, moss, and shredded wood. Sometimes beavers build their lodges against the banks of streams, lakes, and ponds.

In summer, beavers eat soft plant foods, including grasses, bur reeds, duckweed, algae, and the leaves, stems, and fleshy root stocks of cattails and water lilies. They also eat the bark, twigs, and buds of aspen (a favorite high-energy food), as well as maple, willow, birch, alder, cherry, oak, and other trees. A beaver cuts down an average of one tree every two days. Smaller trees are generally cut from one side, and larger trees are attacked from two sides or all around the trunk. A beaver cannot cut a tree to make it fall in a certain direction, and some trees get hung up in other trees, rendering them useless to the beavers who felled them. Beavers cut most of their trees within 300 feet of the water: apparently they feel safest within this zone, and the trees don't need to be dragged a great distance. Beavers dig canals, 1 to 4 feet wide and up to 2 feet deep, from the pond or lake inland, in which they float logs back to the lodge. I canoe up beaver canals on the lake at Black Moshannon State Park, near my Centre County home, to reach prime blueberry-picking spots on shore.

Beavers fell trees to get at the higher, newer, more succulent growth of twigs and buds; after feeding, the beavers gnaw the trees into pieces that are used for building or repairing the dam or lodge. In autumn, beavers cut branches, carry them to the bottom of the pond, and anchor them in the mud. Even after the pond freezes over, the beavers can swim to the storage area, or cache, carry the branches back to the lodge, and feed on them.

Beavers are among the few wild species that pair for life—although some observers believe the male may breed with more than one female. Mating takes place from January into March, peaking in mid-February. After around twelve weeks' gestation, the female bears four or five kits in May or June. The newborns are fully furred, with their teeth erupted and their eyes open; they weigh about a pound. They grow rapidly during summer and may weigh up to 16 pounds by November. Sometimes they seem oblivious to their surroundings: I once canoed right up to a young beaver who was feeding on the tuber of an aquatic plant; he didn't even flee when I rubbed his back with my paddle.

The typical beaver colony consists of an adult male and female and their kits, including yearlings that have not yet dispersed. Colonies do not overlap.

Nictitating membranes protect a beaver's eyes when it swims underwater.

A square mile of habitat may support up to eight colonies. Young beavers stay with their parents for up to two years, as they mature sexually; at that time they leave on their own. Two-year-olds usually travel downstream to look for their own territories, or they strike off across dry land. Beavers have been found many miles away from water.

Coyotes, domestic dogs, bobcats, foxes, and bears may kill some individuals—especially young beavers migrating overland—but on the whole, beavers have little to fear from predators. Some are struck by cars, others are trapped by humans, and a few die when hit by trees they have cut down. The life span in the wild is ten to twenty years.

Habitat. Beavers inhabit streams and creeks that are narrow and slow enough to be dammed. They also live along rivers, on timbered marshy land, and on forest-edged lakes. They prosper in wooded areas with young trees, especially aspens, maples, and willows. Each year an adult beaver may cut up to three hundred trees, most of them less than 3 inches in diameter; under average conditions, an acre of aspen will support a five- or six-member colony for one to two and a half years.

Beaver dams affect many other wildlife species. After a dam is raised, part of a wooded valley is changed to an open pond. Water covers the bases of trees, preventing oxygen from reaching the roots and killing the trees within a few years; these snags may yield homes for cavity-nesting birds. Beaver ponds vary in size from less than 1 acre to many acres. They provide a habitat for ducks, geese, shorebirds, fish, reptiles, amphibians, and insects. Otters, minks, raccoons, herons, ospreys, hawks, owls, and other predators are attracted by the rich variety of life.

After beavers exhaust the supply of winter food in the area, they move on. The dam usually lasts for several years, accumulating silt, leaves, and other organic matter. Finally during a spring thaw or after a hard rain, the dam gives way. Most of the pond's water drains out, leaving an open area. Grass grows in the rich soil; later, berry bushes and shrubs push up, and insects and small rodents thrive. Deer, bear, grouse, turkeys, and songbirds and other insectivorous birds come to the beaver meadow, which provides an opening in the forest. The stream continues to flow through the meadow past standing dead trees. Aspens and willows send up shoots. In time, another beaver colony may find the valley to be a good habitat.

Population. The Pennsylvania Game Commission began restoring *Castor canadensis* to Pennsylvania by releasing a pair of beavers from Wisconsin in a remote valley in Cameron County in 1917. Over the next decade, the pair and its offspring reproduced and prospered. Individuals from that first colony were live-trapped and, supplemented with animals bought from Canadian wildlife agencies, were released on refuges throughout the state. By 1934, the population was large and stable enough to allow a trapping season, with over six thousand taken. Today beavers are found throughout Pennsylvania and are most abundant in the northern tier. Statewide, the population is estimated at thirty thousand.

Rue, L. L. III. *The World of the Beaver.* Philadelphia: J. B. Lippincott, 1964.

MICE, VOLES, AND RATS

They are the rarely seen multitudes, the small, unobtrusive creatures at the base of nature's food chain. Mice, voles, and rats are quick, prolific breeders. In terms of biomass—the total mass of living matter in a given area—they greatly outweigh the many predators who depend on them for food. Pennsylvania has two native species of mice, four voles, a bog lemming, two types of jumping mice, and three imported species. All are rodents, with two pairs of constantly growing, chisel-shaped incisor teeth, one pair on the upper jaw opposing another pair on the lower jaw.

Mice, voles, and rats mainly eat vegetation—nuts, seeds, fruits, leaves, and grasses. Most species collect and hoard foodstuffs to eat at a later time and to subsist on over winter. Most are predators, in a small way, on insects and their larvae, snails, slugs, spiders, and, in some cases, birds' eggs and other mammals. In turn, these rodents are fodder for a vast assortment of creatures, including snakes, shrews, weasels, raccoons, skunks, bobcats, foxes, coyotes, domestic dogs and cats, and even creatures as large as black bears. Many of the hawks and owls prey mainly on mice and voles, and the larger herons take these rodents occasionally.

The various small rodents live in nearly every available habitat, from rocky slopes on forested mountains to boggy lowland meadows to urban streets to the insides of people's houses. Some move about on the surface of the ground, while others keep to thick vegetation, rock crevices, or tunnels. Most feed at night and remain active year-round. Only the jumping mice (two species) hibernate in winter; during bitter cold, the other mice and voles become torpid and sleep for a time in their nests, round masses of leaves and grasses whose inside chambers are lined with plant matter. Some species are social in winter, when small groups huddle together for warmth.

The gestation period for most of the small rodents is around three weeks. The young are born without fur and with their eyes closed; their mother nurses them, and they grow rapidly. Litters are weaned and on their own within a month, and the mother—who has already ovulated and bred again—gives birth within a few weeks. Young from early litters can reproduce during their first year. In one of the most prolific species, the meadow vole, a single female can potentially give birth to nine litters with a total of seventy-two offspring per year. It's not hard to see how a population might explode quickly were it not for constant attrition from predators, parasites, disease, and accidents such as fires and floods.

Deer Mouse *(Peromyscus maniculatus)*. A small mouse with a huge range (the Northeast, Midwest, and the West from Alaska to Mexico), the deer mouse occurs statewide in Pennsylvania. It is 6.5 to 8 inches long, including the tail, which is 3 to 4 inches. A deer mouse weighs 0.4 to 1 ounce. For the first month of its life, an individual is colored gray; then it molts into its brownish gray adult pelt. In both juveniles and adults, the under-surfaces are pure white. The name deer mouse refers to the way that this rodent's coat resembles, both in color and in pattern, the pelage of a deer.

Deer mice inhabit nearly every type of land habitat in Pennsylvania: farm fields, fencerows, grassy berms of roads, brushy land, and deep woods, both dry and damp, pine and hardwood. Some taxonomists recognize two forms of *Peromyscus maniculatus,* the woodland deer mouse and the prairie deer mouse. Deer mice eat the seeds of many plants, cultivated grains, soybeans, corn, berries, buds, nuts, and mushrooms. They consume beetles, grasshoppers, crickets, and caterpillars (including those of the gypsy moth); other invertebrates, such as earthworms, centipedes, slugs, and spiders; and carrion.

Deer mice have sharp hearing. They possess excellent eyesight, and their large eyes are well adapted to night vision. However, they locate most of their food by smell. They can swim if necessary and run at nearly 5 miles per hour for short distances. The tail, covered with fur, acts as a tactile organ and a balancing aid; when climbing, a deer mouse wraps its tail around twigs or branches to gain steadiness. Deer mice weave ball-shaped nests, 6 to 8 inches in diameter, out of leaves, grasses, and other plants, lined with fur, feathers, and shredded plant matter. They nest in hollow logs,

Deer mouse.

stumps, and fence posts, beneath rocks, in underground root channels, and in abandoned squirrels' and birds' nests in trees up to 50 feet above the ground. Deer mice rest in their nests during the day and rear their young there. Nests at ground level may have a nearby burrow with a latrine area and a chamber for storing food.

In winter, if snow covers the ground, deer mice spend most of their time beneath the white blanket, where the temperature is warmer than in the open air. In extreme cold, deer mice cut down on their activity, sometimes sleeping for several days, perhaps huddled in a communal nest with two to four other mice, some of which may be white-footed mice, a different although closely related species.

Deer mice breed from March to October. Females raise three or four litters per year, each with three to seven young. In a year, one female can produce nearly thirty young, although few survive long enough to do so. Young mice, called pups, utter high-pitched squeaks. Males do not help females raise the litters. Deer mice are preyed on by foxes, cats, short-tailed shrews, minks, weasels, hawks, owls, and snakes. Home ranges vary in size from 0.05 to 2.5 acres, with three to thirty-six mice per acre of habitat. Like most other small mammals, deer mice are very abundant in some years and rather scarce in others.

White-Footed Mouse *(Peromyscus leucopus)*. Found statewide, this handsome nocturnal mouse may be the most abundant rodent in Pennsylvania. The coat is reddish brown above and white on the belly and feet. The length is 6 to 7.5 inches, including a 2.5- to 3.5-inch tail. Adults weigh 0.6 to 1 ounce. The white-footed mouse looks much like the deer mouse, except that its tail is shorter in relation to its body. In the white-footed species, the tail is shorter than the head and body combined; in the deer mouse, the tail is as long as or longer than the head and body.

White-footed mice live in shrubby areas, woods, cultivated fields, pastures, rhododendron thickets, fencerows, stream margins, ravines, revegetated strip mines, and farm buildings and houses. Some authorities believe the white-footed mouse prefers a slightly drier habitat than the deer mouse. White-footed mice nest in stone walls and rock crevices, under old boards, and in woodchuck burrows, beehives, tree cavities, and the abandoned nests of squirrels and birds. Like deer mice, white-footed mice do not dig burrows but use the runways of other small mammals. They are very agile and can climb trees. Individual home ranges vary from 0.11 to 0.86 acre, with males' ranges slightly larger than those of females. Typically, one to thirteen white-footed mice inhabit an acre of habitat. Researchers in Virginia found that the mouse population jumped from around one per acre to forty-two per acre following a heavy acorn crop; a year later, with acorns scarce, the population had fallen to five mice per acre.

White-footed mice eat about a third of their body weight daily, or around 0.2 ounce. They eat seeds, nuts (especially acorns), berries, fungi, green plant matter, insects (chiefly ground beetles and caterpillars: white-footed mice prey heavily on gypsy moth caterpillars), centipedes, snails, and small birds and mice. They cache food in autumn, carrying seeds in their cheek pouches to chambers beneath logs and stumps. White-footed mice breed from March through October; the three or four annual litters have three to seven young each. Females can mate when they are two months old. Males sometimes help females rear the young.

Meadow Vole (*Microtus pennsylvanicus*). The meadow vole is a stocky, mouselike creature with a blunt head, beady eyes, very small ears, and a short, scantily furred tail. Its upper parts are a dull chestnut brown, with a darker area along the middle of the back; the underparts are grayish or buffy white. The meadow vole is 6 to 7.6 inches long, including a 1.3- to 2.5-inch tail; the weight is 0.7 to 2.3 ounces. The species, often called a field mouse, lives across northern North America and is the most common vole in the East. In Pennsylvania, it is abundant statewide.

Meadow voles thrive in moist meadows and fields thick with grasses and sedges. They do not live in forests but may inhabit small clearings, bogs, and grassy openings in the woods. They are good swimmers and can run at 5 miles per hour. Meadow voles move about in low, thick grass and weeds that screen them from hawks and owls. I remember one winter when the uncut hayfield next to a friend's house was practically swarming with meadow voles. His dogs spent hours digging the rodents out, pouncing, then gruesomely eating. I was struck by the intricate network of surface runways visible when the grass was parted: the small pathways, approximately the width of a garden hose, branched this way and that and were obviously much used by the voles as they went about feeding on vegetation.

Meadow voles eat grasses and sedges. They cut stalks with seedheads and store them, to be eaten later, in small piles in the runways. They also feed on tubers, roots, grain, and the inner bark of shrubs and trees. Voles sometimes girdle small trees, killing them. Meadow voles are active all year, by night and by day, especially around dawn and dusk. They nest in shallow burrows 3 to 4 inches underground or hidden in the grass. During winter,

*Meadow voles travel in runways through thick grass
and weeds, which screen them from predators.*

voles huddle together in their nests or move about and feed in runways beneath the snow.

In the breeding season, from late March into November, meadow voles vigorously defend individual territories of 0.1 to 0.8 acre. The territories are larger in summer and smaller during peak population years, when up to 166 voles may live on a single acre. Usually a high population crashes to a low level, then builds up again to another high. Females produce from eight to ten litters in a high population year and five or six litters in a year when food is scarce. The average litter has four to seven young. Among the myriad predators that attend to the vole population are herons, crows, gulls, foxes, weasels, opossums, skunks, shrews, bears, house cats, dogs, snakes— even bass and pickerel, in habitats where land and water meet. Many voles are snatched up by hawks and owls, particularly barn owls. The maximum longevity in the wild is around a year and a half.

Southern Red-Backed Vole *(Clethrionomys gapperi).* This rodent is 4.7 to 6.2 inches long, including a 1.2- to 2-inch tail, and weighs 0.6 to 1.3 ounces. A reddish band down the back and a pale gray belly distinguish the species, found in much of upland Pennsylvania. A woods dweller, the red-backed vole favors cool, damp forests with hemlocks, mossy rocks, stumps, and rotten logs. It inhabits deciduous and mixed woodlands with mosses and ferns, rocky outcrops, stone walls, reverting fields, and grassy clearings. When traveling, it uses the burrows of moles and shrews and casts about beneath the fallen leaves. It also climbs into low trees. The species breeds from late March through November, nesting in cavities or the abandoned nests of other species. The southern red-backed vole feeds on nuts, seeds, berries, green vegetation, roots, and fungi.

Rock Vole *(Microtus chrotorrhinus)*. This species of New England and Canada inhabits a limited area of northeastern Pennsylvania. It closely resembles the more common meadow vole, except that the rock vole has a yellowish orange nose. In Pennsylvania, the rock vole lives in cool, damp woods of maple, yellow birch, and hemlock, among boulders and lush ground cover, mainly ferns. Foods include green plants, seeds, leaves, stems, fungi, and insect larvae. Weasels, foxes, timber rattlesnakes, and copperheads prey on rock voles. Females bear two or three litters annually, with each litter having one to seven young. Considered rare in Pennsylvania, *Microtus chrotorrhinus* has been classified as a "vulnerable" species by the Pennsylvania Biological Survey.

Woodland Vole *(Microtus pinetorum)*. Also called the pine vole, this species is found in the Midwest, the East, and New England. In Pennsylvania, it is statewide, with the greatest numbers in the southeastern lowlands. *Microtus pinetorum* is Pennsylvania's smallest vole, with a length of 4.3 to 5.5 inches; tail, 0.7 to 1 inch; and weight, 0.9 to 1.3 ounces. Its soft, glossy fur is chestnut brown on the upperparts and gray on the belly. The woodland vole inhabits wooded bottomlands, hemlock and hardwood forests, old fields, thickets, fencerows, farmland edges, and orchards.

This molelike species burrows beneath the soil just below the leaf litter. The woodland vole breaks up the dirt with its head, incisors, and forefeet, then turns around and shoves the soil out of the tunnel's entry, forming a cone-shaped pile 2 or 3 inches high. Meadow voles, hairy-tailed moles, and shrews use the burrows of *Microtus pinetorum*. Woodland voles seldom leave their burrows, and an individual's home range is about 100 feet in diameter. Foods include roots, stems, leaves, seeds, fruits, and tree bark; in gardens, potatoes and flower bulbs are eaten. Woodland voles kill fruit trees by girdling the bark or damaging the roots. They cache food in storage chambers as deep as 18 inches underground, and they rear their young in nests beneath rocks, logs, and stumps. Woodland voles breed less prolifically than other voles, bearing one to four litters per year, each with one to five young.

Southern Bog Lemming *(Synaptomys cooperi).* The southern bog lemming looks much like the meadow vole, with chestnut brown upperparts
and silver-gray sides and belly, but the fur on the bog lemming is shaggier.
Length is 4.5 to 5.7 inches, including a tail of 0.6 to 1 inch; weight is 0.9
to 1.6 ounces. The species is found in scattered pockets across Pennsylvania,
mainly in wet fields grown up with poverty grass, timothy, broom sedge,
hawthorns, crab apples, and locust trees. Bog lemmings live beneath matted
dead grass in surface runways, which they create by cutting down and feeding on low-growing plants. Bog lemmings eat stems and seeds of grasses and
sedges, along with berries, fungi, and mosses. The species breeds from early
spring to late autumn. Females bear several litters each year, with three to
five young per litter. Southern bog lemmings will often share a habitat with
red-backed voles, meadow voles, white-footed mice, and deer mice.

You can verify the presence of southern bog lemmings by their bright
green droppings: *Synaptomys cooperi* does not digest the chlorophyll in the
plants it eats, and the green pigment is passed on in its feces.

Meadow Jumping Mouse *(Zapus hudsonius).* The meadow jumping
mouse has big feet, long hind legs, and a skinny, tapering, sparsely furred tail
that is longer than the head and body combined. The overall length is 8 to
9 inches, including a 5- to 6-inch tail; weight is around 0.6 ounce. The fur
is yellowish brown, with a dark stripe on the back and orangish sides; the
belly and feet are white. Found in the East, Midwest, Canada, and Alaska,
Zapus hudsonius ranges throughout Pennsylvania.

Meadow jumping mice inhabit moist, grassy and brushy fields, thick
vegetation, and woodland edges. An individual's home range is usually less
than an acre. The name jumping mouse is something of a misnomer, as these
animals do not normally travel by jumping: they prefer to take short hops a
foot or two in length. Active at night, they eat seeds, grasses, berries, nuts,
roots, fungi, earthworms, insects, spiders, and slugs. Meadow jumping mice
breed from May to October, with two litters of three to six young born
yearly, in nests beneath boards, in hollow logs, and in grass tussocks. *Zapus
hudsonius* hibernates in winter; after laying on up to 0.2 ounce of body fat,
in October or November, this mouse retires to a nest about 18 inches below
the ground. The creature curls into a tight ball, buries its nose in its belly,
coils its tail around its body, and sleeps. Its breathing lags, its temperature falls
to near freezing, and its heart rate slows to a few beats per minute. After six
months' suspended animation, the meadow jumping mouse emerges in late
April or early May.

Woodland Jumping Mouse *(Napaeozapus insignis)*. Found in the North-east, New England, and Canada, the woodland jumping mouse lives throughout Pennsylvania except for lowlands in the southeast. It is 8.4 to 9.8 inches long, including a 5.5-inch tail. It has a bright yellowish brown back and sides and a white belly; the tail is tipped with a prominent white tuft. *Napaeozapus insignis* prefers cool, moist hemlock-hardwood forests in the mountains; it lives near streams, rarely in open fields or meadows. I know for a certainty that woodland jumping mice also inhabit dry oak-and-maple woods, because I found one of these handsome creatures dead on the township road at the edge of our land.

Woodland jumping mice eat seeds, berries, nuts, green plants, fungi (particularly subterranean fungi of the genus *Endogene*), insects, worms, and millipedes. An individual's home range includes 1 to 8 acres. Although mainly nocturnal, woodland jumping mice venture out on cloudy days. They use burrows and trails made by moles and shrews. Normally they travel on all four feet, but for greater speed, they hop with their long hind legs and can leap up to 10 feet. They evade predators by taking several bounds, then stopping suddenly under cover. Screech-owls, weasels, skunks, minks, bob-cats, and snakes prey on woodland jumping mice. Like the closely related meadow jumping mouse, the woodland species hibernates from October to late April or early May (about half the year) in an underground nest, singly or in pairs. Females bear three to six young in late June or early July; a second litter may be born in August.

Norway Rat *(Rattus norvegicus)*. The Norway rat is 12 to 18 inches in length, including a naked, scaly, 6- to 9-inch tail. Weight ranges from 10 ounces to more than a pound. This rodent's fur is thin, coarse, reddish to grayish brown above and paler below. The species arrived from Europe aboard ships around 1775. Today it is found statewide, and it ranges across North and Central America.

Norway rats have poor vision, but their senses of smell, taste, hearing, and touch are well developed. Extremely adaptable, they live in and under barns and farm buildings, in city sewers and dumps, along streams and rivers, and in marshes and open fields. They dig burrows about 3 feet long, with several escape holes lightly plugged with weeds or dirt and hidden in grass or under rubbish. *Rattus norvegicus* lives in colonies composed of several family groups that share feeding and nesting areas. Although mainly nocturnal, these rats also move about and feed during the day. They eat anything they can find or subdue, including fish, eggs, vegetables, grain, fruits, nuts, garden crops, carrion, and garbage. They kill poultry, snakes, young rats from neighboring colonies, other small mammals, and wild birds; in some areas, rats may suppress or wipe out native birds, especially the ground-nesting species. In turn, rats are preyed on by dogs, cats, minks, snakes, and large hawks and owls.

Norway rats breed throughout the year, with peak activity in spring and autumn; a female may bear six to eight litters per year, with an average of six to nine young per litter. Rats carry many diseases, including rabies, tularemia, typhus, and bubonic plague. Another introduced Old World rat is the black rat *(Rattus rattus),* found in limited numbers in southeastern Pennsylvania.

House Mouse *(Mus musculus).* Like the Norway rat and black rat, the house mouse is an Old World species inadvertently brought to North America by European settlers. It inhabits Pennsylvania statewide, living in and near houses and on farms. Six to 8 inches long, it has a 3-inch, scaly, nearly hairless tail; its weight is 0.5 to 1 ounce. House mice come in various shades of gray. They eat everything from grain and seeds (their preferred foods) to paper, glue, and household soap. *Mus musculus* is agile and quick, able to run at 8 miles per hour. An adult female produces five to eight litters annually, each with an average of five to seven young. House mice live in colonies and are active year-round.

Elton, C. *Voles, Mice and Lemmings.* New York: Oxford University Press, 1942.

WOODRAT

The Allegheny woodrat *(Neotoma magister)* lives in remote rocky habitats in the mountains of Pennsylvania. The species ranges in a broad band across the Keystone State from the northeast to the south-central region and the southwest, and on through West Virginia and the mountains of Ohio, Maryland, Virginia, Kentucky, Tennessee, and North Carolina. Woodrats of the species *Neotoma floridana* live farther to the south, often in very different habitats, including flatlands and swamps. Until the early 1990s, both of these eastern woodrats—which are nearly identical in appearance—were considered to be *Neotoma floridana*. Recent studies of their chromosomes and skull characteristics have led taxonomists to believe that the two types are indeed separate.

The name woodrat is unfortunately pejorative, because this shy, secretive woodland denizen has little in common with the aggressive Norway rat, other than the fact that both are rodents. The woodrat is as rare as the Norway rat is common; the woodrat is a lover of wilderness, while the Norway rat lives cheek-to-jowl with humankind. The woodrat has vegetarian food habits, and the introduced Norway rat sometimes preys on native wildlife species.

Biology. Adult woodrats are 16 to 17 inches long, including a 7- to 8-inch tail, and weigh from 13 ounces to 1 pound. The coat is grayish brown above and white below. The colors are darker in winter, when the coat is also softer and longer. The fully furred tail is dusky to brown above, white below. The woodrat's prominent rounded ears, long whiskers, and large and slightly bulging eyes all indicate its heightened senses of hearing, touch, and sight, particularly night vision.

Caves, rocky cliffs, ridge crests, overhangs, and boulder fields with deep crevices and underground chambers—these are the places where Allegheny woodrats make their homes, although occasionally they take up residence in abandoned buildings. Woodrats eat leaves; berries (including the pulp and seeds of wild grapes); nuts (acorns are a particularly important food); stalks and fruit of pokeweed; fruits of sassafras, dogwood, mountain ash, cherry, red maple, and shadbush; ferns and other plants; and fungi. Woodrats don't seem to rely on insect food as much as deer mice and white-footed mice do. Woodrats are nocturnal, feeding and shifting about within their home ranges—estimated at half an acre—under cover of darkness. They will forage out to 100 feet and farther from their nest sites.

The woodrat hoards leafy twigs, seeds, nuts, and mushrooms in and near its expansive nest. The creature builds—or accumulates—a nest of twigs, sticks, leaves, moss, and bark scraps, situated out of the weather in a crevice between boulders, on a shelf or on the floor of a cave, or beneath a rock ledge. The nest is open at the top like a bird's nest. Most nests are around 20 inches in diameter, although some, used by generations of woodrats, have been measured at 12 feet long, 6 feet wide, and 3 feet high. Often there will be two living areas, each about 5 inches in diameter and lined with grasses, shredded bark, and fur. A woodrat uses its nest year-round for its entire life. At times, other creatures shelter in woodrat nests: cottontail rabbits, opossums, white-footed mice, snakes, toads, salamanders, insects, and spiders.

As well as stockpiling food, woodrats collect treasures, oddments such as old mammal skulls, feathers, bottle caps, nails, coins, shards of china, spent rifle cartridges, rags, and leather scraps. These objects are hidden in the nest or heaped up outside, sometimes mingled with stored food items. The woodrat gets the name pack rat from its habit of packing off such items; it is also known as the trade rat because, if it comes upon an intriguing object while carrying another article, it may leave its burden behind (leafy twig, mushroom, or the like) and carry off the new item (a camper's spoon or car keys, for instance).

Woodrats leave piles of oval-shaped droppings, each dropping three or four times as large as a mouse's scat, on rock surfaces at latrine sites. The latrines may be used by several individuals. For most of the year, woodrats are solitary and unsociable, guarding their territories and warning off other woodrats by chattering their teeth, thumping their hind feet, and vibrating their tails against the ground. They fight over food and nest sites, rearing up on their back legs and jabbing at one another with their muzzles and front feet. Most adults become scarred, with torn ears, skin wounds, and in some cases, a bitten-off tail.

In Pennsylvania, woodrats breed from midwinter or early spring until autumn, with young arriving from mid-March to early September. The gestation period is about thirty-five days. One to four young are born per litter, and two or three litters are produced each year. Young woodrats are born naked, with their ears and eyes closed; they are about 4 inches long

The Allegheny woodrat sites its nest in a rock fissure out of reach of most predators.

and weigh 0.5 ounce. Their teeth have already erupted. The decurved incisors have an oval hole between them, which, when the jaws close, fits around the mother's nipples. The nursing young hold on tenaciously; when the mother wants them to let go, she will pinch them on the back, jaw, or neck with her teeth and twist them off with her paws. A young woodrat's eyes open in its third week, and it is weaned after about a month. By six or seven weeks, juveniles weigh about 5 ounces.

Allegheny woodrats live up to three years in the wild. They are preyed on by foxes, weasels, skunks, raccoons, bobcats, hawks, barred owls, great horned owls, blacksnakes, and timber rattlesnakes. Uninformed spelunkers sometimes harass and kill woodrats in caves.

Habitat. In Pennsylvania, woodrats seem to be restricted to rocky cliffs, outcroppings at high elevations, and caves. A rocky habitat is important, because Allegheny woodrats like to place their nests in fissures and deep crevices out of reach of most predators. Nearly all of the known historical and current woodrat sites in the state are on the Appalachian High Plateau and on ridges in the state's Ridge and Valley province.

Population. Pennsylvania is near the northern limit of the Allegheny woodrat's geographic range, and populations on the fringe of any species' range are often at risk. Since the late 1960s, woodrats have vanished from sites where they once occurred, particularly in eastern Pennsylvania. During the same period, the species disappeared altogether from New York, where it is now considered to have been extirpated.

Biologists believe several factors have contributed to the woodrat's decline. Over the years, changes in forest composition resulted in fewer oaks and, by the 1940s, in the near eradication of the American chestnut, once a prime source of nuts for woodrats. More recently, gypsy moth caterpillars have defoliated wide areas of oak trees, causing shortages of acorns, a key food item. Woodrats can contract a fatal parasite, the raccoon roundworm, by eating undigested seeds found in raccoon droppings. Human development near woodrat sites—land cleared for farming or homes—has spurred population increases in raccoons, and also in great horned owls, which prey on woodrats.

Since 1986, the Pennsylvania Game Commission has surveyed over 360 sites from which woodrats have been reported. Biologists have identified twenty metapopulations, multiple colonies linked by patches of rocky habitat through which young woodrats can migrate to set up territories of their own and to find mates. Five of these metapopulation areas no longer support woodrats, and seven of them have fewer active colonies than in the past.

The Allegheny woodrat is listed as a threatened species in Pennsylvania and has been proposed as a candidate for the federal endangered species list.

MUSKRAT

Why the name muskrat? *Musk* refers to a strong-smelling substance produced by the animal's perineal glands, under the skin near the anus, and *rat* alludes to its ratlike appearance. The common muskrat, *Ondatra zibethicus,* is a rodent, related to mice, voles, and rats and linked more distantly to tree squirrels and the beaver, woodchuck, and porcupine. A semi-aquatic mammal, the muskrat lives in and near still or slow-moving waters. The species ranges over most of North America, from the Rio Grande to northern Canada and Alaska, and it is common throughout Pennsylvania.

Biology. Adults are the size of a house cat: 22 to 25 inches long, including the tail, and weighing 1.5 to 4 pounds. The muskrat has a stout body, short legs, and an 8- to 12-inch tail, which is flattened vertically, covered with small scales, and nearly hairless. The small ears and eyes are set in a broad, blunt head. The muskrat has the typical rodent incisor teeth, two above opposite two below. Overall, the muskrat looks like a small beaver with a rat's tail.

The muskrat uses its tail as a prop when standing on its hind feet and as a rudder and a scull when it swims. The large, broad, partially webbed hind feet stroke alternately to propel the creature through the water; a muskrat swims at 1 to 3 miles per hour, can travel up to 150 feet underwater, and can stay submerged for fifteen minutes. A muskrat has dense, soft underfur overlain with longer, dark brown guard hairs shading to gray-brown on the throat and belly. The overall pelt color ranges from chestnut brown to almost black. The small forefeet have well-developed claws for digging and burrowing. Like the beaver, the muskrat has a valvular mouth and can close its lips behind its incisors, letting the creature gnaw freely while underwater.

A Louisiana researcher reported that *Ondatra zibethicus* eats about a third of its body weight daily. Muskrats eat the roots and stems of aquatic plants, including cattails (the single most important food item), arrowhead or duck potato, bur reed, bulrushes, pondweed, duckweed, pickerelweed, and water lilies. On land, they consume smartweed, dandelion, grasses, grains, garden crops, and fruits. Muskrats eat a small amount of animal matter, including crayfish, insects, freshwater clams, mussels, snails, fish, frogs, and carrion. Following an intraspecific fight, one muskrat may eat the carcass of another. During winter, muskrats rely on the roots and shoots of plants dug up from the marsh bottom, along with the twigs, buds, and bark of various trees, including willows, cottonwood, and ash. Muskrats hoard food. They cut through the ice from below, then push out roots and other vegetation, forming large piles that act as temporary feeding shelters and breathing spaces.

Although they are mainly nocturnal, muskrats sometimes emerge to feed on cloudy or rainy days. The best way to spot one is to canoe quietly along the bank of a pond or a slow stream, watching for a V-shaped ripple in the water made by a swimming muskrat.

Muskrats do not dam streams as beavers do. However, they build houses or lodges, usually in water about 2 feet deep and often on top of a submerged stump, brush pile, or log. The muskrat piles up mud, cattail stalks, roots, and stems. As the cone-shaped mound rises above the water, the muskrat uses less mud and more plant matter. The creature digs and gnaws from the pond bottom upward through the foundation and into the dwelling; as the lodge settles, the muskrat adds more building material to the outside and continues to gnaw away on the inside. The resulting living chamber is a foot or more in diameter and high enough that its occupant can move about freely. Lodges may be 8 to 10 feet across and 1 to 4 feet above the water, sometimes with two chambers, each housing a separate family group, with walls about a foot thick. Muskrats also burrow into stream banks, earthen dikes, and dams, entering below the water level and angling upward. The tunnels lead to one or more dens containing bulky nests of dried vegetation. Holes to the surface, hidden among brush or thick plant growth, ventilate the dens.

Muskrats work on their lodges throughout the year; new ones found in late summer and early fall are often the work of young muskrats. Muskrats also build feeding huts near their dens, roofed platforms made of aquatic plants where the animals can bring food and eat out of the sight of predators. Feeding huts are circular, smaller than the lodge, and accessed via tunnels from underwater. A muskrat usually stays within about 200 yards of its lodge. From a bank burrow, an individual's home range may extend out into the water several hundred feet and along the water's edge. In spring and fall, muskrats may leave their territories and travel overland up to 20 miles in search of better habitats. Late-summer droughts may prompt muskrats to leave marshes that are drying up; spring flooding can also trigger migrations.

An adult muskrat may eat vegetation equaling a third of its body weight daily.

To attract mates during the breeding season, which runs from March to October in Pennsylvania, muskrats place their pungent-smelling musk on feeding platforms, logs, and lodges. A male may impregnate several females. Males do not help raise the young. Mature females have two, three, and rarely four litters per year, depending on the length of the warm season—more litters in southern Pennsylvania, fewer in the north. After a month's gestation, the female bears five to eight naked, blind, and helpless young, each about 4 inches long and weighing 0.75 ounce. In a month, the well-furred kits are feeding on vegetation and are fairly independent. After two months, they are weaned, and the female drives them off if she is about to bear another litter. A female may overwinter with her final litter of the year. Young muskrats disperse along streams or colonize new sections of marsh. They reach adult size after six or seven months.

Muskrats mature sexually during the year following their birth, but a minority survive to breed. Young muskrats, dispersing juveniles, and migrating adults are vulnerable to minks, the major predator, as well as hawks, owls, river otters, foxes, and coyotes. Young muskrats fall prey to snapping turtles and large snakes; pickerel, largemouth bass, and northern pike also take some. Muskrats are susceptible to sudden cold snaps, and subzero temperatures may kill exposed individuals. Many migrating muskrats are run over by cars. Maximum longevity in the wild is probably three to four years.

Habitat. Muskrats live along marshes, ponds, lakes, and slow-moving creeks; they are less common along swift mountain streams. They inhabit the shores of farm ponds, where their burrowing can undermine and weaken dams. Through their feeding, a population of muskrats can open up areas of densely vegetated marsh, benefiting waterfowl and other aquatic wildlife.

Population. Muskrats are common in Pennsylvania and the Northeast. A normal population density is five to twenty-five individuals per acre of good habitat. The muskrat population does not exhibit marked rises and crashes like that of the meadow vole, a related land-dwelling species. However, the muskrat population in Pennsylvania is thought to be cyclical, with peaks of abundance about every ten years.

Ondatra zibethicus is the most valuable furbearing mammal in North America. In Pennsylvania, around two hundred thousand are trapped and their pelts sold to the fur industry each year.

PORCUPINE

The porcupine, Erethizon dorsatum, is a ponderous, quill-armored rodent with an appetite for tree bark and salt. It lives in forests and often can be seen hunched into a black ball high in a tree. Although it doesn't inhabit all parts of Pennsylvania, the porcupine is one of our best-known and most easily identified wild animals. It is North America's second largest rodent; only the beaver is larger.

The word porcupine comes from two Latin words, porcus, "swine," and spina, "thorn." The porcupine also has a colloquial name: quill pig. In the East, porcupines inhabit Canada and New England; most of Pennsylvania, except for the southeast and southwest; and western Maryland and northern Virginia. The species ranges through the Upper Midwest and Canada to the Pacific Northwest, and south in the forested Rocky Mountains nearly to Mexico and north to Alaska.

Biology. Adults are around 30 inches long, including a 6- to 10-inch tail. Males are larger than females. Porcupines weigh 9 to 20 pounds. An individual has four large incisors, two above and two below, bright orange in color and adapted to gnawing. The skull is heavily constructed; the small, rounded head has a blunt muzzle, small, rounded ears almost hidden in fur, and black eyes. The front and back feet bear long, curved claws. The large hind feet have broad soles covered with small bumps that lend traction for climbing trees. Short-legged and stout of body, a porcupine has a pronounced arch to its back.

Porcupines vary in color from brownish black to pure black, sprinkled on the sides and belly with yellow- or white-tipped hairs. The most distinctive aspect of a porcupine's coat is its sharp, pointed quills—actually specialized hairs—covering the animal's upperparts and sides from the crown of its head to the tip of its tail. Each quill is 1 to 4 inches long (those on the back are longest), yellow or white tipped with black, with a hard exterior and a spongy core. An individual may have thirty thousand quills. When the animal is relaxed, the quills lie smoothly along its body; when it feels threatened, the quills rise and bristle outward. The rest of the pelt consists of long, stiff guard hairs and soft, woolly underfur. Two molts occur each year: in spring, short hairs replace the winter underfur; and in fall, the long, insulating underfur grows back in. Quills are present in the pelage at all times and are replaced as they fall out.

To defend itself, a porcupine turns its back to a potential enemy, tucks its head between its front legs or under a convenient shrub, and flails its quill-studded tail back and forth. It may chatter its teeth and back up toward an adversary. Porcupines cannot throw their quills, but because the quills are loosely attached, they dislodge easily on contact. Each has a needle-shaped tip covered with hundreds of minute, overlapping, diamond-shaped scales, which slant backward and act as barbs. When a quill lodges in flesh, actions of the victim's muscle fibers engage the scales, drawing the quill inward as much as an inch a day. An animal badly impaled in the body and face will suffer intensely; quills may fatally pierce the heart, arteries, or lungs, or they may sever the optic nerves and cause blindness.

A porcupine walks flat-footed and can muster a top speed of only 2 miles per hour. Sensitive facial whiskers help it maneuver through thick underbrush and in the dark. Porcupines are more at home in trees than on the ground; a porcupine scales a tree by digging in with its sharp claws, pressing the soles of its feet against the bark, bracing with its sturdy tail, and hunching its way upward. It descends tail first. A porcupine can see moving objects only at short ranges, and it is almost blind to stationary objects. Its hearing is probably inferior to that of most mammals, but it has a keen sense of smell. A porcupine can swim, its air-filled quills helping to keep it afloat.

Porcupine.

Porcupines do most of their feeding at dusk, during the night, and at dawn. In winter, porcupines climb into evergreens and eat needles, twigs, and small limbs. They eat the inner bark of hemlock, spruce, white and pitch pines, basswood, maple, beech, birch, aspen, ash, cherry, apple, and other trees. As a porcupine strips off bark or foliage, small branches fall to the ground, trimmings that play a minor role in providing winter food for other animals. If a porcupine chews off an isolated section of a tree's trunk, in time the bark will close over the wound; if the rodent girdles the trunk, the tree will die. Trees with upper branches freshly "barked"—the newly exposed wood showing pale against the dark outer bark—advertise the presence of porcupines. In spring, summer, and fall, porcupines consume a variety of vegetation, including grasses, sedges, and the flowers, leaves, twigs, roots, buds, catkins, and seeds of many plants.

Porcupines gnaw on shed deer antlers (to obtain calcium and other elements), wooden buildings, telephone poles, signs, and ornamental trees. Sometimes they eat the tires and radiator hoses of automobiles. Porcupines crave salt, and sniff out and chew on objects that have been in contact with human perspiration, such as axe handles, ropes, work gloves, and leather boots. They lick highway pavement where salt has been spread to get rid of ice; they chew on plywood, apparently for the glue that holds the laminates together.

Porcupines grunt, groan, shriek, bark, and whine; their calls may carry up to a quarter of a mile. They are especially vocal during the breeding season: September, October, and November. Courting porcupines rub noses, chatter their teeth, walk on their hind feet, and perform stylized, weaving body movements; just before mating, the male showers the female with urine. Males are promiscuous breeders and play no role in rearing young. In females, the heat period repeats itself every thirty days until mating occurs or the breeding season ends. Up to half of all adult females may go unmated in a given year.

After a gestation period of 205 to 217 days, the female gives birth in April, May, or June to a single youngster called a pup or a porcupette. Birth usually takes place in a den below the ground. The newborn is precocial: fully furred, with its eyes open, a foot in length, and weighing about a pound. The soft, hairlike quills are a quarter inch long; as they dry, they harden and become functional. A porcupette can climb trees and eat plant food a few days after its birth. It nurses for about seven weeks. After it is weaned, the juvenile receives little attention from its mother; females and young separate for good after about six months. Porcupines mature sexually at fifteen or sixteen months and breed during their second autumn.

Porcupines den in caves, rock crevices, hollow logs and trees, deserted fox and woodchuck dens, brush piles, and abandoned buildings. They have a habit of defecating at the den's entrance, where a large pile of brown, crescent-shaped droppings soon builds up. In winter, porcupines take to their dens for protection from snow and wind; several porcupines may use the same den, together or at different times. Porcupines do not hibernate. From November until spring, a porcupine spends its days sleeping in the den or feeding in a tree, usually a conifer, whose needles blunt the wind. In summer, a porcupine tends to use a large deciduous tree, often an oak, as a daytime rest site.

Thanks to their quill defense, porcupines are not preyed on heavily. Foxes, bobcats, dogs, and owls take some. Coyotes may work in pairs to maneuver a porcupine onto its back, exposing the rodent's quill-free belly. The fisher, a large weasel recently reintroduced into Pennsylvania, frequently preys on porcupines, flipping them over or biting them repeatedly in the head. In Cameron County, I once found a large porcupine whose face had been bitten into a shambles: both of its eyes and the external part of its nose were missing, but the wounds had healed over, and the animal seemed fit and in good health. This was before the fisher had been reintroduced, and I guessed that a coyote or a bobcat had done the damage. I also supposed that if any creature could survive blind in the wild, it would be the stolid porcupine.

Porcupines have a ten- to twelve-year life expectancy. Often people kill them, fearing that the animals will destroy trees or to prevent dogs from being quilled by them. Many others are hit by cars.

Habitat. Porcupines live mainly in forests, at times venturing away from tall trees into brushy land. They prosper in areas of mixed conifers and hardwoods. The winter range includes a den, coniferous feeding areas, and the travel lanes linking them, up to around 20 acres; a porcupine may spend several months feeding in only one or two trees. Summer ranges are larger, between 15 and 65 acres, with an average of 45 acres in the preferred habitat of deciduous woods. The summer range may lie a half mile or farther from the winter range.

Population. In Pennsylvania, porcupines thrive in areas of extensive forests: the rugged mountains of the north-central region, forested land in the northeast and northwest, and wooded sections of the Ridge and Valley region. *Erethizon dorsatum* seems to be holding its own in a time of generally decreasing wildlife populations.

Costello, D. F. *The World of the Porcupine.* Philadelphia: J. B. Lippincott, 1966.
Roze, U. *The North American Porcupine.* Washington, DC: Smithsonian Institution Press, 1989.

COYOTE

On a summer night in 1995, while tenting along Rock Run in Centre County, I woke to a haunting sound: a string of shrill yips followed by drawn-out, mournful wails. I woke my son, and we lay in our sleeping bags listening as the wild chorus echoed through the wooded valley. I had often heard coyotes howling in the western states, but this was the first I'd heard them in Pennsylvania.

The coyote is not listed as a resident species in *Mammals of Pennsylvania,* by J. Kenneth Doutt, published in 1966, although the first coyotes had entered our state by then. It is thought that *Canis latrans* moved into northern and eastern Pennsylvania from New York's Catskill Mountains in the 1960s; from there, coyotes spread south and west across the state, perhaps augmented by individuals migrating east from Ohio. The subspecies found in Pennsylvania and throughout the Northeast is *Canis latrans latrans*, the eastern coyote. The species as a whole inhabits America from Nova Scotia to Florida and from Alaska south to Nicaragua. The name *coyote* (pronounced "ki-O-tee") comes from the Aztec Indian word *coyotl*; coyotes are also called brush wolves. *Canis latrans* probably originated in the plains and deserts of North America, from where it has expanded its range, filling niches left vacant when humans exterminated larger and less adaptable predators such as mountain lions and gray wolves.

Biology. DNA studies show that the coyote interbred with the gray wolf in Canada during its eastward expansion. This hybridization accounts for the larger size of the eastern coyote, compared to its western counterpart. In Pennsylvania, adult coyotes are 48 to 60 inches long, including a 12- to 16-inch tail. Weights range from 35 to more than 60 pounds, with males larger and heavier than females.

Coyotes look like slim German shepherds, with pointed, erect ears and a long, slender nose. A coyote carries its bushy tail held low, near or between its back legs. The fur is coarse, dense, and long; the basically tan coat is sprinkled with rusty brown, black, and gray. Some coyotes are pale in color; others are dark. In most, a dark stripe runs down the back, and dark fur marks the front of each foreleg. The tip of the tail is black.

In the Northeast, coyotes live singly, in pairs, or in packs of three to eight. The usual grouping consists of two adults and some of their offspring, subadults six to eighteen months of age who have not yet dispersed to find territories of their own. Adult males and females pair in a monogamous union for one to several years, sometimes for life.

Coyotes spend their days resting in brushy areas and mountain laurel thickets. They are crepuscular and nocturnal, and their activity peaks at dawn and dusk, although in areas where few humans live, they may become more active during the day. Extremely alert and cautious, coyotes are informed by their keen senses of smell, eyesight, and hearing. They swim capably, run at 25 to 35 miles per hour, and jump with great agility. They use a variety of yips, barks, and howls to communicate with each other, and they mark out territories by sprinkling their strongly scented urine on bushes and stumps and by leaving feces in prominent places. An individual's home range may encompass 1 to 10 square miles. Coyotes in packs defend home territories; lone coyotes and pairs probably do not defend a home range.

Over 90 percent of a coyote's diet is flesh, animals caught and killed or found as carrion: small rodents, rabbits, snowshoe hares, muskrats, woodchucks, deer, domestic dogs and cats, livestock, birds, snakes, frogs, turtles, fish, crayfish, and insects. When coyotes hunt small animals, they creep up quietly, pause while determining the prey's exact location, and pounce. Coyotes kill deer a lot less frequently than they take small rodents. On larger prey, two or more coyotes may chase an animal in relays and finally kill it by biting the neck or throat, severing the jugular vein or causing suffocation. They bury any surplus food and mark the cache with urine. Coyotes wade into marshes to take waterfowl, and they eat fruits and berries in season.

A study of three hundred coyote scats in Pennsylvania found deer to be the dominant food, showing up in 57 percent of droppings. While coyotes may have trouble killing adult deer, they more easily prey on fawns, and they find road-killed deer in abundance. The Pennsylvania study uncovered no evidence of predation on livestock, but coyotes do occasionally kill sheep, goats, and poultry. The droppings included remains of many small mammals, particularly mice and voles, followed in frequency by rabbits and woodchucks. Insect remains showed up in 18 percent of scats and birds in 10 percent; half of the droppings contained plant matter.

Female coyotes have one heat, or estrus, per year. In Pennsylvania, this usually occurs in February. The gestation period is fifty-eight to sixty-three

The eastern coyote is found throughout Pennsylvania and the Northeast.

days. In April or May, the female seeks out a natal den, often on a brushy south-facing slope; she may enlarge a woodchuck, skunk, or fox burrow or choose a hollow log, a hole under a stump or a fallen tree, or a crevice among rocks. Dens are a foot in diameter and up to 20 feet long. The female bears four to eight pups; the average is six. The pups' eyes are closed, and they weigh about 9 ounces and are covered with woolly gray-brown fur. The male—and sometimes other members of the pack—bring food to the nursing female. The pups' eyes open after two weeks. In another week, they begin venturing out of the den. Both parents feed them, finding prey away from the den, chewing and swallowing the meat, and regurgitating it for the young. The pups are weaned at nine weeks. The family leaves the den, and the parents teach their offspring to hunt.

Young coyotes begin to leave the family group in early fall, when they are around six months old. Juveniles disperse 30 to 50 miles, with males traveling farther than females; some go as far as 100 miles. They achieve full size and weight by around nine months. Normally females do not breed until their second winter. Maximum life span in the wild is ten to twelve years; in captivity, coyotes have lived eighteen years.

I am told by other Pennsylvanians that more and more frequently they hear coyotes howling, both at night and during the day. Not long ago, I heard coyotes yipping within a few hundred yards of my house; a siren on the highway a few miles away apparently set them off. One can go looking for coyote tracks in the snow. A coyote's paw measures about 2.75 inches long and 2.5 inches wide. Coyote tracks are similar to those of dogs, but dogs' tracks are usually offset on each side, and coyotes lay their tracks in a straight line.

Habitat. Prime habitat is brushy land with high numbers of rodents, rabbits, and deer. Coyotes also prosper along the edges where wooded areas meet agricultural land. They inhabit deep woods, marshes, and farmland. They have colonized the margins of many cities, including Philadelphia, Pittsburgh, Harrisburg, and Erie.

Population. The coyote population in Pennsylvania has grown rapidly since *Canis latrans* expanded its range into the state in the 1960s. An estimated seven thousand to eight thousand coyotes are killed by hunters and trappers in Pennsylvania each year, perhaps a quarter of the population. Coyotes are found throughout the state, with the largest concentrations in the northern and northeastern counties. Some biologists believe that high concentrations of coyotes suppress red fox populations.

Elsewhere in America, particularly in the West, coyotes are controlled through poisoning, trapping, shooting from aircraft, and destroying the young in the den. Where coyotes are persecuted heavily, they respond by producing larger litters.

Bekoff, M., and M. C. Wells. "Social ecology and behavior of coyotes." *Scientific American* 242 (1980): 130-48.

Boer, A. H., ed. *Ecology and Management of the Eastern Coyote.* Fredericton, NB: University of New Brunswick, 1992.

Van Wormer, J. *The World of the Coyote.* Philadelphia: J. B. Lippincott, 1964.

FOXES

R ed and gray foxes are small, alert carnivores that belong to family *Canidae,* along with the dogs, coyotes, and wolves. Both red and gray foxes are found throughout Pennsylvania; in some areas their ranges overlap, and in other places they occupy different habitats. The red fox is most at home in farmland interspersed with brushy and wooded areas. The gray fox does not favor agricultural land as strongly and thrives in deciduous forest. Both species are intelligent and wary, with sharp senses of sight, smell, and hearing. Both eat plant as well as animal food. Much more is known about the ecology and habits of the red fox than about the gray fox.

The red fox inhabits the East, the South, the Midwest, and parts of the U.S. West, Canada, and Alaska. The gray fox is the more southerly species, ranging from southern Canada and New England to Florida, and from the Upper Midwest, Colorado, Utah, and Oregon south through Central America to northwestern South America.

Red Fox *(Vulpes vulpes).* For many years, the red fox of North America was considered to be a separate species, *Vulpes fulva,* from the Old World red fox, *Vulpes vulpes;* however, recent genetic studies have shown them to be one and the same. Some biologists question whether the red fox is native to North America. In Pennsylvania, archaeological excavations of Indian villages have uncovered remains of the gray fox but not of the red. It seems possible that today's red fox descends from European foxes introduced in the 1700s by colonists wanting to re-create the traditional horseback hunts of England.

The red fox is 35 to 41 inches long, including a 12- to 16-inch tail. Most individuals weigh from 8 to 12 pounds, and an unusually large male may tip the scales at 17. Foxes look heavier than they really are, an impres-

The red fox eats a wide variety of foods, including
small mammals, insects, berries, and fruits.

sion created by their thick fur. A red fox has long, reddish orange fur slightly darker on the back than on the sides; black ears, legs, and feet; and a long, bushy, white-tipped tail sometimes called a brush. Dramatic color variations occur, mainly in the species' northern range in Canada. These include the cross fox, which has a dark stripe extending from the head down the back, transected by another dark stripe over the shoulders, forming a cross; the black fox, a melanistic form; and the silver fox, black with white-tipped guard hairs giving a frosted appearance. A red fox always has a white tip to its tail, no matter what the color phase or the shade of red fur, which also varies among individuals.

Opportunistic feeders, red foxes take whatever comes their way: mice, voles, squirrels, rabbits, muskrats, woodchucks, birds and their eggs, insects (including crickets, grasshoppers, beetles, moths, and flies), earthworms, snakes, turtles and their eggs, frogs, domestic cats, chickens, and carrion. A fox will pad along quietly, detect prey by smelling it or hearing it, sneak closer, and make an arching jump, pinning the creature to the ground with the forefeet and dispatching it with a quick bite. On larger prey, the fox approaches as closely as it can, attacks with a quick dash, and chases after the animal, trying to snag it by a leg or the rump. The fox drags its victim down and kills it with a bite to the neck or head. Foxes often kill more than they can eat and then cache the excess by burying it in loose earth. During autumn, foxes supplement their diets with berries and fruits, including apples, grapes, and wild cherries. A red fox can live on approximately 1 pound of food per day.

Red foxes seem to be less shy of people than gray foxes are, and red foxes often inhabit populated areas. Still, they are rarely seen because they are active mainly at night. They breed in January and February. Males (dogs) locate females (vixens) through barking and by smelling their musky urine. Vixens have a short estrous period of two to four days. Gestation takes fifty-one to fifty-six days. The one to ten (usually four or five) young are born in late March or early April. Vixens give birth in dens, which are usually remodeled woodchuck burrows or hollow logs. They also nest beneath abandoned farm buildings. Sometimes they dig their own dens.

For a few days before and after the birth, the male brings food to the vixen in the den. Fox pups weigh 3 to 4 ounces at birth. By five weeks of age, the pups are frequenting the den entrance, and by early May, they are spending much time above the ground, playing with bones, feathers, skins, and bird wings strewn about near the den. Both parents watch over the pups and bring food to them. The young are weaned at about eight weeks. By the middle of June, the young foxes, now called kits, are two-thirds the size of the adults; they start to follow their parents on hunting forays. At three to four months, the young begin hunting on their own, and in autumn they disperse.

Males leave earlier and go farther than females, usually traveling more than 15 miles; if food is abundant, females may stay in the area of their birth. The following year they may help their parents raise the next litter by bringing food to the young. Both males and females mature sexually at ten months and can breed during their first winter. Many, although not all, red foxes are monogamous and remain paired for life. Home territories can cover 150 to 1,500 acres.

Red foxes seldom shelter in holes or dens during the winter, preferring to sleep in the open with their tails curled over their noses and feet to keep warm. I once spotted a sleeping fox before it saw me; the creature was lying in 6 inches of snow on a slight knoll next to a cornfield. It lifted its head about every thirty seconds, looked around, then resumed its catnapping.

Red foxes are subject to several diseases, including rabies, distemper, tularemia, and mange. They are hunted and trapped by humans, but because fur prices are low at present, hunting and trapping have little effect on the population. Red foxes live three to seven years in the wild.

Gray Fox *(Urocyon cinereoargenteus)*. The gray fox measures 35 to 41 inches in length, including a 12- to 15-inch tail, and weighs 7 to 13 pounds. The gray fox has a blunter muzzle and shorter legs than the red fox does. The claws on the forefeet are curved, an adaptation for climbing trees. The coarse, dense fur is a grizzled gray; rusty brown or buff coloration shows on the neck, chest, sides of the belly, and undersurface of the tail. The cheeks, throat, belly, and insides of the ears are off-white. The gray fox has a prominent dark spot on each side of the muzzle, narrow black rims around the eyes, and a black tail tip.

The gray fox lives in deciduous or mixed woods, swampy lands, and rugged, mountainous, and often rocky terrain with abundant brushy cover. Like the red fox, the gray fox is an adaptable predator and may live in areas not considered to be prime habitat. Gray foxes eat many of the same foods that red foxes do. They prey frequently on cottontail rabbits, mice, and voles; they also take shrews, birds and their eggs, and reptiles. They eat insects, including many grasshoppers and crickets. Important vegetable foods are wild fruits, nuts, corn, and grasses.

The gray fox is most active at dusk, night, and dawn. An individual can run at 20 miles per hour. A gray fox climbs a tree by scrabbling up to a limb, then jumping from branch to branch. Gray foxes climb to catch prey,

Curved claws on it forefeet help the gray fox climb trees.
A favorite daytime resting place is a broad limb.

feed on grapes, and escape from dogs. Often they spend the daylight hours resting in a tree crotch or on a broad limb. Although they are more secretive and warier of humans than red foxes, gray foxes are considered to be less cunning and easier to trap. They are more aggressive than red foxes, and where the two species overlap, the gray fox is the dominant animal.

Gray foxes probably mate for life. They breed from mid-January to May, with a peak in March. Gestation averages fifty-three days. The single yearly litter has two to seven pups, usually three to five. Gray foxes give birth in dens in hollow logs or trees (sometimes as high as 50 feet up), or underground in converted woodchuck burrows. Through late summer, gray foxes live in family units consisting of the paired adults and their kits. In autumn, when the young are almost fully grown, they disperse as far as 50 miles from the den.

Home ranges vary in size from 200 acres to 1.2 square miles. Individuals' home ranges may overlap. An area of prime habitat may support three to ten gray foxes per square mile. *Urocyon cinereoargenteus* has few natural enemies; coyotes may take some, and the young are vulnerable to hawks and owls. In Vermont, one researcher found the remains of a gray fox in the stomach of a large bobcat. Gray foxes live five to six years in the wild, although many perish at a younger age. In captivity, they may live for fifteen years.

Like the red fox, the gray fox has expanded its range during the last century, moving into areas it had not previously occupied (or from which it had been wiped out), including parts of New England and Michigan.

Henry, J. D. *Red Fox, the Catlike Canine.* Washington, DC: Smithsonian Institution Press, 1986.
Rue, L. L. III. *The World of the Red Fox.* Philadelphia: J. B. Lippincott, 1969.

BLACK BEAR

The black bear is the largest carnivore in Pennsylvania. Its scientific name, *Ursus americanus,* means "bear of America," and North America is the sole place where the species is found. The black bear inhabits all or parts of forty of the lower forty-eight states; Alaska; much of Canada; and the mountains of northern Mexico. Forests are the prime habitat requirement: the black bear is absent from the treeless central plains and the northern tundra. At one time, the biologists believed that black bears needed large wilderness tracts in which to forage and reproduce. Pennsylvania has little true wilderness, and yet bears in the state thrive in both remote wooded areas and in mixed forested and farming country.

Biology. Adult black bears are 50 to 70 inches in length, including a stubby 3- to 5-inch tail. They stand about 30 inches at the shoulder and weigh 110 to 400 pounds, with some extremely large individuals weighing 650 pounds. Males, or boars, are larger and heavier than females, or sows. A bear has long, thick, soft fur. Most Pennsylvania bears are black to brownish black, although a few are cinnamon colored. The muzzle is tinged with tan; often a bear will have a small white mark on its chest and in some cases a prominent V. The sexes are colored alike.

Bears walk with a shuffling, flat-footed gait. They are agile and fast, with a top speed of 30 miles per hour. They swim capably: I once saw a bear cross the Allegheny River, paddling like a dog, with its head tilted up and out of the water; on the far bank, it got out, streaming water and looking surprisingly skinny. A bear has five toes on each foot, with curving claws that help the animal climb trees. Along Wallace Run in Centre County, my wife and I happened onto a young bear, scaring it up a tree. It climbed the trunk in a spectacular, rushing bound: back legs pumping, claws hooking, front legs hugging, breath whooshing—in about four seconds, 20 feet up.

Black bears have adequate eyesight, and their senses of smell and hearing are acute. They stand erect on their hind feet to see around them and to sniff the air. Although normally silent, sometimes they growl or *woof* at other bears or, when provoked, at humans. They bellow, bawl, or make a sobbing sound when hurt. Sows communicate with their cubs by using grunts, huffs, and mumbles. Bears are mainly nocturnal, although they sometimes feed and travel by day. Alert and wary, they tend to avoid open areas. Individuals are solitary. Although most bears will run from a human, a female with cubs should be avoided: on rare occasions, in Pennsylvania and elsewhere, sows have attacked people who have gotten too close to their cubs. The more accustomed to humans bears become—as in a park or a place where people put out food for them—the less likely they are to flee and the greater their potential danger. In wilderness areas such as northern Canada, where black bears have never seen people, bears sometimes treat humans as prey.

Bears find food mainly by scent. Opportunistic feeders, they have a largely vegetarian diet that includes skunk cabbage, wild fruits (blueberries, blackberries, and grapes are favorites), mast (acorns, hickory nuts, beechnuts), the succulent leaves of hardwoods, the inner bark of trees, the roots of many plants, grasses, corn, and apples. They eat insects (adults, eggs, and larvae), reptiles and amphibians, mammals as small as mice and as large as beavers, fish, carrion, and garbage.

An obese bear is a healthy bear: it won't starve during winter, and if a female, the more cubs it will produce and the stronger its young will be. An occasional bear runs afoul of humans by preying on pigs, goats, or sheep; by raiding campers' food stores; and by breaking into honeybee colonies to eat the honey and the protein-rich larvae. Bears drink water frequently, and in hot weather, they wallow in streams.

In autumn, bears eat heavily to fatten themselves for winter, accumulating fat up to 4 inches thick. They begin denning in November and December. Females den up before males, and pregnant females den earlier than females who are not pregnant. The winter den may be a hollow tree or log; a hole dug beneath tree roots; a rock crevice or cave; or a nest on top of the ground, under fallen trees, in a pile of brush, on a hummock in a swamp, or in a drainage culvert. Some bears line their dens with bark, grasses, leaves, or bitten-off blueberry stems. Females often select more sheltered sites than males. Males den alone, as do pregnant females, who give birth in the den. Females with year-old cubs den with their young.

For about seven months, bears lapse into a deep sleep, without eating, drinking, defecating, or urinating. There has been a certain amount of debate among scientists as to whether this deep sleep actually constitutes hibernation. An individual's body temperature falls from around 100 degrees Fahrenheit to 94, and the heart slows from forty beats per minute to eight beats. Bears live off their fat and may lose a quarter to a third of their weight over winter. If disturbed, they waken readily. Bears in poor physical condition den for shorter periods than do those in better (fatter) shape.

In some areas, bears wear down trails with their traveling. They often use their claws or teeth to scar prominent trees: these "bear trees" mark an individual's territory or signal availability during mating season. Other signs of bears include their large paw prints (3.5 inches wide by 7 inches long), logs and stumps overturned when searching for insects or amphibians, bushes broken down for their fruit, and large droppings, discrete or in loose piles, loaded with seeds, insect parts, animal hair, and nutshells.

Bears mate from June to mid-July. The act of copulation causes the sow to ovulate. The fertilized eggs remain dormant in the uterus until late fall; if, by that time, the sow has achieved an adequate hibernation weight, the eggs implant in the uterine wall and develop over the next ten weeks.

The female gives birth to one to five cubs in her winter den, from late December through early February. Newborns are covered with fine dark hair through which their pink skin shows. They are 6 to 9 inches long and weigh 6 to 12 ounces. Blind, the cubs crawl toward the warmth of their

Black bear.

mother's mammaries. After four weeks, the cubs' eyes open. They begin to walk at about sixty days. They leave the den with the female when they are three months old, become weaned by seven months, and by the fall may weigh 60 to 100 pounds. Groups of bears seen in the autumn are usually females and their cubs.

Cubs are playful, romping in water and wrestling with their littermates. The female protects them, sending them up trees if danger threatens. (Adult males have been known to kill and eat cubs.) The family group disbands during the second summer, when the female is again ready to breed. A sow raises a litter every two years. In Pennsylvania, most sows breed for the first time when they are three and a half years old.

The size of a bear's range varies with the animal's gender, the time of year, and the quality of the habitat. In Pennsylvania, males range over territories 10 to 15 miles in diameter. Females' home ranges are 3 to 5 miles across. The population shuffles around during the summer, as mature males search for mates, and males a year and a half old leave their mothers' home ranges and look for vacant territories in which to settle. A young female may inherit a part of her mother's range.

Habitat. Black bears live mainly in wooded country. In spring and summer, they frequent openings to feed on fresh vegetation and berries; in fall, they occupy brushy regenerating clear-cuts and mountain laurel thickets, and they seek out areas with abundant acorns on which to gorge. In northeastern Pennsylvania, bears favor dense swamps tangled with rhododendron, blueberry, and spruce; they also inhabit mixed hardwood forests, especially where the underbrush is thick. The loss of habitats for bears is a problem in Pennsylvania and in other parts of the East. Although bears show a remarkable ability to live close to humans, their numbers must decline as their habitats dwindle.

Population. Thanks to an abundant food supply, and particularly to large acorn crops, Pennsylvania's bears are among the heaviest, fastest-growing, and most fecund black bears on the continent. Females breed at younger ages in Pennsylvania than in other places; litters average 3 cubs here, compared to 2.4 or fewer cubs in most other states. In Pennsylvania, the population is estimated at 10,000 bears statewide. The highest populations of bears are found in the north-central and northeast regions; less concentrated populations exist elsewhere in the state. The Pennsylvania Game Commission controls and manages the bear population through licensed sport hunting. An average of 1,750 bears were taken annually by hunters during the years between 1990 and 2000.

Van Wormer, J. *The World of the Black Bear.* Philadelphia: J. B. Lippincott, 1966.

RACCOON

The raccoon is a medium-size mammal with the taxonomic name *Procyon lotor. Procyon* means "before dog," implying that the raccoon is less advanced than the dog from an evolutionary standpoint, and *lotor* means "washer," referring to the raccoon's fabled habit of washing or dunking its food in the water before eating it. The common names raccoon and coon are anglicized versions of a Virginia Algonquian word. The raccoon is a New World animal, ranging from southern Canada south through most of the United States and Central America to Panama; the species has been introduced in Europe. Unlike some wild animals, the raccoon readily adapts to an environment changed by humans, and today it is abundant.

Biology. Adults are 28 to 38 inches long, including a 10-inch tail; most weigh 15 to 18 pounds, with very large males weighing 30 pounds or more.

(The record is 62 pounds.) A raccoon's fur is long, soft, and colored a grizzled black-brown. Four to seven black rings, alternating with lighter fur, mark the bushy tail. Broad cheeks, a long, slender muzzle, erect ears, and a black masklike band across the cheeks and eyes give the raccoon an alert appearance. The fur on the feet is light gray, and the soles of the paws are hairless. Raccoons molt in April and in October; the summer fur is thinner and lighter than the winter pelt.

Raccoons live in wetlands, moist forests, farmland, and wooded neighborhoods in towns and cities. Nocturnal creatures, they start foraging around sunset, with most activity peaking before midnight, although many individuals continue feeding until dawn. A raccoon spends the day lying on a tree limb or curled up in the crotch of a tree, resting in a mashed-down leaf nest of a squirrel, or hidden in a hollow log, a woodchuck burrow, or a rock crevice. In marshy areas and fields, raccoons sleep on the ground in thick vegetation.

Raccoons are curious, intelligent, and have good memories. They are agile climbers and strong swimmers. On land, their normal gait is an ambling, flat-footed walk, but when pressed, they can run speedily for short distances. A cornered raccoon defends itself fiercely. A raccoon makes a variety of sounds, including barks, hisses, a wailing tremolo, a *churr-churr* noise often produced while the animal is feeding, and a piercing scream that the creature gives when alarmed or in pain.

Raccoons have excellent senses of hearing, sight, and smell and possess a highly developed sense of touch in their forefeet, letting them catch fish and other small, quick prey. Dextrous digits and long, sharp claws help them anchor slippery food items. Does *Procyon lotor* "wash" its food? According to one expert, only captive raccoons dunk their food, and the behavior has nothing to do with cleanliness or moistening the food. "Dabbling" is the fixed motor pattern a raccoon uses in searching for aquatic prey in the wild, and "washing" food is simply a substitute for a normal natural behavior that has no other outlet in captivity.

Raccoons forage mainly by following the edge of a river or creek or the margin of a lake or pond. They eat a wide variety of vegetable and animal food. They consume wild fruits (cherries, grapes, blackberries, persimmons), apples, melons, corn, nuts (beechnuts, hazelnuts, acorns, walnuts), grain, grasses, leaves, buds, and fungi. Animal foods include earthworms, crickets, grasshoppers, beetles, insect larvae, crayfish, frogs, fish, snakes and snake eggs, turtles, small perching birds and their eggs, duck eggs and ducklings, mice, shrews, young muskrats, squirrels, rabbits, carrion, and garbage.

In spring and early summer, raccoons eat mostly animal matter; in late summer, fall, and winter, they rely more heavily on fleshy fruits and seeds.

By late autumn, raccoons have built up a thick layer of fat. Unlike woodchucks, raccoons are not true hibernators: their temperature and heart rate do not fall markedly. During winter, raccoons den up (in woodchuck burrows, hollow trees, caves, rock crevices, abandoned buildings, rarely in muskrat lodges) and become torpid, sleeping soundly for days or even weeks when temperatures fall to 25 degrees Fahrenheit or lower. If the weather warms to 30 degrees, they leave their dens to forage, even when the ground is snow-covered. From late November to late March, a raccoon will typically lose half of its body weight.

Raccoons breed from late January to mid-March, with a peak in February. Mature males typically mate with several females. The female chooses a natal den, usually in a hollow tree. Following a two-month gestation period, she bears her young in April or May. The single annual litter has three to seven cubs, with an average of four. If for some reason a female doesn't breed in late winter, she will become receptive in the spring and will give birth in summer; small raccoons found in the fall are a result of late breeding.

Cubs weigh 3 ounces at birth, are covered with yellow-gray fur, and have faintly banded tails. After nineteen days, their eyes open, and they are weaned at about sixteen weeks. The female often moves her cubs to a

Raccoon.

ground bed after seven or eight weeks. Soon the cubs begin following their mother on short feeding forays. By the time they are three to four months old, juvenile raccoons are large and independent enough to be on their own, but generally they stay within their mother's home range and continue to sleep with her. In winter, the family may den together in the same tree cavity or in a group of trees close to one another.

Most yearling females breed after leaving the winter den; males of the same age generally do not breed until another year passes. In dispersing from the areas where they were born, young raccoons may shift only a mile or two, or they may travel long distances; records exist of males moving more than 150 miles. The typical home range for an individual might stretch along a watercourse and measure 1 by 3 miles. In an Ohio suburb, ranges were much smaller, around 12 acres per animal. Raccoons exhibit a social hierarchy, and older males and females with young are the dominant animals. Individuals generally do not defend fixed territories against other raccoons, however.

Captive raccoons have lived to age eighteen, but the life expectancy in the wild is around six years. Humans and dogs are the chief predators, and young raccoons are also taken by bobcats, coyotes, fishers, and great horned owls. Many raccoons are killed by cars. Among the diseases afflicting *Procyon lotor* are coccidiosis (caused by a parasitic protozoan), canine distemper (an infectious viral disease), fox encephalitis (an inflammation of the brain caused by a virus), and rabies.

Habitat. Because they are so adaptable, raccoons live in many settings. Prime habitat is moist woodland interspersed with farm fields, not far from marshes, streams, rivers, lakes, and ponds, along whose margins good hunting can be found. Raccoons favor relatively open, mature hardwood stands, where trees provide nuts and fruits and often have cavities for denning. Raccoons are less common in dry woods, and they are rare in dense evergreen forests.

Population. Local populations may fluctuate because of severe weather, food scarcities, development of rural land, hunting and trapping pressures, and disease outbreaks. Researchers have estimated one raccoon per 0.63 acre of excellent habitat and one raccoon per 2 acres of good habitat. Population densities are highest in bottomlands, marshes, and suburban areas.

Rue, L. L., III. *The World of the Raccoon.* Philadelphia: J. B. Lippincott, 1964.

FISHER

Over a century ago, the fisher, *Martes pennanti,* was wiped out in Pennsylvania by extensive logging that destroyed its habitat, combined with unregulated trapping. The species is currently being reintroduced into the Keystone State. In North America today, the fisher inhabits forested parts of Canada's southern tier; New England; northern Minnesota, Wisconsin, and Michigan; and the Pacific Northwest.

A member of family Mustelidae, the fisher is closely related to the weasels, river otter, mink, and skunks. The fisher is sometimes called pekan or wejack, from two Canadian Indian names. It does not rely on fish for its food; the name fisher was probably bestowed by early settlers who noted the creature's resemblance to the European polecat, which is sometimes called fitch, fitchet, or fitchew, stemming from a Dutch term.

Biology. The fisher is about the size of a fox. Adult males are 36 to 48 inches long, including a 12- to 16-inch tail, and weigh 8 to 12 pounds. Females are smaller and lighter than males, measuring 30 to 38 inches and weighing 4.5 to 5.5 pounds—about half the weight of most males. *Martes pennanti* has a long, muscular body with short legs; small, rounded ears; dense fur; and a bushy, tapering tail. The overall color is dark brown, sometimes with a white patch on the throat; tricolored guard hairs on the face and shoulders give those areas a grizzled or a bronzed appearance. The fisher's toes end in semiretractile claws, which help the animal climb trees. A fisher has extremely mobile ankle joints and can turn its hind feet nearly 180 degrees, letting the animal maneuver among branches and descend trees headfirst.

Although they sometimes move about during the day, fishers are most active at night and at dawn and sunset. True woodland denizens, they avoid places that lack overhead cover, choosing routes to keep away from gaps in the forest and running quickly if they must cross open spaces. Fishers swim capably. They den high up in hollow trees and rest in abandoned hawk and owl nests. They leap from one tree to the next, springing as far as 9 feet, and may jump down into the snow from as high as 20 feet above the ground.

Fishers hunt for prey in trees, but they do most of their foraging on the ground. Solitary predators, they travel widely in search of food, slipping into and out of brush piles and darting between logs and overturned stumps. They do not try to run down or stalk prey but rely on a quick, rushing attack. They seize shrews, mice, voles, squirrels, chipmunks, muskrats, rabbits, and snowshoe hares, killing the larger prey animals with a bite to the back of the head. Fishers feed on the carcasses of deer that have died of wounds received during hunting season and on gut piles left by successful deer hunters. They also eat ruffed grouse, small birds, amphibians, insects, fish, nuts, and berries.

One of few carnivores to selectively hunt porcupines, the fisher may travel from one known porcupine den to another. Should a fisher find a porcupine in a tree, the predator will try to force the rodent out to the end of a branch, where it may lose its grip and fall. On the ground, the fisher repeatedly circles the porcupine, trying to bite the animal's face, which is not protected by quills. The porcupine keeps its back to the fisher and may try to shield its face with a tree trunk. If the porcupine backs up toward the fisher, flailing its tail, the fisher will dart around to the front and inflict another bite. The attack may go on for thirty minutes or longer; if successful, the fisher finally turns the dead or dying porcupine onto its back and begins feeding on the unprotected belly, consuming the heart, lungs, and liver, and eventually eating everything but the skin, intestines, large bones, and feet. Should the fisher ingest a few quills, they will be expelled in its feces.

Male fishers have home ranges of about 30 square miles, and females' ranges are approximately 12 square miles. Males move about more than females, especially during the spring breeding season. Male fishers defend their territories against other males, and females defend against other females. Within its range, a fisher will use several dens: in hollow trees and logs, under stumps, in rock crevices, and in brush piles. Fishers remain active year-round, although they wait out severe winter weather in their dens.

Females are sexually mature when one year old; males do not generally breed until they are two years old. *Martes pennanti* has a relatively low reproductive rate, bearing only one litter per year, with an average of two to three cubs. Biologists believe that fishers are polygamous and that the males take no part in rearing the young.

Fishers breed in spring. After fertilization, the blastocyst (an early stage of the embryo) remains in the female's uterus but does not immediately attach to the uterine wall. After ten to eleven months, the blastocyst becomes implanted, and after another 30 to 60 days' gestation, the young are born in March or April. Altogether, the gestation period lasts 327 to 358 days. This process is known as delayed implantation, and it is common among the mustelids, or weasels. About 10 days after giving birth, a mother fisher will temporarily leave her den, find a male, mate, and return to her litter.

*The fisher is a superb climber. It is one of the few
predators to regularly kill porcupines.*

The only fisher maternity dens that have been found were in hollow trees. At birth, fishers are blind, helpless, and partly covered with fine gray hair; they weigh about an ounce and a half. Their eyes open after fifty-three days; they begin to walk in their ninth week and to climb during their tenth week. By ninety-six days they weigh around 28 ounces. They continue to nurse into the fourth month, by which time they are also eating meat brought to them by their mother. The young have learned to kill prey by the time they're four months old, and they disperse during their fifth month.

There are records of fishers living for seven years in the wild. Kits may be taken by black bears, the larger hawks and owls, foxes, bobcats, and coyotes, but the adults are large and fierce enough to have few natural enemies. The fisher is easily caught in traps. Today *Martes pennanti* is a protected species in Pennsylvania.

Habitat. Fishers need unbroken woodland. They thrive in coniferous forests and also do well in habitats where conifers and hardwoods mix. Many acres that were logged years ago have now grown back into mature woods in northern Pennsylvania; in the same region, state agencies manage large forested tracts that should ensure an adequate, secure habitat for reintroduced fishers.

Population. The last indigenous fishers in Pennsylvania were killed in the early 1900s. Between 1994 and 1998, biologists turned loose 189 New York and New Hampshire fishers at five sites in Pennsylvania, from the Allegheny National Forest in the northwest to remote state game lands in Sullivan and Wyoming counties in the northeast. The Pennsylvania Fisher Reintroduction Project was a cooperative venture between Penn State University and the Pennsylvania Game Commission. Biologists hope that the population will expand across Pennsylvania's northern tier. In the East, *Martes pennanti* has also been reintroduced in New York, Massachusetts, and West Virginia, and West Virginia fishers have begun to colonize western Maryland.

Powell, R. A. *The Fisher: Life History, Ecology and Behavior.* Minneapolis: University of Minnesota Press, 1993.

SKUNKS

Two kinds of skunks inhabit Pennsylvania. The striped skunk is common statewide. The eastern spotted skunk is a southern and western species whose range may edge northward into Bedford, Fulton, and Franklin counties in south-central Pennsylvania. The word *skunk* comes from the Algonquian Indian name for the animal, *seganku*. Other names include polecat and woods pussy.

A skunk has a small head, with small ears and eyes and a pointed nose; short legs; and a broad rear end. The bottoms of the feet are hairless, like those of bears and raccoons. As with those two other mammals, skunks walk in a plantigrade manner, on the soles of their feet with their heels touching the ground. The claws of a skunk's forefeet are long and sharp, adapted for digging. Skunks eat a range of plant and animal food.

Skunks are well known for a unique self-defense weapon: pungent musk sprayed from their anal glands. As members of family Mustelidae, skunks are related to the weasels, river otter, fisher, and mink, species that also have prominent musk-producing anal glands but are not able to spray their musk. Most mammals have subtle colors, to blend into their surroundings; skunks, with their eye-catching white-on-black pelage, advertise both their presence and their potent defense. Most smaller creatures would flee when confronted by a predator, but a skunk stands its ground, displaying unconcern or a bold fearlessness.

Alarmed, a skunk may stamp its feet, snarl, or chatter its teeth; it may arch its back, raise its tail, and walk in a stiff-legged manner toward its adversary. When about to spray, a skunk aims its posterior at its enemy. Strong muscles surrounding the anal glands contract, squirting a stream of scent out of two nozzlelike ducts that protrude through the anus. The scent stream is accurate to 12 to 15 feet. A skunk can spray in any direction and can even discharge its musk when hoisted by the tail. Skunk spray looks like skim milk with curds; its active ingredient is mercaptan, a sulphur-alcohol compound. The spray is vile smelling and nauseating and can cause temporary blindness. If your dog comes home reeking of skunk, you'll never forget it—and you can hope your pet won't forget the lesson, either. (Try soaking the unfortunate beast in tomato juice, which reduces the stench.)

Striped Skunk *(Mephitis mephitis)*. Adult striped skunks are about 2 feet long, including a 7- to 11-inch tail. They weigh 3 to 12 pounds, with males averaging 15 percent larger and heavier than females. A white blaze marks the forehead, and often the tail is tipped with white. The fur is long, thick, and soft, black with a white patch running down the back of the head, forking at the shoulders, and continuing toward the tail as two prominent stripes. The amount of white in the coat varies from individual to individual. The fur industry gives the highest grades to skunk pelts having the least amount of white.

Skunks are mainly nocturnal. (A skunk moving around during the day should be strictly avoided: it could be sick with rabies, a deadly disease that shows up fairly frequently in skunks.) By nature, skunks are placid and somewhat sluggish. They move at a deliberate walk, a slow trot, or a clumsy gallop; their top speed is about 10 miles an hour. They can swim but do not enter the water readily. The striped skunk does not climb trees. A skunk's senses of sight, smell, and hearing have been judged poor to fair compared with those of other wild mammals. Skunks have an acute sense of touch in their dexterous forepaws.

In summer, striped skunks feed heavily on insects, both adult and larval forms, including ants, grasshoppers, crickets, moths, and beetles. They dig out bumblebee and yellowjacket colonies and scratch at the entrances of beehives, catching and eating any honeybees that come out. Often they leave evidence of their feeding: small, cone-shaped holes in the soil, pine needles, leaf duff, or lawns, where they've dug out grubs. Other foods include spiders, toads, frogs, snakes, mice, chipmunks, and the eggs of turtles and ground-nesting birds. Skunks steal the eggs of wild ducks, and they raid hen houses. They consume garbage and carrion. They eat berries and fruit (wild grapes, Virginia creeper, cherries, nightshade, and others) and corn. Individuals build up a layer of fat to sustain them over the winter.

Skunks den beneath buildings and stumps, in wood and rock piles, and under overhanging creek banks. A skunk may use a burrow abandoned by a fox or a woodchuck, or it may dig its own hole. The burrow has a central chamber 12 to 15 inches in diameter, about 3 feet underground and connected to the surface by one or more tunnels 5 to 15 feet long. The central chamber is lined with leaves and dry grass. In spring, summer, and early fall, a skunk may shelter in several different burrows; in winter, it tends to use just one den.

Before releasing a stream of pungent musk, a skunk aims its posterior at its enemy.

Skunks do not hibernate, although they stay dormant underground for much of the winter, emerging during mild spells to look for food. In winter, a sleeping skunk falls into a torpid state, its body temperature dropping from 98 to 88 degrees Fahrenheit. Striped skunks lose from one-quarter to one-half of their weight between November and March.

Striped skunks become increasingly active in February, when the snowy ground may be crisscrossed with the tracks of males searching for mates. Mating takes place from mid-February to mid-April, peaking around the middle of March. A male may share a winter den with a female, and the pair may mate in the den. Skunks are not monogamous, and the male does not help the female raise the young.

Following a gestation period of around one month, the female bears four to eight young. Kits are born naked and helpless, weighing less than an ounce. Young skunks have musk in their scent glands at birth and can deliver it after eight days. They nurse for six to eight weeks and then emerge from the den with their mother, following behind her in single file during nightly hunting forays. Family groups usually break up in August or September, although some mothers winter with their offspring. People have found communal dens holding twelve or more skunks, generally females and young. Skunks are able to breed in their second year.

Striped skunks usually stay within a half mile of their dens. Females may have a daily activity area about 700 feet in diameter. Skunks may visit the same feeding area several nights in a row, then switch to a new den to exploit some other food source. Prime habitats include mixed woods and brushland, weedy fields, fencerows, wooded ravines, and rocky outcrops in or near farming areas. In Pennsylvania, the lowest populations occur in the heavily forested mountains. A radio-tracking study in Illinois found home ranges averaging 578 and 701 acres for juvenile females and males, respectively, and 934 and 1,264 acres for adult females and males.

Great horned owls, which lack a well-developed sense of smell and apparently are not bothered by mercaptan, kill many skunks. Dogs, foxes, bobcats, and coyotes take a few, but the potent musk warns off most predators. Skunks die from diseases, including pneumonia, distemper, tularemia, brucellosis, and rabies; in traps; and when struck by cars. Biologists believe they live an average of two years to forty-two months in the wild. Captives have lived for ten years.

The closest I ever came to a skunk was in McKean County, where I was working at planting trees on a state game land. Our efforts apparently disturbed a striped skunk, which we first noticed when she was loping toward us at a distance of about 20 feet, carrying one of her young between her teeth by the scruff of its neck. We left the field in due haste and watched from afar as the mother moved her kits to another den.

Spotted Skunk *(Spilogale putorius).* The spotted skunk is smaller than the striped skunk: 16 to 24 inches in length, including a 7- to 11-inch tail, and weighing 1 to 2 pounds. Its coat is glossy black overlain with four broken white stripes along the neck, back, and sides. In Pennsylvania, the spotted skunk has occasionally been found in dry, rocky, forested mountains of mixed hardwoods and softwoods in the south-central part of the state. In Maryland and farther south, the spotted skunk lives in a range of habitats, including oak and hickory woods with dense tangles of wild grape; weedy cultivated fields; and woodlots. The species tends to avoid wetlands and deep forest.

The spotted skunk, quicker and more agile than the striped skunk, catches more small mammals, including baby rabbits, mice, and voles. Spotted skunks can climb trees. They do not seem to defend territories but move about from one area to another while feeding. When a female bears a litter, she maintains a more permanent den. The spotted skunk has a fifty- to sixty-five-day gestation period and produces two to four young per litter. When menaced, a spotted skunk will stand on its front feet, elevate its rear quarters, and hoist its plumelike tail aloft. Should a predator advance, the skunk will drop to all fours and twist its body into a U shape, rump and face directed toward the enemy and tail held to one side, ready to let loose with its powerful spray.

Verts, B. J. *The Biology of the Striped Skunk.* Urbana, IL: University of Illinois Press, 1967.

WEASELS

Three weasel species inhabit Pennsylvania: the long-tailed weasel, or New York weasel; the short-tailed weasel, also called the ermine, stoat, or Bonaparte's weasel; and the least weasel, sometimes referred to as the mouse weasel. All belong to genus *Mustela,* along with the mink. Other related creatures include the fisher, otter, and skunks.

Weasels are brown above and white or pale yellow on their underparts. They have sharp teeth, excellent senses of sight, hearing, and smell, and lightning-quick reflexes. They are active year-round, mostly at night but also during the day. Consummate hunters, they lope along on their short legs with their heads held low and their hindquarters elevated, pausing to listen and to sniff at the air (often while standing on their hind legs), and insinuating their long, lithe bodies into rodent burrows. They attack in a rush, sinking their claws into their prey, using their strength to wrestle the creature down, and delivering a fatal bite to the back of the skull. Weasels do not suck blood, as is sometimes asserted, but they may lap blood from a fresh kill. To achieve a balanced diet, a weasel eats the organs and the partially digested vegetation in the stomach and bowels of its prey, as well as the muscle tissue.

Weasels eat mice and voles, insects, ground-nesting birds and their eggs and nestlings, and mammals as large as small rabbits, which may weigh more than the weasels themselves. The diminutive least weasel preys on the smallest creatures, the medium-size short-tailed weasel takes somewhat larger prey, and the long-tailed weasel homes in on even larger animals, including smaller weasels. Weasels are important components in many ecosystems; they help control rodents, which, without constant predation, would soon overpopulate an area and damage the habitat. Weasels are themselves preyed on by hawks, owls, bobcats, foxes, coyotes, domestic cats and dogs, and snakes.

With its fast metabolism, a weasel needs a lot of food and must eat frequently. If confronted with several prey animals, a weasel may kill them all—not because the weasel is ruthless or cruel, but because its instincts direct it to do so: excess kills are cached in the den to be eaten later. Weasels drink frequently, and they usually live near water sources.

Weasels make three kinds of vocalizations. They sound a high, trilling note during play, mating, and hunting. They screech when antagonized by another weasel or threatened by a predator. And they squeal when wounded or distressed.

Of the three Pennsylvania species, the short-tailed weasel turns white in winter; the least weasel does not turn white; and less than half of all long-tailed weasels turn white, while the majority remain brown year-round. Weasels that stay brown have a markedly paler coat in winter, however. The white coloration blends with the normally snowy background, making it harder for the weasel to be seen, both by its prey and by larger creatures that kill weasels. Other than the snowshoe hare, the weasels are the only Pennsylvania animals to turn white in winter. Weasels that turn white are called ermines in the fur trade.

Weasels do not hibernate in winter, nor do they sleep for long periods. They hunt beneath the snow in what is termed the subnivean environment, in tunnel systems where rodents are active. A weasel's slender body gives it access to rodents' runways, but there's a downside to the slim shape: in winter, a weasel can't curl up into a ball the way a mouse or a vole can, which means that the weasel has a greater exposed surface area relative to its body mass and thus loses heat more quickly. To stay warm when resting, weasels insulate their nests heavily, using dried leaves and the fur and feathers of their prey.

Except during their breeding season, weasels are solitary, which has the effect of spreading them out over a given habitat. Males are larger than females, and the territory of one male will usually overlap the territories of several females. Members of both sexes use musky secretions from their anal glands to mark out home ranges. Males exclude other males from their homes, and females bar other females. Within its territory, a weasel will use more than one den; the typical den is an abandoned rodent burrow with a central chamber filled with insulation.

Short-tailed and long-tailed weasels exhibit delayed implantation, a breeding adaptation assuring that the young, called pups, are born in early spring, a time when their mother can find ample prey while nursing and when the developing young can hone their own hunting skills on immature creatures that are less wary and easier to kill. In delayed implantation, the female mates soon after she bears her litter in the spring. The developing blastocysts (early stage embryos) move from her ovaries to her uterus, where their development is suspended because they do not attach to the uterine wall. During the following winter, the blastocysts become attached to the uterine wall and resume developing. The least weasel, the smallest of the three species in the Northeast, does not have delayed implantation.

Long-Tailed Weasel *(Mustela frenata)*. The long-tailed weasel ranges from Nova Scotia and southern Canada south through Central America to Peru; it is common across Pennsylvania in open woods, brushy land, fencerows, streamside thickets, and wetlands. *Mustela frenata* is 12 to 17 inches in length, including a 3.2- to 6.3-inch tail. In most individuals, the tail makes up more than a third of the overall body length. Weights vary from 2.5 to 9.3 ounces.

The long-tailed weasel is similar to the short-tailed weasel in proportion, color, and markings, although the long-tail is larger and its tail is longer. The dark brown tail is tipped with black. The long-tailed weasel normally becomes white only in the northern sections of its range. In Pennsylvania, considerably fewer than half of all long-tailed weasels become white, and

An adult long-tailed weasel may be 17 inches from its nose to the tip of its tail.

none do south of a line near the border between Pennsylvania and Maryland. The change back to brown starts in March and takes six to ten weeks.

When hunting, a long-tailed weasel darts from one rodent burrow to another. Long-tailed weasels kill shrews, moles, mice, voles, chipmunks, woodrats, Norway rats, young rabbits and squirrels, small birds, and snakes; they also eat insects and earthworms. They sometimes climb trees when pursuing their prey.

Long-tailed weasels den in shallow burrows below the ground, often under stumps. After a gestation period of nine to ten months (owing to delayed implantation), females bear six to eight young in April or May. Newborns are about 2.5 inches long and weigh 0.11 ounce. The young grow quickly; they become independent in midsummer and are almost fully grown by November. Young females breed when they are three to four months old and bear their first litter the following spring. Males do not mature sexually until they are a year old. Biologists estimate a maximum density of one long-tailed weasel for every 7 acres of good habitat. A male's home range is 25 to 60 acres and includes the territories of more than one female. Males travel an estimated 600 feet on nighttime hunting forays; females travel about 300 feet.

Short-Tailed Weasel *(Mustela erminea).* The short-tailed weasel is about the size of a chipmunk, although much slimmer. It is found across Pennsylvania, except for the southwest. Its range stretches across New England, Canada, Alaska, the Pacific Northwest, and the Rocky Mountains. The short-tailed weasel inhabits agricultural lowlands, woods, meadows, and mountains. Total length is 7.5 to 12 inches, including a short tail (1.6 to 3.2 inches), which generally makes up less than a third of the total body length. Weights range from 1.6 to 3.7 ounces.

A short-tailed weasel's pelt consists of short, soft underfur and long, coarse, glossy guard hairs. The sexes are colored alike, and immatures are similar to adults. The tail has a distinct black tip; in winter, the tail tip and the dark eyes are the only parts of the visible body that do not turn white. A short-tailed weasel molts twice a year, in spring and fall. The molts, which take three to five weeks, are triggered by the amount of light per day.

Short-tailed weasels are good swimmers and sometimes pursue their prey through water. They can climb trees but spend most of their time on the ground. The normal gait is a series of bounds, each covering about 20 inches; a short-tailed weasel can leap up to 6 feet and run 8 miles an hour for short distances. *Mustela erminea* preys mainly on mice, voles, chipmunks, and shrews; it also eats frogs, lizards, small snakes, birds, and insects.

Males do not breed until their second year, but females mate during the year of their birth. They ovulate in June and then on a monthly basis until they are bred. An adult female and her female offspring will be bred by the male whose territory encompasses the den site. (Males' territories usually cover 25 to 100 acres.) During mating, the female remains passive, and the male drags her about by the scruff of the neck. Even very young, immature females may be bred; in some cases, the female's eyes have not opened by the time she is impregnated. After the fertilized eggs develop for about two weeks, they remain dormant in the uterus for the next nine to ten months, with active gestation resuming the following spring.

In mid-April, in an underground nest, females bear four to nine pups. The newborns are blind, naked, and weigh about half an ounce. The pups' eyes open after five weeks, and they begin feeding on prey brought to them by their mother; a favorite foodstuff is earthworms. After six to eight weeks, the youngsters begin playing outside the nest, and they start killing prey at ten to twelve weeks. Females reach adult size after about six months, males after a year.

Least Weasel *(Mustela nivalis).* The least weasel is the world's smallest carnivore. It is found in Europe, in northern Asia, and in North America, where it ranges across Alaska, Canada, and the Upper Midwest; it is absent from New England. In Pennsylvania, *Mustela nivalis* inhabits the Allegheny Plateau and the south-central mountains; it ranges south in the Appalachians to North Carolina and Tennessee. The least weasel lives in deep forests, mixed grasslands, fencerows, and along pond edges. It does not seem to be common in any part of its range.

Males are about 7.5 inches long, and females are about 7 inches, including a very short tail that is rarely longer than an inch. Least weasels weigh 1 to 2 ounces. Coloration is brown above and white below; the chin and toes are white, and the brown tail has no black tip. White winter coats have been reported from Pennsylvania, Ohio, and Indiana, but most least weasels in the eastern United States turn a pale shade of brown during winter.

Least weasels are as aggressive and predatory as the larger weasels and kill their prey in the same manner. They take mice, voles, small birds, insects, earthworms, and small amphibians. Their hearing is sensitive toward high-frequency sounds such as those made by rodents. Least weasels spend most of their time hunting; they require nine or ten meals per day and consume food equaling 40 percent of their body weight.

Delayed implantation does not occur in the least weasel. The gestation period for *Mustela nivalis* is thirty-four to thirty-seven days. A female produces two or more litters per year, each with one to six young. Least weasels breed year-round, with a peak in spring and summer; in Pennsylvania, lactating females with young have been found in October, January, and February, in addition to the spring and summer months. The young are born blind and naked, but they develop rapidly. Hair covers the body by day four; the first set of canine teeth appears at eleven days; the eyes open at twenty-six to thirty days; and the young are weaned and on their own after seven weeks. Juveniles reach an adult length after about eight weeks and achieve an adult weight when they are twelve to fifteen weeks old.

Biologists have estimated males' home ranges at 17 to 37 acres and females' ranges at 2 to 10 acres. Least weasels may have several dens in their home range. They take over nests and burrows of mice, moles, and voles, lining them with fine grass and fur; in winter, the fur lining may be an inch thick and matted like felt. Hawks and owls take many least weasels, as do foxes and the other mammalian predators, including long-tailed weasels. Most least weasels live for less than a year, although a few may survive to age three.

King, C. M. *The Natural History of Weasels and Stoats.* Ithaca, NY: Cornell University Press, 1989.

MINK

The mink, *Mustela vison,* is a medium-size aquatic weasel found through-out much of Canada and the United States, except for the desert Southwest. The species is statewide in Pennsylvania. Although few people are lucky enough to glimpse a mink in the wild, *Mustela vison* is not a par-ticularly rare animal. On wild Lushbaugh Run in Cameron County, I watched a mink hunting through boulders and downed trees, darting into and out of crevices in perfect weasel fashion. I've also found mink tracks in the snow along several central Pennsylvania streams and in the marsh that fringes Black Moshannon Lake in Centre County.

Biology. A mink is almost as large as a house cat, although more slender. It has a long body and neck, short legs, and a short head tapering to a pointed muzzle. The overall length is 18 to 26 inches, including a bushy 8-inch tail; males are about 10 percent longer than females. Adults weigh from 1 to 2.5 pounds. The fur is a dark chocolate brown, with white on the chin and sometimes on the throat. The mink's dense, oily underfur repels water; longer, coarser guard hairs overlaying the underfur give the pelt its sparkling luster.

Minks have excellent senses of sight, hearing, and smell. On land, they travel in a leisurely, arch-backed walk or a bounding lope, at 6 to 8 miles per hour, that they can keep up for miles. They swim and dive with ease; a webbing of stiff hairs between the toes of the hind feet help propel a mink through the water. Minks can dive as deep as 18 feet and swim underwater for 300 feet. They are most active in the evening, at night, and early in the morning, but their near-constant appetite often has them on the move in broad daylight as well.

Minks are solitary except during the breeding season. They den in ground cavities beneath the roots of large streamside trees, in abandoned lodges or bank dens of muskrats, and in areas hollowed out beneath logs and stumps. Minks sometimes dig their own burrows. They line their nests with leaves, grass, and the fur and feathers of their prey. Minks are active year-round, although they may curl up in their dens and sleep for several days during storms and cold snaps in winter. Males in particular may use several dens in their extensive home ranges. Like otters, minks sometimes play, skidding down muddy or snowy banks on their bellies.

Quick, agile, and fierce, minks wrap their bodies around their larger prey and kill with a bite to the back of the skull. They take rabbits, mice, shrews, fish, frogs, crayfish, insects, snakes, turtles, waterfowl (mainly flightless young and molting adults) and other birds, and the eggs of birds and turtles. Minks kill many muskrats, even though an adult muskrat is a tough and formidable foe. A mink is an opportunistic feeder, eating whatever is most easily caught or found; it might avoid a scrap with a healthy adult muskrat if, for instance, crayfish were abundant nearby. At times, minks kill more than they can eat, caching the carcasses and eating them later. In winter, they rely heavily on mammals, and in summer, they turn to fish, frogs, and crayfish, especially when streams are low and aquatic prey are concentrated in shallow pools.

From February to April, minks use a potent scent from their anal glands to attract mates. Males search widely for females, and both sexes make a chuckling call at this time. The male grasps the female by her nape when mating with her. Copulation stimulates the female to release eggs; the eggs are fertilized, develop for a short period, and then become quiescent for nine to forty-six days before the embryos attach to the female's uterine wall and resume their development. This lapse in the reproductive cycle is known as delayed implantation. Females give birth in May, twenty-eight to thirty-two days after the embryos become implanted.

Litters have one to ten young, with four the average. Newborn minks are 3.5 inches long, blind, hairless, and weigh 0.2 ounce. In two weeks,

A mink may have a territory covering 8,500 feet of stream bank.

they are covered with reddish gray hair. They are weaned at five weeks; the female brings them food. By the time they're seven to eight weeks old, the young are hunting with their mother and killing prey. The family breaks up in early fall.

Females' ranges are 20 to 50 acres; males' ranges are larger. An adult male may hunt along 8,500 feet of stream; a female may be active on 5,900 feet. The ranges of males do not overlap, nor do those of females. The population density in good habitat is estimated at one mink per 30 acres. When juveniles disperse, most move less than 3 miles. Minks have few enemies: fishers, foxes, bobcats, coyotes, and great horned owls occasionally take them. *Mustela vison* matures sexually at ten months. Minks have lived up to ten years in captivity; life expectancy in the wild is probably two or three years.

Habitat. Minks live along streams and rivers with scattered downed trees, in cattail marshes, near lakes, in bogs and swamps, in tidal areas, and in wooded bottomlands. When water freezes over, minks spend more time in and near forests, where they may den in rabbit or woodchuck burrows. Minks do best where water is unpolluted, for here exist the greatest concentration and diversity of prey.

Population. Minks are more common than the casual observer might think. They are, however, highly susceptible to mercury and polychlorinated biphenyls (PCBs), pollutants found in many wetlands. Pesticides can accumulate in the bodies of minks, weakening or killing them. A fair number of minks are killed on highways. Over a recent four-year span, Pennsylvania Game Commission biologists estimated that trappers took between eighty-six hundred and fourteen thousand minks per year.

RIVER OTTER

In Pennsylvania, the river otter, *Lutra canadensis,* is found mainly in the northeastern counties, and it has been reintroduced in other parts of the state. The river otter belongs to the mustelid family and is closely related to the mink and the weasels. It ranges through parts of New York and New England, the Upper Midwest, much of Canada and Alaska, the Pacific Northwest, and the South.

Play in animals is considered a measure of intelligence; if so, the river otter must be extremely intelligent, for it is one of the most playful of all wild creatures. Otters slide on ice or snow; shoot down muddy banks into creeks; juggle food, sticks, and stones; and chase and wrestle with each other. The last time I saw otters in the wild was in Minnesota's Boundary Waters wilderness. At the end of a portage, I set down the canoe and watched two otters combine work with play as they looped and twisted gracefully through the water, touched noses, splashed, dived—and came up every few moments with a fish. Nowadays, while canoeing on Pennsylvania's waterways, I'm on the lookout to see something similar.

Biology. An adult otter weighs 12 to 20 pounds and is 35 to 50 inches long, including a 12- to 20-inch tail; height at the shoulders is about 10 inches. Females are slightly smaller than males. An otter's head is broad and flattened, and its eyes protrude slightly. The body is streamlined, muscular, and solidly built. The feet are wide and webbed between the toes. To give purchase on wet rocks and logs, the soles of the hind feet have rough protuberances that work like the studs on snow tires. An otter's tail is long and tapered, thickest where it joins the body and furred along its entire length.

The pelt is a rich dark brown, lighter on the underparts. The throat and chin are grayish, and the nose is black. The two layers of fur—dense, oily, waterproof underfur and longer guard hairs—combine with a subcutaneous layer of fat to insulate the body. In autumn, the normally thick fur grows in even more densely.

An otter's hearing is acute, its eyesight is adequate above water and excellent underwater, and it has a keen sense of smell. A set of long, stiff whiskers serve as feelers when the otter searches for food in murky water. Vocal animals, otters utter low chuckles, shrill chirps, growls, grunts, and hissing barks.

A river otter can swim as fast as 7 miles per hour, dive to 60 feet, travel underwater for a quarter mile, and stay down for four minutes. When an otter dives, valvelike structures seal its ears and nose, and its pulse rate drops, slowing blood and oxygen circulation and making long submersion possible. Underwater, the feet and tail are used mainly for steering, with propulsion coming from an up-and-down body motion, as opposed to the side-to-side flexing of a swimming fish. An otter can tread water, sticking its head high above the surface to look around. On land, otters run at a maximum speed of 18 miles an hour in the bounding gait typical of weasels.

Otters eat mainly fish: minnows, sunfish, suckers, chubs, carp, catfish, trout, and others. They consume crayfish, clams, frogs and toads, tadpoles, salamanders, snails, turtles, small mammals (especially voles), earthworms,

River otter.

snakes, and birds. A study in New York's Adirondack Mountains found fish in 70 percent of otters' stomachs, and only 5 percent of the fish were trout. Otters deposit their scats on prominent streamside rocks and logs that biologists call "haul-outs"; fish scales and crayfish shells predominate in otter droppings. Otters keep tabs on one another by checking for scents left in fresh droppings. Males mark their territories with their scats and, in the mating season, prowl about looking for sign left by receptive females.

Otters den on the banks of lakes, rivers, and streams, on islands, and on hummocks in marshes. They use hollow stumps, voids in logjams, and natural openings among tree roots or in brush piles; sometimes they enlarge the burrows or lodges of woodchucks, beavers, and muskrats. The den entrance may be above or below the water level; the floor is bare or covered with leaves, grasses, and aquatic plants. An otter may use its den for several years.

Males are polygamous, and several may follow a female in estrus and fight among themselves to mate with her. Otters copulate in the water and on land; the female may caterwaul when mating. Otters breed mainly in March and April. As with many other mustelids, *Lutra canadensis* exhibits delayed implantation: after fertilization, the eggs remain dormant in the female's uterus until the following winter, when they attach to the uterine wall and resume developing. Some two months later, from February to April, the female gives birth to one to five pups (usually two or three). Baby otters are born fully furred, with their ears and eyes closed, and weigh 4 to

5 ounces. They open their eyes at five weeks and are weaned at ten to eleven weeks, when they begin foraging with their mother.

Mother otters help their young learn to swim. The mother carries or pushes a pup into the water; she submerges, keeping close as the youngster tries to swim and letting it climb onto her back when it tires. After several sessions, young otters begin to enter the water on their own and eventually play and hunt in the aquatic setting. By autumn, they are nearly adult size. They disperse in the fall or stay with their mother until the following spring, when she is ready to bear another litter. Otters are sexually mature by age two. A few females may breed when they are fifteen months old.

Otters have large home ranges. A study in Idaho found one breeding adult male for each 12 to 18 miles of waterway. In spring, summer, and fall, otters range along the shores of streams and lakes, often crossing overland from one drainage to another. In winter, they travel many miles looking for unfrozen entrances to water so they can hunt for fish. They do not store food in winter and must forage continually. Young otters, both males and females, may move up to 125 miles when seeking territories of their own. Some kits fall prey to coyotes, domestic dogs, foxes, bobcats, and black bears; adult otters are strong and fearless fighters, and other predators leave them alone. Longevity is estimated at eight to ten years in the wild.

Habitat. Otters live in swamps, freshwater marshes, and along rivers, streams, and lakes. In coastal parts of its range, *Lutra canadensis* uses brackish streams and saltwater marshes. For denning, otters require undisturbed streamside areas with ample forested cover. Pollution from strip-mine runoff, industrial wastes, and sewage has rendered many Pennsylvania waterways unfit for aquatic wildlife, otters included. As antipollution laws are enforced and as streams cleanse themselves, otter habitat should expand.

Population. In Pennsylvania, the river otter has been protected from hunting and trapping since 1952. The remaining native population is concentrated in the northeast, in Susquehanna, Wayne, Lackawanna, Pike, Monroe, and Carbon counties. Otters in the Lower Susquehanna drainage enter Pennsylvania from Maryland. Both New York and Maryland have substantially more otters than exist in Pennsylvania; their populations are stable, and trapping is permitted in both states. In New York, the population is centered on the Adirondacks, and a reintroduction project is under way in the western part of the state. Maryland is reintroducing otters in the western panhandle in the Youghiogheny River drainage.

In Pennsylvania, biologists have released 110 river otters in six drainages: Loyalsock Creek, Pine Creek, Kettle Creek (with otters moving south to the West Branch of the Susquehanna), Tionesta River, Allegheny River, and Youghiogheny River. Pennsylvania's otter population numbers at least 500 and is growing.

Chanin, P. *The Natural History of Otters.* New York: Facts on File, 1985.

BOBCAT

The bobcat, *Lynx rufus,* inhabits wild sections of the Northeast and southern Canada south through the United States to Mexico. In the north, the range of the bobcat may be limited by the presence of the closely related Canada lynx *(Lynx canadensis),* a larger cat better adapted to hunting in deep snow. Also called the wildcat or bay lynx, the bobcat is Pennsylvania's only wild feline predator. It lives mainly in the forested mountains but may occupy the fringes of farming country where swamps, forested bottomlands, or wooded hills provide prey and offer secluded denning sites.

Biology. Adults are 28 to 47 inches long (averaging 36 inches), including a stubby 6-inch tail whose "bobbed" appearance gives the cat its name. Bobcats are about twice the size of house cats. They weigh 18 to 24 pounds, with the rare individual as heavy as 35 pounds. Males are larger than females. A bobcat is rangy and muscular; its back legs are longer than its front legs, giving the creature a high-tailed, bobbing gait when it runs.

Most Pennsylvania bobcats are tan to grayish brown with dark spots on the body, dark horizontal bars on the fronts of the forelegs, and three or four dark bands on the tail. Its spotted coat blends a bobcat into the dappled woods background of light and shadow. A bobcat's ears are pointed and tipped with black tufts, and its underparts, lips, and chin are off-white. One dull December day, I was hunting deer near my Centre County home when two bobcats—one small and the other large—passed in file. The cats' coats were the same drab gray of deer in winter, and what struck me was the dazzling visibility of the white fur on the insides and backs of the ears and the undersides of the tails—visual cues no doubt used by bobcats for locating and communicating with each other.

Bobcats have keen vision and hearing and a decent sense of smell. Agile climbers, they ascend trees to get away from dogs, pursue prey, and rest on branches and in the crotches of limbs. Bobcats do not particularly like the water, but they will swim when it is necessary; in the snow, their tracks show how they will go out of their way to cross a stream on a fallen log. Bobcats are most active during twilight. They move about from three hours before sunset to midnight, and again from just before dawn until about three hours after sunrise.

Quick, strong, and fierce, bobcats are equipped with retractable needle-sharp claws for gripping their prey. They hunt by lying, crouching, or standing still in areas where prey is abundant, and then pouncing when a target wanders near. They also stalk through cover, pausing to look and listen, creeping within 20 or 35 feet, and attacking in a rush. Bobcats take mice, voles, shrews, squirrels, chipmunks, birds (including ruffed grouse and turkeys), rabbits, and snowshoe hares. Less frequently, they kill minks, muskrats, skunks, porcupines, foxes, and house cats. They also eat fish, frogs, insects, and carrion.

Bobcats also kill deer, both fawns and adults. A bobcat is more likely to attack a deer that it spots bedded down on the ground: the cat stalks close, rushes in, grabs the deer by the neck, and bites the throat or the base of the skull. More deer are killed in winter—when other, smaller prey is scarce—than during summer. After killing a deer, a bobcat will eat its fill, cover the carcass with leaves or snow, and return several times to feed.

Bobcats breed in February and March. A male may travel with a female, mating with her several times. Later, the male leaves and plays no role in rearing the young. In April or May, after about sixty-two days of gestation, one to six (usually three) kittens are born. The female gives birth in a den, in a small cave or rock overhang, beneath a stump, or in a hollow log insulated with dry leaves. The kittens are born blind and helpless; they are fully furred, and their coats have discernible spots. They open their eyes after eight or nine days, and they are weaned within two months. Juveniles stay with their mothers into the fall and sometimes into the winter; because of the pronounced difference in their sizes, I'm fairly sure that the two bobcats I observed together when deer hunting were a mother and one of her kittens from the previous spring. Females have one litter per year. Females mature sexually within a year of their birth. A female's first litter ususally yields a single kitten.

Individuals typically move from 2 to 7 miles per night, hunting and traveling along habitual routes. A bobcat has a definite home range, which it marks with feces, urine, scent from its anal glands, and by clawing on prominent trees. A bobcat will have a number of shelter dens scattered

Bobcats do most of their hunting at dawn and dusk.

though its territory, under rock ledges and in brush piles, thickets, and hollow logs. Males' home ranges in summer are around 16 square miles, and females' ranges are less than half that. In winter, individuals may expand their territories. The home ranges of males seem to overlap, but females seldom trespass on other females' territories.

Studies in Pennsylvania state show that some bobcats live fifteen years in the wild, and many die during their first or second winter before they have mastered hunting skills. Other surveys suggest that bobcats live an average of six to eight years. *Lynx rufus* has no major predators; automobiles, starvation, accidents, and disease are the main mortality factors. Foxes, owls, coyotes, and fishers sometimes kill kittens.

Habitat. An ideal habitat is wooded land broken up by brushy thickets, reverting fields, and south-facing rock outcrops. The general range of the bobcat in Pennsylvania follows a broad band from the southwest through the northern tier, an area dominated by wooded uplands. Bobcats also live in remote sections of southeastern and south-central Pennsylvania. Bobcats may be more adaptable than we suppose. On the ridge above my house are a series of rock overhangs where I have found bobcat fur; I believe bobcats breed there, at the top of a steep slope that people rarely climb, less than a quarter mile from a road dotted with houses.

Population. Bobcats live in low densities even in prime habitat. Most of Pennsylvania's forests were logged around the turn of the century, and for years the resulting brushy land supported high populations of small rodents, rabbits, and other prey for bobcats. As second-growth forests matured, the number of prey animals decreased, and the bobcat population fell. Since 1970, the bobcat's population has increased, and *Lynx rufus* has expanded its range. An estimated 3,500 to 8,000 bobcats live in Pennsylvania.

Van Wormer, J. *The World of the Bobcat*. Philadelphia: J. B. Lippincott, 1963.

WHITE-TAILED DEER

The white-tailed deer *(Odocoileus virginianus)* gets its name from the snowy white hair covering the underside of its tail: when the deer runs, it holds its tail erect or flags it back and forth, so that the white sur-face is strikingly visible. The white-tailed deer belongs to the Cervidae family, which in North America includes elk, moose, caribou, and mule deer. Cervids are split-hoofed mammals lacking incisor teeth in the upper jaw; they are ruminants, which means they have four-chambered stomachs and chew a cud.

Odocoileus virginianus ranges from southern Canada south through the United States and Mexico to northern South America; it is absent from northern Canada and parts of the U.S. Southwest. The white-tailed deer is one of the most ubiquitous and well-known wild animals in Pennsylvania, and its large population has a huge effect on other kinds of wildlife and on the natural environment as a whole.

Biology. In Pennsylvania, the average adult male, or buck, weighs 140 pounds and stands about 33 inches at the shoulder. A typical deer is 70 inches from the tip of its nose to the base of its tail; the tail itself adds another 11 inches. Females, or does, are smaller and weigh less than bucks.

Deer weights vary considerably, depending on age, sex, diet, and season of the year. Breeding-age bucks may weigh 30 percent less at the end of the autumnal mating season, or rut, than at the beginning. Does and yearlings lose weight in winter when food is scarce.

Both sexes are colored alike. In adults, the belly, throat, underside of the tail, and the areas around the eyes and nose are white. In summer, a deer's upperparts are a rich reddish brown, and in winter, they turn a dull gray-brown. The summer coat is shed in August and September, and the winter coat in May and June. Summer hairs are short, thin, straight, and wiry; winter hairs are long, thick, hollow, and slightly crinkled. The winter coat provides such effective insulation that snow will lie unmelted on the resting animal's back.

Deer have several scent-producing skin glands. The tarsal glands lie inside each hind leg at the hock joint; they give off a musky scent used by a deer to advertise its presence. The metatarsals, on the outside of each hind leg between the hock and the foot, discharge odors conveying excitement and fear. Interdigital glands between the toes of each foot scent-mark a deer's trail.

Deer can sprint at 35 miles per hour for short distances and can maintain speeds of 25 miles per hour over longer stretches. When fleeing from a person or a predator, a deer will take three or four long, swift strides followed by one or more high bounds. Excellent jumpers, deer can clear obstacles up to 8 feet high or 25 feet wide. They are strong swimmers, and the air-filled hairs of the coat help to buoy them up.

White-tailed deer do not identify motionless objects well, but they are instantly alerted by movement. Their eyes, positioned on the sides of the head, take in a wide field of view. A deer's hearing is very sharp, and its sense of smell is keen: a deer can tell whether a plant is palatable by sniffing it; bucks find does by following their scent trails; and deer can smell predators from a great distance. Although usually silent, deer sometimes bleat, grunt, or whine. When alarmed or suspicious, they make a loud *whiew* by blowing air out of their nostrils and often strike the ground with a forefoot. A mother deer whines to summon her fawn, and the fawn bleats to call its mother.

Male deer have antlers, twin growths of bone projecting from the skull in front of the ears. Each antler consists of a main beam, up to 2 feet long and curving forward, from which two, three, four, or more tines stick up. The antlers begin developing in March or April. They are covered by a layer of skin, the velvet, richly supplied with nutrient-carrying blood ves-

White-tailed deer are social animals that often live in family groups.

sels. At first the antlers are soft and subject to injury (bent and twisted tines result from damage to the growing bone). Later the bone hardens. In August or early September, antler growth stops. The velvet is shed or rubbed off by the buck scraping his now-hardened rack against trees and shrubs. The buck carries his polished antlers throughout the fall breeding season. In winter, when the buck's testosterone level dwindles, a separation layer forms between the antlers and the skull. The antlers then fall off or are knocked off when they strike a branch or some other obstruction.

A buck grows a new rack of antlers each year. Under normal conditions, each new rack will be larger than the preceding year's growth until the deer reaches old age, and then its rack will decline in size each year. The age of a buck cannot be determined by the spread of the antlers or the number of tines: nutrition plays a greater role than age. A year-old buck with plenty of food may sport large antlers, while an older buck, if living in a nutritionally poor setting, may grow narrow single points, or spikes.

Deer are strictly plant eaters. A Pennsylvania study of food items in the rumens (stomachs) of road-killed deer identified ninety-eight different plant species. Deer eat leaves and twigs from a vast assortment of woody plants, including aspen, ash, beech, birch, dogwood, maple, oak, willow, witch hazel, pine, and hemlock. In contrast to rabbits, which snip stems neatly with their paired incisors, deer—because they lack the upper incisors—must grasp and rip off twigs and stems, leaving a ragged end. Deer grub out the corms of ferns, nibble on lichens, strip bark from trees, and consume lily pads and pond plants. Much of their summer diet consists of grasses and the leaves of various plants. They eat garden vegetables, wild mushrooms, fruits such as apples and pears, and crops, including soybeans, corn, and alfalfa. Acorns are a favorite food, and deer consume them in great quantities when putting on fat for winter. A deer will eat 5 to 9 pounds of food daily.

Social animals, deer gather into two basic kinds of groups. The family group consists of an adult doe and her fawns, who stay together for nearly a year; sometimes three or four generations of related does form a single family assemblage. The second type of group is a "bachelor" band of two to four bucks; membership constantly changes until this association breaks up before the autumn rut.

Deer breed from October to January. Bucks establish dominance by sparring with each other or locking antlers and engaging in shoving matches. The strongest bucks with the largest racks do most of the breeding. Using their antlers, bucks thrash shrubs and small trees, leaving "buck rubs," gashes in the bark that advertise their presence. They also paw the ground into bare patches called scrapes, which they check frequently; when a receptive doe finds a scrape, she urinates in it, and the buck seeks her out by sight or scent. The rut peaks in mid- to late November, and most adult females have been bred by the end of December.

After she is mated, a doe remains with her family group until the following spring, when she leaves for a short period to give birth. Does bear their fawns from late May to early June, after approximately two hundred days of gestation. Year-old does may have one fawn, and older does generally drop twins or, rarely, triplets.

Fawns are born fully furred and with their eyes open. The reddish coat is dappled with white spots in a pattern that helps blend the young animal in with forest shadows and sunlit foliage. Fawns weigh 4 to 8 pounds at birth. They nurse almost immediately and can walk within an hour. Does leave their newborns in secluded places and return to nurse them several times a day. The fawns instinctively keep still in their beds. Once, in a patch of mountain laurel, my springer spaniel pounced on a very young fawn and mouthed it without hurting it; the fawn didn't move a muscle during what must have been a terrifying experience. When two weeks old, fawns begin playing with each other. At three to four weeks, they start eating grass and, with their mother, rejoin the family group. They are weaned at around four months. Fawns begin losing their spots in September, when the spotted coat is gradually replaced by the gray-brown winter coat. By early fall, male fawns typically weigh 60 to 80 pounds.

In the past, gray wolves and mountain lions preyed on deer, but both of those predators have been wiped out in Pennsylvania. Today, domestic dogs, coyotes, bobcats, and foxes kill deer—particularly fawns. Many more deer die when struck by automobiles, and others perish in accidents. In the winter, freezing rain can turn snowy mountains into steep, deadly slides for deer. *Odocoileus virginianus* is the most important big-game animal in Pennsylvania, and hunters take 350,000 to 400,000 each year. In winters with deep snow, deer die of starvation, especially in areas where they have severely browsed back the vegetation. Although few survive beyond age four or five, whitetails may live twenty years in the wild.

Except for does that have just dropped fawns, deer are not territorial. They live out their entire lives in the same home range, about 40 acres in good habitat to over 300 acres in marginal habitat. Mature bucks usually have larger home ranges than those of does and younger deer.

Habitat. An ideal habitat is brush-stage forest with a wide variety of tree and plant species. Creatures of the forest edge, deer use thickets interspersed with open, sunny glades and abandoned fields. White-tailed deer are highly adaptable and live in many habitats, including woodlots in farming country, suburbs, and deep woods.

Population. It is hard to believe it today, but when the twentieth century began, deer were rare in Pennsylvania. Vast areas of forested land had been logged off, and deer had been hunted almost into oblivion. In 1895, the Pennsylvania Game Commission was established. The agency bought and released deer from Michigan, Maine, and other states and set hunting seasons and bag limits. Today the state's deer population is estimated at 1.4 million. Other states in the Northeast also have large, burgeoning deer populations.

The deer herd in the Northeast has had a tremendous impact on the habitats of many other wild animals. By severely browsing back tree seedlings, deer open up brushy habitats and suppress forest regrowth. They affect the composition of the forest by selectively browsing certain tree seedlings. Today, many oak woods that are logged off do not come back in oaks, which deer browse heavily, but regenerate as stands of red maple, black birch, and other species that are less attractive to deer—and less commercially valuable to foresters. Severe overbrowsing has caused some timber stands to fail to regenerate, and the only plants that have come up are bracken ferns. Deer also damage forest ground-cover plants, including orchids and many wildflowers.

Because deer no longer have large natural predators, humans must control deer numbers, and hunting—of bucks, does, and juveniles—is the most logical and effective way to keep the population in check. Today in certain parts of Pennsylvania—particularly in the southeast and southwest, in areas bordering farmland—deer populations are so large that the herd is dangerously out of balance with the environment. In other parts of the state, brushy second-growth woods have become mature forest offering less browse to deer; there, populations are much lower.

Halls, L. K., ed. *White-Tailed Deer: Ecology and Management.* Harrisburg, PA: Stackpole, 1984.
Putnam, R. *The Natural History of Deer.* Ithaca, NY: Cornell University Press, 1989.

ELK

Before European settlers arrived in Pennsylvania, the eastern elk *(Cervus elaphus canadensis)* inhabited much of the state. By 1867, this large woodland deer had been extirpated; ultimately it became extinct throughout its range, which had included New York, New England, and southern Canada. Today a small herd of elk lives on a limited range in north-central Pennsylvania. The animals are descendants of Rocky Mountain elk *(Cervus elaphus nelsoni,* another subspecies) released by the Pennsylvania Game Commission between 1916 and 1926.

The word elk comes from a German name for the European moose. The elk is also called wapiti, a Shawnee Indian word meaning "white deer," perhaps referring to the animal's light-colored rump. The elk is the second largest member of the deer family in North America; only the moose is larger. Many western states and several Canadian provinces support thriving elk populations, and in those places, elk are popular big-game animals. Michigan, Wisconsin, Arkansas, Kentucky, and Minnesota have also reintroduced elk.

Biology. Elk are much larger and heavier than white-tailed deer. A mature male elk, or bull, stands 50 to 60 inches at the shoulder and weighs 600 to 1,000 pounds. Females, or cows, weigh 500 to 600 pounds. In summer, an elk's coat is short, thin, and reddish brown in color. In winter, long, coarse guard hairs overlay woolly underfur; the color is tawny brown or brownish gray, with the neck, chest, and legs dark brown. Buffy or whitish fur covers the rump and the 4- to 5-inch tail.

Strong, muscular animals, elk can run at 30 miles an hour. They jump well and swim readily. Their senses of smell and hearing are acute. A cow elk barks and grunts to communicate with her calf, and the calf makes a

A small herd of elk occupy a range of about 200 square miles in northern Pennsylvania.

sharp squealing sound. The best-known elk call is the bull's bugling: a low bellow that ascends to a high note, held until the animal runs out of breath and followed by guttural grunts.

Each year a bull grows large, branching antlers that sweep up and back from the head. Yearlings usually grow spikes 10 to 24 inches long, and older bulls produce racks with main beams 4 to 5 feet in length and with five or six tines to a side. Bulls carry their antlers into late winter or early spring, then shed them.

Elk are primarily grazers, eating a variety of grasses and leafy plants. In winter, they paw through the snow to reach grass; they also eat the twigs, buds, and bark of trees. In Pennsylvania, elk browse on aspen, red maple, fire cherry, oak, striped maple, black cherry, shadbush, witch hazel, and blackberry.

The mating season is in September and October. Bulls bugle invitations to cows and challenges to other bulls. The bulls fight with each other, joining antlers and pushing and shoving. Battles rarely end in serious injury; the weaker bull usually breaks off the confrontation and trots away. In the West, bulls may amass harems of fifteen to twenty cows; in Pennsylvania, four or five cows is the norm. Most harems are controlled by large, mature bulls, although younger males hang around on the fringes of the groups and sometimes get a chance to breed.

In May or June, about eight and a half months after mating, a cow will leave the herd and, in a secluded area, give birth to a single calf, rarely to twins. An elk calf weighs about 30 pounds and can stand up when only twenty minutes old. Within an hour, it starts to nurse, and it begins grazing on vegetation when less than a month old. Calves are dappled with spots, a camouflaging pattern.

In spring and summer, bulls go off by themselves and live alone or in small groups. Cows and calves tend to stay in family units composed of a mature cow, her calf, and immature offspring from the year before. Sometimes several family groups band together; an older cow will lead the group, giving a barking alarm call and guiding the band away from intruders. In hot weather, elk bed down in the shade of dense timber.

The potential life span for an elk is twenty years. Pennsylvania elk die from old age, collisions with cars, disease, and poaching; occasionally a black bear takes a calf.

Habitat. Elk are attracted to clear-cut forests coming up in brush, revegetated strip mines, open stream bottoms, and farms. The major elk range covers 200 square miles, including 80 square miles in state game and state forest lands. Elk frequent agricultural areas near the town of St. Marys and dwell among revegetated strip mines and aspen clear-cuts north of Benezette.

Population. The Game Commission released 177 elk between 1913 and 1926, most of them in northern Pennsylvania. The population dwindled to fewer than 40 in the early 1970s but has increased greatly since that time. Today around 500 elk live in a mountainous, mainly wooded area in Clearfield, Elk, Cameron, Clinton, and Potter counties. Sixty-eight percent of adult cows give birth to calves each year, and about 70 percent of the calves survive. The population has been expanding by about 12 percent each year. In the late 1990s, to relieve crowding in the core elk range, the Pennsylvania Game Commission translocated around 90 animals to the Kettle Creek drainage in western Clinton County. The Game Commission has considered setting up a controlled annual hunt to minimize damage to farmland and to keep the herd in balance with its habitat, with a goal of 1.5 elk per square mile of occupied range.

Murie, O. J. *The Elk of North America*. Harrisburg, PA: Stackpole Books, 1951.
Thomas, J. W., and D. E. Toweill. *Elk of North America: Ecology and Management*. Harrisburg, PA: Stackpole Books, 1982.

BIRDS

In birds, every aspect of the body seems dedicated to the ability to fly. Birds have streamlined shapes and thin, hollow bones. Their skulls weigh little: evolution has whittled away excess ounces by eliminating teeth, heavy jaws, and jaw muscles. A bird's lungs and heart are large relative to its body size. Connected to the lungs is a system of air sacs, which supplement lung capacity and help cool the speedy avian metabolism. Rapid, skillful flight requires sharp eyesight, quick reflexes, and superb coordination, and so the nervous system is complex in birds. The brain of a small perching bird weighs ten times that of a lizard having the same body weight. In birds, the parts of the brain that control vision and muscle coordination are particularly well developed.

Birds' abilities to fly and to survive in dramatically differing environments have let them colonize almost every part of the globe. Like mammals, birds are warm-blooded, able to regulate their own internal temperature so they can remain active when it is hot or cold. Birds evolved from an early line of reptiles, and their feathers are believed to have developed from reptilian scales, the feathers at first functioning as heat-conserving insulation and only later being used for flight.

Birds lay eggs, out of which hatch young that are precocial (able to move about and feed themselves almost immediately) or altricial (requiring several weeks of parental care and feeding before they become independent). Most birds eat insects and plant matter, particularly seeds and fruits; some are predators of other birds and of mammals, amphibians, and reptiles. Birds' bills vary in shape and may be dedicated to exploiting certain foods.

Scientists recognize around 10,000 species of birds worldwide, with new species still being discovered, mainly in tropical forests. About 900 species inhabit North America. Pennsylvania has around 187 breeding species, and many other birds pass through the state when migrating during spring and fall.

As of 1999, eleven species of birds were classified as endangered in Pennsylvania: the American bittern, least bittern, great egret, yellow-crowned night-heron, bald eagle, peregrine falcon, king rail, common tern, black tern, short-eared owl, and loggerhead shrike. Five species were classified as threatened: the osprey, upland sandpiper, yellow-bellied flycatcher, sedge wren, and dickcissel. The passenger pigeon, an extinct species, occurred in Pennsylvania until the early twentieth century. Since European settlement, five species have become extirpated in the state, although they survive in other places: the greater prairie chicken, piping plover, olive-sided flycatcher, Bewick's wren, and Bachman's sparrow.

Brauning, D. W., ed. *Atlas of Breeding Birds in Pennsylvania*. Pittsburgh: University of Pittsburgh Press, 1992.

McWilliams, G. M., and D. W. Brauning. *The Birds of Pennsylvania*. Ithaca, NY: Cornell University Press, 2000.

Santner, S. J., D. W. Brauning, G. Schwalbe, and P. W. Schwalbe. *Annotated List of the Birds of Pennsylvania*. Pennsylvania Biological Survey Contribution Number Four, 1992.

Poole, A. and F. Gill, eds. *The Birds of North America*. (Informational Series). Philadelphia: The Birds of North America, various dates.

Kaufman, K. *Lives of North American Birds*. Boston: Houghton Mifflin, 1996.

Eastman, J. *Birds of Forest, Yard, and Thicket*. Mechanicsburg, PA: Stackpole Books, 1997.

Eastman, J. *Birds of Lake, Pond, and Marsh*. Mechanicsburg, PA: Stackpole Books, 2000.

Stokes, D. W. *A Guide to Bird Behavior*, Vol. II. Boston: Little, Brown, 1983.

Weidensaul, S. *Living on the Wind: Across the Hemisphere with Migratory Birds*. New York: North Point Press, 1999.

HERONS, EGRETS, AND BITTERNS

Have you ever hiked along the edge of a quiet stream and startled a big, long-legged bird that flapped up from the water, leaving only a widening ripple? Chances are good that the bird was a heron.

Herons are wading birds with long, slender legs, long necks, and heavy bills tapering to a point. Their wings are broad and rounded, their tails short. Most herons, especially the larger ones, are graceful in form and movement. They inhabit both freshwater and saltwater areas; in Pennsylvania, they are found on lakes, ponds, rivers, and woods streams, and in bogs, marshes, and swamps. Herons stand at the water's edge or walk slowly through the shallows; they also perch in trees near or over the water. Some sixty-five species of herons live throughout the world. They are most common in the tropics. They are closely related to storks, ibises, spoonbills, and flamingos.

Herons feed on animal life—fish, frogs, snakes, crayfish, insects, other invertebrates, and small rodents—found in the zone of shallow water and shoreline. They swallow their food whole and later regurgitate pellets of indigestible matter. Several adaptations help a heron wade about and catch prey. The most obvious are its long legs, which elevate the bird above the water. A heron's toes are long and flexible, letting it keep its balance and stand on mucky ground. The attenuated muscular neck delivers a quick blow with plenty of force to penetrate the water and seize a fish. (A heron uses its bill more often to grasp rather than to impale its prey.) Herons have well-developed "powder down," small feathers whose tips continually disintegrate into powder, which the bird's preening helps distribute about its body; the powder absorbs and removes fish slime and pond scum, keeping the plumage clean and dry. Herons preen using a serrated middle claw called a comb toe.

Herons fly with the head and neck drawn back in a compressed S-shape and the legs held straight to the rear. Strong fliers, they propel themselves with deep, steady wing strokes.

During the breeding season, males fight with one another (rarely causing serious injury), sound harsh calls, and go through elaborate courtship movements, such as raising the wings, stretching the neck, or erecting a head crest. Some put on showy flight routines. In many species, bright colors appear on the bill or legs or in the bare skin around the eyes. Often the male begins building a nest to attract a mate; if suitably impressed, the female takes over construction, with the male bringing sticks and twigs to her. Mated herons defend a zone immediately around their nest against the intrusion of other birds. Some species nest in colonies, called rookeries or heronries, while others are solitary nesters. Herons may nest in mixed colonies—great egrets, black-crowned night-herons, and yellow-crowned night-herons nesting in the same grove of trees, for instance. In parts of their range, herons share nesting habitats with cormorants, pelicans, and ibises.

After breeding, the female lays three to six unmarked bluish, greenish, or brownish eggs in a nest of sticks in a tree (the herons and egrets) or of grasses on the ground (the bitterns). Both parents help incubate the clutch for two and a half to four weeks, depending on the species. Some species begin incubating after the first egg is laid, so that the young hatch at intervals and differ in size at any given time. Newly hatched herons are sparsely feathered and incapable of feeding themselves; they remain in the nest for several months. At first, their parents regurgitate predigested liquid food to the nestlings. Later they present partially digested food, and finally whole prey. Using its bill, a young heron will seize the base of its parent's bill in a scissors grip and wrestle with it, triggering an impulse in the adult to drop or regurgitate the food.

Where several species of herons share a habitat, specialized feeding patterns may develop. The great blue heron wades in deeper water, looking for larger fish. Great egrets hunt for smaller fish closer to shore. The green heron waits motionless for its prey on a half-submerged log or on the bank. Bitterns snatch frogs and tadpoles among the reeds. On drier ground, cattle egrets forage for insects stirred up by livestock. Black- and yellow-crowned night-herons patrol the shallows in late evening and at night.

Herons may become prey for foxes, mink, raccoons, hawks, and owls, although few predators dare to tackle adults of the larger species. Crows, grackles, and tree-climbing snakes rob unguarded nests. In the late 1800s and early 1900s, people slaughtered herons and egrets for their plumage, which was used to decorate women's hats. Today these birds are protected by state and federal laws. However, they are harmed by loss of habitat: nesting and feeding areas that are drained for development, covered by waters backed up behind dams, or polluted by human activities.

The following species nest in Pennsylvania: great blue heron, green heron, great egret, black-crowned night-heron, yellow-crowned night-heron, least bittern, and American bittern. Two others are rare breeders in Pennsylvania: The snowy egret, *Egretta thula,* has bred along the lower Delaware and Susquehanna rivers. The cattle egret, *Bubulcus ibis,* is an Old World species that has colonized parts of North America; cattle egrets nested on the Susquehanna in the 1970s and 1980s but have abandoned those areas and are thought not to breed in Pennsylvania at this time. The little blue heron, *Egretta caerulea,* is a more southerly species occasionally spotted in Pennsylvania; it is more common during spring in the western half of the state, and in autumn in the eastern half.

Great Blue Heron *(Ardea herodias).* This is the largest of the North American herons and one of Pennsylvania's largest breeding birds, standing up to 4 feet tall and having a 6-foot wingspread. The head is mostly white, with a black, feathery crest and bright yellow eyes. The underparts are dark gray, and the back and wings are blue-gray. When hunting, a great blue heron stalks slowly through the water or stands in wait with its head hunched back on its shoulders. Great blue herons eat fish (up to a foot in length), snakes, frogs, crayfish, mice and voles (particularly in winter), small birds, and insects. Individuals are solitary or forage in loose flocks. The call consists of three or four hoarse squawks.

Great blue herons nest in colonies with as many as several hundred breeding pairs. Pairs reuse a nest or build a new one, a platform of large sticks anchored in a tree crotch or on a limb and lined with twigs, with an outside diameter of 2 to 4 feet. The male brings nesting material to the female, who does most of the actual building. The female lays three to six (usually four) pale blue eggs. Incubation, shared by both parents, takes twenty-eight days. Both mother and father feed the young, which can fly after about sixty days and leave the nest after sixty-five to ninety days.

In spring, the great blue heron is a common migrant in March and April. In summer, it nests throughout Pennsylvania, often in wooded swamps or remote mountain forests many miles from feeding areas. The great blue heron breeds across the northern United States, southern Canada, and

*The great blue heron stands up to 4 feet tall
and has a 6-foot wingspread.*

southeastern Alaska. In Pennsylvania, the greatest concentration of nests is in the northwest. From July to November, many great blue herons pass through Pennsylvania heading south. The species winters along the Atlantic coast, in the southern states, and in Central and South America. A few individuals winter in Pennsylvania along creeks and other open water.

The most common heron in the Northeast, the great blue heron has increased in numbers over the past twenty years. It is often found on beaver ponds, and the resurgence of the beaver may be partly responsible for the species' proliferation. Logging operations that destroy woodland rookeries are the greatest threat to local populations.

Green Heron *(Butorides virescens)*. This small heron is found in ponds and along shaded riverbanks and wooded streams, often in quiet side channels and backwaters. Its length is 18 inches, its wingspread 24 inches. The bluish green back and wings give the bird its name. The underparts are dark, the neck and head are a reddish brown, and the crown is black. The green heron may appear all dark from a distance, especially in dim light. Immatures are streaked with brown and resemble American bitterns.

A green heron flies with deep wingbeats. Its call is a sharp, descending *kyow,* often given as it flushes ahead of a human interloper. Among its common names are fly-up-the-creek and shipoke—literally "shit bag," describing the bird's habit of squirting out copious white feces when launching into flight. Green herons hunt close to stream banks and feed mainly on minnow-size fish. They also take insects, earthworms, newts, frogs, tadpoles, and snakes. Green herons have been seen dropping objects such as feathers, insects, and twigs into the water to attract fish, making *Butorides virescens* one of the few avian species for which tool using has been documented.

Green herons nest in trees or shrubs overhanging the water and sometimes in orchards and dry woods away from any water source. Most pairs nest by themselves, but some form loose colonies with other herons. The male picks the nest site and starts building, and the female finishes the task. The nest, a platform of twigs and sticks lined with finer material, is about a foot across. Some nests are so shallow and flimsy that the eggs can be seen through the bottom. The parents take turns incubating the three to five pale greenish blue eggs, which hatch after twenty-one days. The young can fly after about three weeks. Their parents continue to feed them for a while longer. Some pairs raise two broods over the summer.

Green herons are common April and May migrants. In summer, they breed in suitable habitat throughout Pennsylvania. (The species breeds across the eastern United States, the Midwest, and parts of the Southwest.) Green herons are common July-to-September migrants, with stragglers until November if ponds and streams stay ice-free. Green herons rarely winter as far north as Pennsylvania. Breeding bird surveys indicate that the population in the Keystone State is currently stable.

Great Egret *(Ardea alba).* The great egret—also called the common or American egret—was nearly extinct by the early twentieth century. For years the birds had been killed for their long, white body plumes, or aigrettes, used to feather women's hats. Strong conservation laws saved the species, which has begun to repopulate its former range. The National Audubon Society has chosen the great egret as its symbol.

A great egret's plumage is pure white, the bill is yellow, and the legs and feet are glossy black. The bird is 38 inches long (not counting the plumes) and has a 54-inch wingspread. Egrets eat fish, small mammals, amphibians, and insects. Egrets inhabit swamps, brushy lake borders, ponds, islands, and mud flats. They nest solitarily or in colonies with other herons, usually 10 to 50 feet up in trees. Built of sticks and twigs, egret nests are 2 feet across and may be lined with leaves, moss, and grass. The three to four eggs are blue or greenish blue. Both sexes incubate the eggs, which hatch after twenty-three to twenty-six days.

Currently, great egrets breed in only one site in Pennsylvania, Wade Island in the lower Susquehanna River near Harrisburg. Egrets also nest farther south along the Atlantic coast and in the Eastern Hemisphere. They migrate through Pennsylvania from July to October. In some years, individuals winter on the John Heinz National Wildlife Refuge at Tinicum, in Delaware and Philadelphia counties. Because of its rarity and the vulnerability of its nesting habitat, *Ardea alba* is listed as an endangered species in Pennsylvania.

Black-Crowned Night-Heron *(Nycticorax nycticorax)*. Night-herons have heavy bodies and short, thick necks. The black-crowned night-heron is 26 inches in length, with a 46-inch wingspread. Adults have glossy greenish black backs, pale undersides, and yellow-orange legs; three 6-inch white plumes extend back from the crown of the head. Immatures are heavily streaked with brown. The call, a single *quok,* is given most often at night.

In flight, black-crowned night-herons resemble slow, light-colored crows. During the day, they sit hunched and motionless in trees, often roosting in groups; they begin to feed at dusk, usually in loose flocks. Some biologists believe that night-herons feed in the dark because they are dominated and driven off by other herons and egrets during the day. Black-crowned night-herons eat mainly fish, some of it as carrion; they also take insects, earthworms, crayfish, and small rodents. *Nycticorax nycticorax* breeds in a variety of habitats: fresh, salt, and brackish water; forests and thickets; and city parks. Pairs nest close together in small to large colonies in trees, shrubs, or on the ground in cattails. Some solitary pairs nest in areas totally removed from colonies. The nest of sticks, twigs, or reeds is sometimes lined with finer material; both sexes build the nest, whose construction takes two to five days. Females lay three to five pale green eggs, which hatch after twenty-four to twenty-six days.

In spring, black-crowned night-herons are uncommon to common migrants in April and May. In the summer, they breed mainly in creek valleys feeding into the lower Susquehanna in Lancaster, Dauphin, Cumberland, and York counties. They migrate south in August and September. Banded birds from eastern North America have been recovered in Mexico, Central America, and the West Indies.

Yellow-Crowned Night-Heron *(Nyctanassa violacea).* Similar in size and body configuration to the much more common black-crowned night-heron, the yellow-crowned has slightly longer legs, a yellow patch on the head, a black and white face, and a gray body. The call, a strident *kwawk,* is pitched higher than that of the black-crowned night-heron. Yellow-crowned night-herons hunt mainly at night but also at times during the day. They eat frogs, fish, salamanders, crayfish, and insects. Most yellow-crowned night-herons nest colonially, placing their stick nests in trees and shrubs. The three to four eggs are a pale bluish green.

Yellow-crowned night-herons migrate through Pennsylvania in April and early May. In summer, they breed in the southeast, mostly in the Susquehanna and Conestoga river valleys, including at the mouth of Conodoguinet Creek at West Fairview in Cumberland County. In the fall, they are rare migrants, headed for wintering territories in the southern United States, Central America, and South America. Pennsylvania is on the northern fringe of the yellow-crowned night-heron's breeding range, and *Nyctanassa violacea* is classified as an endangered species in the state.

American Bittern *(Botaurus lentiginosus).* The American bittern is about 2 feet long, with a 45-inch wingspread. Its plumage is a cryptic mix of dark and light brown, with a black streak on each side of the neck. The bird flies in a slow, deliberate manner, showing distinctive black flight feathers.

Shy and seldom seen, American bitterns inhabit the tall vegetation of freshwater marshes, bogs, and swamps, often among cattails and bulrushes. They are active at dusk and at night, when they prey on frogs, salamanders, snakes, and insects, and, in drier habitats, mice and voles. Individuals hunt by standing motionless and waiting for prey to pass by. The American bittern hides from predators by freezing with its bill pointed upward. Its breeding call, a hollow, booming *oonck-a-tsoonck,* has earned it the name "thunder pumper." Solitary nesters, these bitterns build platforms 10 to 16 inches across, using cattails, reeds, or grasses, on dry ground among tall vegetation. The female lays three to seven eggs (usually four or five) that are buffy to olive in color. Incubation, mainly by the female, lasts twenty-four days.

American bitterns breed in the northern United States and in southern Canada. Habitat loss and perhaps a general decline in the number of amphibians (a favorite prey) have caused the population to fall over the last fifty years. In Pennsylvania, American bitterns are uncommon migrants in April and early May. In summer, they are breeding residents, although many marshes where the species once nested have been destroyed, particularly in southeastern Pennsylvania. American bitterns breed most regularly in large wetlands in Crawford County. Pennsylvania has classified *Botaurus lentiginosus* as an endangered species.

Least Bittern *(Ixobrychus exilis)*. The smallest of our herons, the least bittern is 11 to 14 inches in length, with a 17-inch wingspread. It has large, buffy wing patches; a black crown, tail, and back; and yellow legs. This shy bird hides among tall grasses and sedges. A weak flier, it would rather run from danger or stand motionless with its bill pointed skyward like another reed or stick. The least bittern clambers about in cattails and reeds, clinging to the stems with its long toes. It eats small fish, such as minnows, sunfishes, and perch, and large insects, including dragonflies. It also takes crayfish, salamanders, frogs, tadpoles, and small snakes. The mating call, given by the male, is a low *coo* repeated several times.

Pairs nest on the ground in marshes and bogs. Nests are 6- to 10-inch-wide platforms of dead plant matter interwoven with living vegetation, built in thick cattails, tall grass, or under bushes, 1 to 8 feet from the water's edge. The female lays four or five pale bluish green eggs. During incubation, which lasts seventeen to twenty days, the adults do not fly directly to the nest: they land nearby and approach quietly through the ground cover.

Least bitterns are rare spring migrants in April and May. They are rare breeding residents, with nests reported from various parts of the state, most frequently from the tidewater Delaware Valley and the northwestern counties. In fall, they are rare August-to-September migrants. Most least bitterns winter in Florida, Texas, and Central America. *Ixobrychus exilis* is protected as an endangered species in Pennsylvania.

Eckert, A. W. *The Wading Birds of North America*. Garden City, NY: Doubleday, 1981.
Hancock, J., and H. Elliott. *The Herons of the World*. New York: Harper & Row, 1978.

SWANS AND GEESE

During their spring and fall migrations, these large waterfowl regularly stop on Pennsylvania's lakes, reservoirs, and rivers to rest and feed. The introduced mute swan does not make long-distance migrations; it breeds in Pennsylvania, as does the abundant, well-known Canada goose (treated in a separate chapter). Swans and geese rarely dive. They do most of their feeding by tipping their rear ends into the air and extending their long necks underwater. To take off, they must run along on the surface of the water, beating their wings to build up sufficient speed. At times, they venture onto dry land to feed.

Tundra Swan *(Cygnus columbianus).* Formerly called the whistling swan, this graceful white bird is 48 to 55 inches long, has a wingspread of 85 inches, and weighs 13 to 15 pounds. It breeds in the Arctic and migrates—both during the day and at night, usually in V-shaped formations—to wintering areas in Atlantic coastal estuaries. Adults are pure white with black bills and feet; juveniles have dusky brown plumage and pinkish bills. The voice is a high-pitched, whistling call, *kow-wow,* somewhat like a dog barking and not as harsh as a Canada goose's honking. Wary birds, tundra swans are not apt to let an inquisitive canoeist get too close before taking to the air. When migrating, they feed on pondweeds, wild celery, smartweeds, spike rushes, and wigeon grass; they can reach 3 feet underwater with their long necks. Lancaster County is a key wintering and staging area, where thousands of tundra swans may gather. Tundra swans pick up corn, rye, and soybeans left in fields after the harvest and pull up shoots of winter wheat. On the coastal wintering range, they eat small clams.

Mute Swan *(Cygnus olor).* This is the familiar elegant, white swan of farm ponds, parks, and estates. Adults are 5 feet long and have a 7-foot wingspan. On the water, the mute swan holds its neck in an S shape with its head pointed downward; the bill is orange, with prominent black knobs at the base. Despite its name, the mute swan makes some sounds, mainly grunts and hisses. Mute swans were brought to North America from Europe. A small feral population now breeds in parts of the Northeast and the Great Lakes region; in winter, birds from our area make short migrations to the open waters of Atlantic coastal bays and marshes.

Mute swans eat roots, stems, leaves, and seeds of pondweed, wigeon grass, eelgrass, and other plants, as well as floating algae and duckweeds; an adult may consume up to 8 pounds of food per day. Mute swans feed by ripping out aquatic vegetation by the roots, often destroying the plants. Pairs begin breeding activities as soon as ice melts. They seek out small islands, narrow peninsulas, and clumps of plants on which to nest. Their nests are 5 to 6 feet wide, built of cattails, reeds, and other marsh plants. The female, also called a pen, lays four to seven pale gray to bluish green eggs; each egg is 4.5 inches long and 3 inches wide. Incubation takes about five weeks. Mute swans defend their nests, attacking other swans, dogs, and even people who get too close: blows from the big birds' wings can cause painful injuries. After hatching, the young, called cygnets, stay in the nest for about twenty-four hours, brooded by the pen and watched over by the male, or cob. Very young cygnets may be carried across the water, riding on the backs of the adults. Soon they begin to swim on their own and to feed on insects and crustaceans.

Mute swan populations are increasing. Biologists suspect that these aggressive swans may force other waterfowl into poorer habitats where they do not breed as successfully. Although few predators can threaten an adult mute swan, the cygnets are sometimes taken by snapping turtles, minks, and great horned owls.

Snow Goose *(Chen caerulescens)*. This white goose is a bit smaller than a Canada goose; its wings are tipped black. The call, given in flight and on land, is a high-pitched yipping. Snow geese breed in arctic Canada and winter in various parts of the United States. They migrate in large flocks (from one hundred to more than one thousand individuals) and forage in even greater numbers.

The subspecies seen in Pennsylvania is the greater snow goose; it migrates through the St. Lawrence estuary and overwinters along the Atlantic coast. The population of greater snow geese has increased from around three

Snow geese have learned to feed on farm crops,
and their population has burgeoned in recent years.

thousand in 1900 to more than six hundred thousand at the close of the century, with numbers still rising. Grazing snow geese have damaged some coastal marshes, such as Bombay Hook National Wildlife Refuge in Delaware, as well as staging areas north of the St. Lawrence. An even more serious ecological crisis centers on populations of lesser snow geese that winter along the Mississippi River and the Gulf coast. Those birds have learned to feed on farm crops, particularly rice, and their population has exploded to over five million; they are now denuding their fragile arctic nesting areas, as biologists struggle to find acceptable ways of reducing their numbers.

Brant *(Branta bernicla)*. Brant look like small, dark-headed Canada geese. At 23 to 30 inches in length, they are only slightly larger than a mallard. The call is a throaty *krr-onk*. Brant eat aquatic plants, particularly eelgrass. They are uncommon migrants passing through Pennsylvania from late February to early June and again in October and November. They are seen mostly on Lake Erie; elsewhere in the state, they touch down mainly after storms. They breed north of Hudson Bay and winter along the coast from Connecticut to North Carolina. Migrating brant may cruise at altitudes of several thousand feet; some fly nonstop from southern James Bay to the Atlantic estuaries.

Johnsgard, P. A. *Ducks, Geese, and Swans of the World.* Lincoln, NE: University of Nebraska Press, 1978.

CANADA GOOSE

The Canada goose *(Branta canadensis)* is the large goose commonly seen in suburban and farming areas. Biologists have identified eleven races, or subspecies, in North America, which differ in size and color. The smallest race is the cackling Canada goose, about the size of a mallard (it may ultimately prove to be its own separate species); the largest race is the giant Canada goose, weighing 11 to 13 pounds. Most geese in Pennsylvania belong to the *interior* race *(Branta canadensis interior)*. As a group, Canada geese are known as honkers.

Few wildlife species have proven to be as adaptable to human-caused changes in the environment as the Canada goose. Unregulated market hunting almost wiped out *Branta canadensis* in the 1800s, but aided by wetlands set aside as refuges and by restocking projects conducted by wildlife agencies, the species has come back—to the point that today's large and still-growing goose population is causing problems for humans.

Biology. Canada geese are sturdy, plump birds with long necks, short wings, and short legs. Their legs are set farther forward than those of many other waterfowl, letting them walk easily on land. The feet are webbed between the three front toes. Geese have large amounts of down, fluffy feathers growing close to the body that create an insulating layer, keeping the birds warm in cold weather. Both sexes are colored alike. The bill, head, neck, legs, feet, and tail are black; there is a broad, white "chinstrap" marking on the throat and cheeks; and the upper body is a mottled brownish gray. Adult males, or ganders, of the *interior* race average 3 feet in length and weigh about 9 pounds. Females and immatures are slightly smaller.

Canada geese feed almost exclusively on plants, eating roots, shoots, stems, blades, and seeds. In shallow water, geese tip their bodies forward, dip their heads under, and use their round-tipped bills to pull up aquatic vegetation. They eat wigeon grass, eelgrass, pondweed, spike rush, American bulrush, cordgrass, glasswort, and algae. On land, geese graze on grasses and clover, and they glean grain that is not recovered by farm equipment during the harvest: wheat, millet, corn, barley, and rye. Sometimes they feed on cultivated crops, including soybeans and the shoots of fall-planted wheat. Goslings eat insects, crustaceans, and snails.

Each day around dawn, groups of geese leave the safety of the water where they have spent the night—river, pond, lake, or reservoir—and fly to feeding areas that may be a few hundred yards away to more than 20 miles distant. They feed for two or three hours; at least one member of the group always seems to have its head up, checking for danger. The geese fly back to the water and rest there through the middle of the day, then fly out again and feed for a few hours in the evening.

A goose usually runs along on the surface of the water or ground to gain lift before taking off, although when startled, it can vault into the air almost as directly as the puddle ducks do. Airborne, a goose may look as if it is flying slowly, perhaps because of its deep wingbeats and large size; in fact, a Canada goose can reach speeds of 45 to 60 miles an hour. In flight geese sound a distinctive honking call, *ka-ronk, ka-ronk;* when feeding they make a gabbling sound; and when angry they hiss.

In the spring, Canada geese are among the first waterfowl to breed. When vying for females, unmated males approach each other with their heads lowered and their necks extended, hissing loudly, pecking, and flailing with their powerful wings. Once mated, a pair stays together as long as both geese remain alive; if either dies, the other finds a new mate. Geese nest in places that afford an open view, such as islands in rivers and lakes, the

*While the female Canada goose incubates her eggs,
the gander stays close by to defend his mate and their nest.*

tops of muskrat houses in marshes, rocky cliffs, abandoned osprey and heron nests, and grassy fields near water. The female selects the site and builds the nest while the male stands guard. The nest is a depression in the ground lined with sticks, cattails, reeds, and grasses. The female pads the central cup with down plucked from her breast. Nests are 1.5 to 4 feet wide and 3 to 6 inches deep; the diameter of the central cup is about 10 inches.

The female lays four to ten eggs (usually five or six). The eggs are creamy white and unmarked, becoming stained over time. The gander does not sit on the eggs but stays close by, defending the nest and the surrounding territory, which usually encompasses 5 or more acres of open water close to grazing areas. On a lazy stream feeding a bog in northern Pennsylvania, I canoed past a nest. The goose sat flattened on her eggs, her neck outstretched and her chin resting on the ground; she didn't move a muscle, even though I paddled by less than 5 feet away.

Incubation lasts an average of twenty-eight days. The goslings are precocial: their eyes are open, they are covered with a fine, brown fuzz, and they are able to walk and swim soon after they hatch. They leave the nest from several hours to one day after hatching. Both parents stay with the young, and the female broods them nightly for about a week. Of every one hundred nests, about seventy produce young while thirty fail because of flooding, nest desertion by parents, or predation. Raccoons, opossums, and skunks eat eggs temporarily left unguarded. Foxes, coyotes, hawks, and owls prey on immatures and adults. Snapping turtles and other predators take goslings.

When the young are half grown, their parents begin to molt. After the adults lose their wing feathers, they are flightless for four to six weeks; during that time, the goslings are growing their own flight feathers, so that parents and young are both able to fly at about the same time. As the days shorten, family groups gather and combine into multifamily flocks, leave the breeding grounds, and fly south. Migrating geese travel by day or night, traveling until they tire and then landing to rest and feed. Where food is plentiful and the water doesn't freeze over, geese may not migrate but may winter in the area where they bred.

Geese fly in Vs or in single diagonal lines. Some biologists believe the flights are led by dominant mature ganders. A trailing goose encounters less air resistance and must expend less energy, thanks to turbulence set up by the bird flying just ahead. In overcast weather, geese may fly only a few hundred feet above the ground; under fair skies, they may tower up almost a mile. During the fall migration, most geese fly between 750 and 3,500 feet.

In March and April, geese return in family units to the previous year's breeding sites. A mated pair reestablishes a territory, and the juvenile birds generally leave and travel as far as several hundred miles, often in a northerly heading. Most geese do not breed until their second or third year. On average, Canada geese live five or six years in the wild.

Habitat. Canada geese need expanses of water in which to rest, along with grasslands or farm areas for feeding. On an Illinois wildlife refuge, cornfields attracted 41 percent of all geese observed, small grains drew 24 percent, pasture brought in 22 percent, soybeans 9 percent, and wheat stubble 4 percent. Geese generally will not land close to fencerows, woodlots, houses, and barns. To maximize the efficiency of large, modern farm machinery, many farmers have dug out fencerows, creating the expansive fields that geese favor. As more acres have been put into agricultural production, more potential goose habitat has been created.

Population. By the 1940s, the Canada goose was so rare that it was thought not to breed in Pennsylvania. Federal and state wildlife biologists began trapping wild geese in areas where they remained abundant and translocating them to secure habitats. A system of refuges, combined with strictly controlled hunting seasons, slowly brought the population back.

Today Canada geese probably breed in every county in Pennsylvania. There are essentially two populations of geese in eastern North America: a permanent resident population and a migratory population. The permanent residents, which far outnumber the migratory birds, are the ones causing problems for humans. These geese graze in open suburban settings such as parks, golf courses, cemeteries, and airports, and rest on lakes and reservoirs. Their copious droppings annoy people and pollute water. The birds damage crops, particularly soybeans and winter wheat.

As a migratory species, *Branta canadensis* comes under the jurisdiction of the U.S. Fish and Wildlife Service, which monitors the population and establishes hunting seasons and bag limits. Today hunting is used to reduce the number of local birds, with the dates of hunting seasons set to protect migratory geese. In recent years, hunters have taken over one hundred thousand Canada geese in Pennsylvania. At the end of the twentieth century, an estimated ninety thousand pairs were breeding statewide.

PUDDLE DUCKS

The two major groups of ducks—diving ducks and puddle ducks—differ in several ways. Diving ducks inhabit large lakes and rivers, coastal bays, and inlets. Puddle ducks prefer the puddle-deep shallows of lakes, rivers, and freshwater marshes; they also frequent brackish and saltwater wetlands, especially during migration. Diving ducks are adept at diving and get most of their food by swimming well below the surface. Puddle ducks stay on top of the water or close to it; often they stretch their heads underwater and feed upended with their tails in the air. As a group, the puddle ducks (also called dabbling ducks) are not accomplished divers, but adults dive occasionally and ducklings do so more frequently.

Puddle ducks ride higher in the water than their diving cousins and launch themselves directly into the air when taking flight: they do not need to run across the water to build up speed for takeoff as the heavier-bodied diving ducks must. Puddle ducks are fast and agile fliers. Their wings often display a speculum, an iridescent panel of feathers close to the body on the trailing edge of each wing. The color of the speculum varies from species to species and may serve as a flashing signal to keep flocks together. Puddle ducks feed along the fringes of islands and shorelines, in flooded bottomlands, and on dry land. They eat mainly vegetation, including grasses; the leaves, stems, and seeds of underwater plants; farm crops; and nuts. They also consume a few mollusks, fish, and insects.

In almost every species, the male, or drake, has bright, colorful plumage, while the female, or hen, is drab. In fall, winter, and spring, drakes wear their showy feathers; in early summer, after breeding, they molt into a drab "eclipse plumage" and resemble the hens for several months before another molt restores their bright coloration.

Puddle ducks mate for the first time when a year old. During courtship, drakes chase after hens and engage in posturing, stylized movements, and calling. After mating, the drake either leaves immediately or stays with the hen while she is laying eggs and then departs before the eggs hatch. Pair bonds are weak, and a different mate will be chosen each year. The hen builds a nest of grasses, leaves, and reeds, usually hidden in thick vegetation, and lays a clutch of seven to thirteen eggs, depending on the species. Her dull plumage helps hide her as she incubates the eggs.

Ducklings are precocial. They are covered with soft, fuzzy down; most are a pale brownish color, streaked with darker markings that break up the body outline and conceal them from predators. Minutes after hatching, ducklings can swim and feed themselves. Over the next several weeks, they follow their mother about in the marsh, feeding and resting. The young first fly at about two months of age.

In autumn, puddle ducks fly south. Waterfowl begin migrating through Pennsylvania in late August, including birds that have bred farther to the north; the movement peaks in October and ends in December. Some puddle ducks stay in Pennsylvania during the winter, but most spend the cold months in the southern United States, Central America, and the West Indies.

Raccoons, foxes, coyotes, minks, hawks, and owls prey on ducks. Raccoons, skunks, and crows eat eggs, and snapping turtles and large predatory fish take ducklings.

In addition to the species described in this chapter, the mallard *(Anas platyrhynchos)* and the wood duck *(Aix sponsa)* are dabbling ducks. They are treated separately.

American Black Duck *(Anas rubripes)*. The black duck is 21 to 26 inches in length and weighs about 2.5 pounds. This is our only puddle duck in which the plumage is almost identical for both sexes. The black duck—also called "black mallard" or "red leg"—looks much like a hen mallard, only darker. The body is a dark mottled brown, the underwings are white, and the speculum is violet-blue. When visibility is good, the contrast between the light brown head and the dark brown body is noticeable. The hen voices a loud *quack;* the drake, a lower-pitched *kwek-kwek.*

Black ducks eat eelgrass, wigeon grass, and the seeds of sedges, bulrushes, wild rice, pondweeds, smartweeds, and millet. On land, they consume acorns and waste corn, flying up to 25 miles from watery resting areas to a reliable food source. Animal foods, more important in winter, include clams, mussels, and snails.

The black duck is the third most common breeding duck in Pennsylvania after the mallard and wood duck. *Anas rubripes* breeds most commonly in the Poconos, in the northeastern part of the state. Black ducks nest in brushy and wooded wetlands, on lake and stream margins, and on the edges of beaver

Above, a drake black duck; the hen mallard, below, looks very similar but has a paler overall plumage.

dams. The hen builds her nest in a clump of vegetation, on a stump or a dead snag, and occasionally in a tree cavity. The seven to eleven eggs, creamy white to greenish buff, hatch after about a month. Black ducks are hardy birds that often winter in the state.

Black ducks remain abundant in some areas, but the population in general has declined throughout the species' range over the past fifty years. When wooded land is cleared for agriculture, the more adaptable mallard prospers. The mallard can hybridize with the black duck and may suppress its numbers through genetic swamping: essentially, the mallard has been absorbing the black duck into its population.

Gadwall *(Anas strepera).* Gadwalls are 19 to 23 inches long and weigh 2 pounds. Males in the subtly beautiful breeding plumage have brown heads, gray bodies, and black tails; females are similar, except browner. The legs are yellow. This is the only puddle duck with a white speculum. Gadwalls feed almost exclusively in the water, and they dive more often than other puddle ducks do. They eat leafy parts of pondweed, naiad, wigeon grass, and water milfoil; algae; and seeds of pondweeds, smartweeds, bulrushes, and spike rushes. After breeding, hens seek out dense, dry, weedy cover that will hide the nest from above and on all sides. They lay about ten eggs, which hatch in twenty-six days.

The species breeds mainly in the western United States, Canada, and Alaska, and has been extending its breeding range into the East during the last several decades. In Pennsylvania, gadwalls are uncommon; although they are considered nonbreeding residents, they have nested here in the past and breed in several adjoining states. In spring and fall, gadwalls are sometimes seen with pintails and wigeons; rarely do they congregate in large flocks.

Northern Pintail *(Anas acuta)*. Length, 20 to 29 inches; average weight, around 2 pounds. This slender, trim duck is also known as a sprig. Among the most beautifully marked of our ducks, the male pintail has a chocolate brown head, a white neck and breast, and a gray back and sides. The female is grayish brown. The speculum is a dull brown or bronze; far more noticeable in flight are the long, narrow wings and pointed tail. Pintails are graceful, fast fliers, apt to zigzag down from great heights before leveling off and settling onto the water. The drake's call is a wheezy whistle, and the hen sounds a low *kuk*.

In summer and fall, pintails feed on seeds and vegetative parts of pondweeds and wigeon grass, and on the seeds of bulrushes and smartweeds. Nesting females eat snails, crustaceans, and aquatic insects. Sometimes pintails land in harvested fields to glean waste corn and wheat. The species breeds across Canada and the Pacific Northwest into Alaska, and also in Europe and Asia. The northern pintail seldom breeds in Pennsylvania, although nests have been reported from Crawford County and along the Delaware River. Pintails often nest in dead weedy cover of the past year's growth, which may offer little concealment. The nest is usually within 100 yards of water. The female lays six to ten eggs, which hatch after twenty-one to twenty-five days. A few pintails winter in southeastern Pennsylvania, but most fly to the southern United States and Central America.

Green-Winged Teal *(Anas crecca)*. The smallest of our ducks, the green-winged teal is scarcely larger than a pigeon. Its length is 13 to 16 inches, and it weighs from 0.5 to 1 pound. The male has a dark reddish brown head, a curved green streak over each eye, and a vertical white stripe on each side. The female is primarily brown. The speculum shows green in both sexes.

Green-winged teal fly swiftly, often in small, tight flocks that turn and twist in unison. Drakes whistle and have a chittering call; hens make a faint *quack*. The wings produce a whistling sound in flight.

Green-winged teal prefer small, shallow freshwater ponds near thick cover. They feed on seeds of grasses, bulrushes, and smartweeds, and on stems and leaves of pondweeds. They eat mollusks, snails, other crustaceans, and insects. Invertebrates make up 80 to 90 percent of the ducklings' diet. Most pairs have already mated when they arrive on the breeding grounds. The female hides her nest in a dense patch of shrubs and weeds, or in tall grass at the edge of a lake or slough. She lays six to eleven eggs and incubates them for twenty-one to twenty-three days. Pennsylvania is on the southern edge of the species' breeding range. Some green-winged teal winter here, but most go farther south.

Blue-Winged Teal (*Anas discors*). Length, 14 to 16 inches; average weight, 0.75 to 1 pound. The drake has a brown body and a slate gray head; in front of the eyes is a distinctive white crescent. The hen is primarily brown. Both sexes have a blue patch on the forewing and a green speculum. Blue-winged teal are fairly common waterfowl, found on ponds and marshes, often in the company of other puddle ducks. Their small, compact flocks fly swiftly, low over the marsh, twisting and dodging around trees and emergent shrubs; the birds sound a twittering flight call. Drakes also make a whistling *tseet tseet tseet,* and the hens *quack* softly.

Blue-winged teal forage in very shallow water, eating the seeds and vegetation of aquatic plants, especially pondweeds, wigeon grass, duckweed, and millet. *Anas discors* breeds most commonly in the northern prairies of North America; although uncommon in Pennsylvania, it does breed here, with nests found scattered throughout the state. Hens nest in the drier zones of marshes, in grass on the edges of ponds, and in nearby fields. Ten to thirteen eggs are laid in a basketlike

The male blue-winged teal has a white crescent marking each side of his head.

nest; surrounding vegetation may arch over the nest and conceal it. Incubation lasts around twenty-four days. Blue-winged teal are among the first ducks to migrate south in fall. Many winter in South America after flying long distances over the open ocean.

American Wigeon *(Anas americana)*. Wigeons are 18 to 23 inches long and weigh 1.5 to 2 pounds. The drake has a buffy neck and head with a white stripe from the forehead to the middle of the crown, earning it the name baldpate, and an iridescent green patch coming back from the eye. The body is pinkish brown, the speculum dark green bordered with black. The hen's coloration is similar but duller. A flying wigeon can be identified by its white belly and forewings.

Wigeons fly swiftly in compact flocks, wheeling and turning in unison. Males have a three-syllable whistle with the middle note the loudest; hens utter a low, guttural *kaow kaow.* Wigeons feed on aquatic plants; they also come ashore for grass shoots, seeds, and waste grain. They breed in the northwestern United States, Canada, and Alaska, nesting in dry upland sites grown with grasses, weeds, or low brush. The seven to ten eggs are incubated about twenty-three days. The American wigeon rarely breeds in Pennsylvania.

Northern Shoveler *(Anas clypeata)*. Also called the spoonbill for its long, spatulate bill, the northern shoveler is similar in size to the mallard, for which it is sometimes mistaken. The male has a green head, white breast, and chestnut sides. The female is a mottled brown. The best identification marks are the outsize bill, held downward as the bird floats on the water, and in flight, blue upper wings and white underwings. Females sound a typical quacking, and males utter a *took-took* call. Shovelers usually travel in flocks of five to ten birds.

Shovelers eat caddisfly larvae, dragonfly nymphs, beetles, bugs, duckweeds, and seeds of pondweeds and bulrushes. They strain food items out of shallow, murky water, swimming slowly forward with the bill submerged and swinging from side to side. The species breeds mainly in the northwestern United States, western Canada, and Alaska; only a few nests have been reported in Pennsylvania. Females nest in grassy cover, sometimes well away from water. The eight to twelve eggs hatch in three or four weeks. Shovelers migrate through Pennsylvania in March and April and again from mid-September to late November.

WOOD DUCK

The wood duck's taxonomic name, *Aix sponsa,* can be loosely translated as "waterfowl in wedding dress," and the male of this species is indeed a brilliantly colored, beautiful bird. Shy and retiring, wood ducks inhabit lakes, ponds, marshes, and sluggish streams bordered by woods. Nicknames include Carolina duck, summer duck, woodie, and squealer. Most authorities place *Aix sponsa* with the dabbling ducks, a group distinguished by their habit of feeding on and near the surface of shallow water rather than diving for their food.

Wood ducks range across the United States and southern Canada. They are absent from the Great Plains and Rocky Mountains. Most individuals winter in the southeastern states, from the Carolinas south to the Gulf of Mexico and west to eastern Texas. In Pennsylvania, wood ducks are common migrants in March and April; they are summer breeding residents; they're common migrants in September, October, and early November; and occasionally some individuals remain as winter residents, mainly in the southeastern and southwestern corners of the state.

Biology. Adults are 18 to 20 inches long, have a 2-foot wingspan, and weigh 1.5 pounds. The drake's coloration is showy and exotic. His head is iridescent green shading into blue and purple, with a slicked-back crest of feathers and white chin markings. The eyes are bright red, the bill reddish orange, and the legs yellow. The breast, a rich chestnut speckled with white, is separated from the golden-yellow sides by vertical black and white bars. The hen's plumage combines gray, brown, and white. She has a small head crest and a white circle around each eye. A wood duck in flight can be identified by its pale belly, broad wings, long, rectangular tail, and the head bobbing up and down.

Wood ducks do not quack. The hen, more vocal than her mate, squeals a shrill *hoo-eek, hoo-eek,* both as a warning call and to establish contact with the drake. The drake whistles an ascending, finchlike *jeeb jeeb.* Wood ducks are excellent fliers. In the open, they wing along at 30 miles per hour; in forested swamps they twist and turn between trees and limbs: scientists speculate that their large eyes give them the visual acuity needed to dodge obstructions. Wood ducks are accomplished swimmers. On dry land, they can run at speeds up to 7 miles per hour. The nails on their webbed feet help them grip tree branches; the wood duck is one of few waterfowl to perch in trees.

Wood ducks feed along the shores of woodland streams, slow rivers, and ponds. They take food from the surface and by tipping their heads into shallow water, probing for vegetative parts and seeds of pondweeds, duckweed, wild rice, and water lilies. They leave the water readily and forage on land for wild grapes, berries, seeds, and nuts—mainly acorns, but also hickory nuts and beechnuts, swallowed whole and crushed by the gizzard into digestible bits. Up to twenty large acorns have been found in a wood duck's esophagus. Insects and spiders constitute about 10 percent of the adults' diet; ducklings eat a larger percentage of these high-protein animal foods. In winter, if natural foods are scarce, wood ducks may land in fields and eat waste corn.

Wood ducks choose new mates each year. Most pairs form during the southward migration in autumn and on the wintering areas. To attract a mate, the male preens behind his wings and spreads them to show off their iridescent sheen; he tucks his chin, erects his crest, fans his tail, and performs ritualized drinking motions. When the birds return north, the hen homes in on her last year's nest—or, if she is a yearling, on the same general area in which she was hatched. Wood ducks have been found nesting in Pennsylvania as early as March, but most nest from mid-April into June. Although paired wood ducks do not guard territorial boundaries, the male will defend his mate from the attentions of other males. Several breeding

Wood ducks choose new mates each year. The drake is brightly colored, and the hen's plumage is a camouflaging mixture of gray, brown, and white.

pairs may share the same pond, with nesting concentrations largely determined by the number of available tree cavities.

The mated hen seeks out a cavity in a tree; the male follows her on these search flights, but the hen picks the exact spot. Wood ducks prefer to nest in trees at the water's edge but will settle for sites up to a mile away from water. Tree cavities average 25 feet up, with some as high as 60 feet. Wood ducks do not excavate cavities but use rotted areas behind branch stubs or, less frequently, abandoned woodpecker cavities. They also nest in boxes put up for them by humans. The hen lays nine to fourteen eggs (one per day) on wood chips covered with down plucked from her breast. The eggs are dull white and unmarked. The hen starts incubating with the last egg laid. The drake remains near his mate into her incubation period, but he leaves before the eggs hatch.

After twenty-five to thirty-five days of incubation, all of the eggs hatch on the same day. The hen usually keeps her brood in the nest overnight, and in the morning she flies out and lands on the ground or on the water below and starts calling a soft *kuk, kuk, kuk*. The ducklings clamber up the inside of the nest cavity, perch on the rim of the opening, and jump out; they tumble down like cotton puffs and usually land unharmed. The hen leads them to a lake or a stream. As the ducklings follow her, and in the days to come, they are fair game to a host of predators. It's not uncommon for a fox or a raccoon to kill an entire brood of newly hatched wood ducks. If her first clutch or brood is destroyed, the female may lay a second clutch. A few hens raise two broods per year, mainly in the South; in the North, most hens raise one brood annually.

Hens generally rear their broods on wetlands or ponds that are at least 10 acres in area, and where food and cover are plentiful. Ducklings—and adults—are preyed on by minks, otters, hawks, and owls. Snapping turtles seize ducklings on the water and drag them down. In Maryland, biologists found that up to half of all young wood ducks were killed during their first month. The broods begin to break up after six weeks, and the young can fly when two months old.

After leaving his mate, the drake joins other male wood ducks in dense cover in a swamp or a tract of flooded timber. There he molts into an "eclipse plumage": dull feathering that resembles the drab plumage of the hen. For three to four weeks, molting wood ducks—at first the drakes and later the hens—lack fully formed wing feathers and cannot fly. In late summer and early fall, a continuation of the molt restores the flight feathers and returns the drake to his usual splendor.

Wood ducks are among the earliest of the ducks to move south in autumn. Most depart in late September and early October, with adult birds migrating ahead of the juveniles. On the wintering grounds, wood ducks may group in flocks of less than a hundred to several thousand. Some studies indicate that drakes vary their migratory pattern from year to year: if a drake pairs with a hen hatched in a southern swamp, he may move only a few miles away from the wintering area in springtime when he follows his mate, who instinctively returns to her natal area; the next year, if the drake bonds with a hen hatched farther to the north, he may migrate hundreds of miles to her area of origin.

Habitat. Wood ducks favor slow-moving streams, lakes, swamps, beaver ponds, and backwaters of creeks and rivers. They rest in thick growths of water lilies, smartweeds, and other emergent aquatic plants; hens hide their ducklings in vegetation, under overhanging banks, and among fallen, partly submerged trees.

Wood ducks nest in cavities in mature trees, including sycamore, maple, oak, beech, elm, and black gum. Where large trees are scarce, hens readily use artificial nest boxes. The erecting of predator-proof wood duck boxes has helped *Aix sponsa* become plentiful; the boxes also provide nesting space for kestrels, screech-owls, and mergansers. The resurgence of the beaver, a species that creates wetlands by damming streams, has provided much excellent habitat for wood ducks and other waterfowl.

Population. By the early 1900s, the wood duck was nearing extinction. Most of the Northeast's mature forest had been logged off, market hunters had killed many ducks, and wetlands had been damaged or destroyed by draining and pollution. In 1918, the United States and Canada signed a treaty setting limits on the hunting of migratory birds. Conservation officials quickly banned wood duck hunting, and game departments put up thousands of nest boxes. The wood duck population grew steadily. In 1941, limited hunting was again permitted. In 1976, waterfowl scientist Frank Bellrose reviewed many local studies and concluded that the adult population of wood ducks was about 1.3 million before each year's breeding. Others estimated the annual post breeding population at 2.5 to 3.5 million. Today most biologists believe the population is stable or is still increasing.

Aix sponsa is the second most common breeding duck in Pennsylvania; only the mallard is more abundant. Field surveys for the *Pennsylvania Breeding Bird Atlas,* conducted from 1983 through 1989, documented wood ducks breeding in all Pennsylvania counties, most abundantly in the northeast, southeast, and northwest. In 1998, waterfowl surveys suggested that the state had over fifty thousand breeding pairs.

MALLARD

The mallard, *Anas platyrhynchos,* is the most common duck in the United States, North America, and the Northern Hemisphere. The species possesses a huge breeding range, nesting across Canada and Alaska south to California, New Mexico, Kansas, Ohio, and Virginia. Taxonomists recognize seven races. The mallard may have been the first domesticated bird, and from it have sprung almost all of the strains of domesticated ducks.

The mallard is known as a puddle duck or a dabbling duck. It frequents shallow water, where it picks up plant and animal food on and near the surface by dabbling with its bill; to reach the bottom, it hoists its tail in the air and stretches its neck and head underwater. Rarely does a mallard dive for food. Like all puddle ducks, the mallard springs directly into the air when taking off.

Biology. Adults are 2 feet long and weigh 2.5 to 2.75 pounds. The male, or drake, is easily recognized by his iridescent green head, the narrow white ring around his neck, and his dark chestnut breast. He has a black rump (topped with curling feathers), white outer tail feathers, whitish underparts, gray sides, and a brownish back. The female, or hen, has a buff-colored head and a straw brown body streaked and mottled with many shades of brown. The feet are orange in both sexes. The speculum, a patch of feathers on the trailing edge of the wing close to the body, is violet-blue bordered with white. The male's bill is yellow, and the female's bill is orange with dark spots.

Mallards are among the most vocal of waterfowl. The hen makes a variety of quacks, and the drake utters reedy quacking sounds and, during the mating season, a sharp single- or double-note whistle. Mallards fly in small groups or in V- or U-shaped flocks, usually with ten to twenty members but sometimes as many as several hundred. Mallards are swift fliers and excellent swimmers. They often feed and rest in the company of other puddle ducks, including pintails and black ducks.

The mallard's bill has a serrated edge: when dabbling, the duck picks up food in its bill, forces water out through the serrations, and ends up with a mouthful of edibles. Mallards eat the seeds of bulrushes, pondweeds, millet, sedges, smartweeds, and wild rice; the stems and leaves of many aquatic plants; and acorns. About 10 percent of the adults' diet is aquatic insect larvae and mussels. Occasionally, mallards eat tadpoles, frogs, earthworms, and small fish. Ducklings feed on insects, particularly mosquito larvae, and on crustaceans, in addition to plants. Wild mallards prefer natural foods, but when marshes and ponds freeze over, they head for dry land. Perhaps more frequently than any other duck, the mallard feeds in harvested fields, gleaning waste grain among the stubble. Mallards will travel up to 25 miles to a dependable food source. They make two feeding flights daily, one at dawn and the other in late afternoon; during midday they rest on shallow water where they are screened by vegetation.

Mallards mature sexually in their first year. A period of social display begins in the fall and continues through winter into spring. To attract a mate, the male dips his bill in the water and then rears up, sounding grunts and whistles while settling back down. He may pump his head and preen. The hen stimulates the courtship with calls and body movements. Most pair forming takes place on the water; chase flights are also prominent courtship rituals. Most hens have chosen—and been accompanied by—mates by the time they arrive back in the areas where the hens themselves were reared. Mallards breed and nest from late March through April. The male selects a shoreline breeding and waiting area, up to a quarter acre in extent, which he defends against other mallards. After flying low over marshes and fields in the company of her mate, the female picks a nest site. Mallards nest around freshwater lakes, ponds, marshes, and reservoirs; they may also nest in farm fields, grassy areas, and suburban neighborhoods.

The hen usually nests within 100 yards of water, on the ground in a shallow bowl of plant material lined with down from her breast. The nest is concealed by tall grass, reeds, leatherleaf, brush, alfalfa, or clover. A few individuals nest in stumps, tree cavities, and the crotches of trees. The hen lays one egg per day until she has a clutch of seven to ten—rarely as many as fifteen. Some hen mallards lay eggs in the nests of other ducks. (Ducks of many species practice this "nest parasitism.") The eggs are whitish to olive buff. Only the female incubates the eggs; the male soon leaves her or gets tired of hanging around unvisited on his breeding and waiting area. Incubation takes twenty-three to thirty days, and all of the eggs in the clutch hatch at about the same time.

Mallards normally raise one brood per year, but if a skunk, crow, raccoon, or opossum destroys her first clutch, a hen may nest again. Renesting

The drake mallard, left, has an iridescent green head;
the female is an overall straw brown.

usually results in fewer eggs, six to eight on average. Nests are also lost to
farm activities, such as hayfield mowing, and to flooding. In addition to the
predators listed above, foxes, snakes, largemouth bass, muskellunge, and
snapping turtles take mallard ducklings. Within a day of hatching, the
young follow their mother to the water—up to a mile away, if areas that
were flooded in the spring have dried up by the time the brood hatches. In
the lake margin or marsh, the young are shepherded about by the female
but feed themselves. They can fly after seven or eight weeks, and the family group breaks up soon after.

When the drakes leave their mates in May or June, they fly to secluded
wetlands, where they undergo their annual molt. The birds seek out food-
rich areas, since the replacing of feathers demands considerable energy. A
complete, simultaneous wing molt leaves them temporarily flightless; at this
time they are in a drab "eclipse plumage," which resembles the females'
protective coloration. The hens undergo a similar molt after their ducklings
fledge. The wing feathers take three or four weeks to grow back.

In fall and winter, most mallards fly south when ice and snow lock
away their feeding and resting areas. Of the puddle ducks, the mallard and
the closely related black duck are among the latest fall migrants. Many of
the mallards seen in Pennsylvania during the fall have come from breeding
areas in Ontario, Quebec, and the Great Lakes region. The mallard is one
of the earliest ducks to return north in the spring: in Pennsylvania, mallards
are common migrants in late February, March, and early April. Feral or
semidomesticated mallards—those that rely on human handouts and live in

parks and towns—may be permanent residents, but all wild mallards in North America are believed to be migratory.

The maximum life span of the mallard is seven to nine years, with half of all individuals failing to reach two years of age. Ducks perish from predation, accidents, hunting, and diseases such as botulism, fowl cholera, duck virus enteritis, and aspergillosis.

Habitat. Ideal habitat combines shallow-water foraging areas and thick vegetation for nesting. The mallard prefers open country to woodlands. It breeds on farm ponds, edges of freshwater lakes, sloughs, reservoirs, and marshes. Wetlands in extensive forested tracts will have fewer breeding mallards and more black ducks. Mallards winter on marshes, lakes, rivers, and bays.

Most waterfowl move away from areas frequented by humans and have been driven out of their habitat by expanding cities and towns, rural development, and vacation homes. The mallard, less wary of people, has occupied these altered habitats. A rise in the number of farm ponds also has benefited the adaptable mallard.

Population. In North America, the densest population of mallards centers on the northern prairies of the Great Plains—Montana, North Dakota, and the Canadian provinces of Saskatchewan, Alberta, and Manitoba—with nearly half of the continent's mallards breeding there. The species winters in Canada, Alaska, Mexico, Central America, and most of the United States, with heavy concentrations in the Mississippi River drainage. Along the Atlantic coast, many mallards winter in the Carolinas and on the Chesapeake Bay.

Anas platyrhynchos has bred widely in Pennsylvania for at least several hundred years, and during this period the population has fluctuated, sometimes dramatically. In the late 1940s and 1950s, the mallard population began to grow as continental birds extended their range eastward. Today mallards breed throughout the state, with the greatest numbers in the southeast. Recent estimates put the number of breeding pairs at over ninety thousand statewide, with a total population topping two hundred thousand.

DIVING DUCKS, LOON, GREBES, AND CORMORANT

The diving ducks are chunky, heavy-bodied waterfowl that tend to stay farther out from shore than the puddle ducks do. In Pennsylvania, diving ducks can be found on streams, rivers, lakes, and marshes as they migrate through the state in spring and fall. The best places to look are the larger lakes, such as Raystown Lake in Huntingdon County, Blue Marsh Lake in Berks County, Lake Wallenpaupack in Pike and Wayne counties, Sayers Lake in Centre County, Pymatuning Reservoir in Crawford County, and Lake Erie, as well as the larger rivers—although diving ducks will also rest and feed on smaller bodies of water, such as the lakes at state parks. (I've spotted everything from loons to canvasbacks on the 250-acre lake at Black Moshannon State Park, a short drive from my home.)

The best time to search for these birds is after a storm in March or April. When driven down by a storm, birds may land on wet or ice-glazed roads or parking lots, which they mistake for open water. Because they can't take off from land, they may die if people don't help them back to the water. (Human rescuers should wear heavy gloves and clothing to protect against a strike from one of the sharp-billed species, such as a grebe or a loon.)

Diving ducks eat more animal than plant food. They feed on fish, mollusks, crustaceans, insects, and other invertebrates, and on the seeds and vegetative parts of aquatic plants. To reach their food, they dive underwater, propelled by their large, broad feet, which are fully webbed and have strongly lobed hind toes. Their legs are spaced wide apart and located well back on the body, improving diving efficiency at the expense of mobility on land. Diving ducks' bodies are compact, and their wings have relatively small surface areas, an arrangement that helps them dive and swim but hinders their ability to get airborne. Instead of springing out of the water into flight as puddle ducks are able to do, diving ducks must run along on the water's surface to build up speed for takeoff.

Two species of mergansers (which, although not technically diving ducks, nevertheless are usually grouped with the divers) breed in Pennsylvania. The other diving ducks nest in New England, eastern Canada, the prairies of the Upper Midwest and central Canada, the Pacific Northwest, and Alaska. The pied-billed grebe and the double-crested cormorant also breed in the Keystone State. Some diving ducks winter on inland waters, but most stay along the coast in large bays and estuaries as far south as Mexico. Several species of diving ducks inhabit both the Eastern and Western hemispheres.

In the spring, males vie for females. Courtship may include ritualized drinking and preening motions, posturing, and calling. Males and females form monogamous pairs that generally stay together until the female begins incubating eggs; then the male leaves the area and joins a band of other males. There's no evidence that the same males and females pair up every year.

Nesting habits and habitats vary from species to species. Females lay five to fifteen eggs in nests hidden in vegetation, tree cavities, or rock crevices overhanging or near the water. Because females do not start incubating until they lay the last egg in their clutch, the young all hatch at about the same time. The down-covered ducklings are patterned with shades of yellow and brown to break up the body outline and provide camouflage. Their eyes are open, and they can swim and feed themselves soon after hatching. Broods usually stay together until the juveniles can fly, around eight to ten weeks after hatching.

Adults go through a post-breeding molt, when they shed their old, worn feathers and grow new plumage. Males molt first; their bright nuptial plumage is replaced by a drabber, less-conspicuous eclipse plumage. While their flight feathers are growing in, ducks cannot fly; they stay hidden or float in the middle of broad waters during this time of vulnerability. Later, after the males' bright breeding plumage has filled in, the birds migrate south.

Ducks fall prey to raccoons, foxes, coyotes, minks, hawks, and owls; the young are taken by snapping turtles. Crows, raccoons, and skunks rifle nests and eat eggs. The populations of some species, such as the canvasback and the redhead, have generally declined, because the prairie marshes in which they nest have been drained for farming; other species, including the ring-necked duck and the oldsquaw, are believed to have populations that are stable or increasing.

This chapter includes information on loons, grebes, and cormorants, birds that also dive for their food and sometimes share habitats with the diving ducks.

Canvasback *(Aythya valisineria).* This big, wary duck is about 20 inches long and weighs 2.5 to 2.75 pounds. Its body plumage is black and white; the head is brick red in the male and brown in the female. Canvasbacks can fly at 70 miles per hour in calm skies, faster with a tailwind. The species breeds in prairie marshes and winters on saltwater bays.

The species name, *valisineria,* refers to the taxonomic name for wild celery, an aquatic plant that is a favorite food. The canvasback's diet is mostly plant matter picked up in water only a few feet deep. Canvasbacks eat the

Canvasbacks breed in prairie marshes and winter along the Atlantic coast.
They migrate through the Northeast in spring and fall.

leaves, roots, and seeds of various water plants, including pondweeds, sedges, and grasses. They also consume mollusks, crustaceans, insects, and small fish. Canvasbacks begin to pair up on their northward migration. In March and April, males display for females by jerking the head back and then pushing it forward while sounding cooing and clicking calls. When migrating, canvasbacks fly in flocks of five to thirty birds, often quite high and in V-formation. In Pennsylvania, the canvasback is an uncommon migrant in spring and fall. Along the Atlantic, wintering canvasbacks concentrated on Chesapeake Bay make up almost half of the North American population.

Redhead *(Aythya americana)*. The redhead is about 19 inches long and weighs 2 to 2.5 pounds. Its plumage is black and gray, with the head red in the male and brown in the female. Redheads feed in shallower waters than do most other diving ducks, eating the seeds, leaves, stems, and roots of plants, along with aquatic insects and mollusks. In Pennsylvania, redheads are uncommon migrants in spring and fall. They breed mainly in the northern United States and southwestern Canada and winter across the southern states and in Mexico. Hens may lay eggs in the nests of other ducks, of their own and at least ten different species, leaving them to be incubated by the nest owners. As well as parasitizing nests, female redheads usually lay a clutch of their own. The current population of *Aythya americana* is significantly reduced from original levels, probably because much prairie nesting habitat has been lost.

Ring-Necked Duck *(Aythya collaris)*. Also known as the ringbill, the ring-necked duck is 16 to 17 inches long and weighs from 1.25 to 2 pounds. The plumage is black and white in the male, brown and white in the female. The male has a faint brown ring around the neck, not easily seen in the field, and both sexes have a pale ring near the tip of the bill. They fly swiftly in flocks of up to twenty. Ring-necked ducks feed in shallow waters on seeds and vegetative parts of pondweeds and other water plants, and on insects and mollusks. Look for *Aythya collaris* on small, tree-lined ponds, flooded cropfields, and large lakes; generally ring-necked ducks do not set down on saltwater bays. Common migrants through Pennsylvania during spring and fall, these ducks breed across southern Canada and the northern United States. Some winter in Pennsylvania, although most go farther south.

Greater Scaup *(Aythya marila)* and **Lesser Scaup** *(Aythya affinis)*. These two similar species are 16.5 to 18 inches in length and weigh 1.5 to 2.5 pounds. They are also called broadbills or bluebills. The males are black and white, the females brownish and white. The bill is blue in both sexes and both species. Greater scaup inhabit large bays, sounds, and inlets of both coasts and the Great Lakes; although mainly found in these more exposed situations, they occasionally land on smaller inland waters. The lesser scaup is the species generally seen in Pennsylvania. Scaup eat mollusks, insects, crustaceans, and aquatic plants, feeding both during the day and at night. In Pennsylvania, they are common migrants in spring and fall. The two species breed in Canada and Alaska and winter in coastal areas.

Oldsquaw *(Clangula hyemalis)*. The oldsquaw is so named because of its whistling call, freely given (mostly by the male) and imagined to resemble a person wailing or keening. Adults are 16 to 20 inches in length and weigh 1.75 to 2 pounds. The plumage is a striking mix of black and white; the male has a long, needlelike tail. Oldsquaws may dive to 100 feet when foraging. They eat crustaceans, mollusks, insects, and fish. They are uncommon spring and fall migrants through Pennsylvania. Occasionally they winter in the state, but more often they rest and feed in large flocks along the coasts and on the Great Lakes. They breed in western and northern Canada and in Alaska.

Black Scoter *(Melanitta nigra),* **Surf Scoter** *(Melanitta perspicillata),* and **White-Winged Scoter** *(Melanitta fusca).* All three scoter species are basically black, with varying amounts of white in the plumage. Lengths range from 18.5 to 22 inches, weights from 2 to 3.5 pounds. These sea ducks fly in long, undulating lines, in irregular groups, or in V-shaped flocks. They eat mollusks, crustaceans, aquatic insects, and plants. When traveling over land, scoters tend to fly at high altitudes and for long distances. They are rare to uncommon migrants through Pennsylvania, traversing the state in March and April and again in October and November. Scoters breed in Canada and Alaska. They winter on the Great Lakes and along the Atlantic and Pacific coasts.

Common Goldeneye *(Bucephala clangula).* This species is 17 to 19 inches long and weighs about 2 pounds. The goldeneye is also called a whistler for the sound of its wingbeats. The plumage is black and white in the male (with a large, round, white spot between the eye and the bill), gray and white in the female. Goldeneyes dive for crustaceans, insects, mollusks, and fish. They breed across Canada and Alaska on lakes, bogs, and rivers. Hens nest in tree cavities 5 to 60 feet above the ground. In Pennsylvania, golden-eyes are common migrants in spring and autumn.

Bufflehead *(Bucephala albeola).* This smallish diver is 13 to 15 inches long and weighs around 1 pound. Its name derives from "buffalo-head," for the male's peculiar puffy head shape. The plumage is mostly black and white in the male and brown and white in the female; the male has a large white patch on the back of the head. Buffleheads are fast fliers with rapid wing-beats. In summer, on the breeding grounds in Canada and Alaska, buffle-heads eat mostly aquatic insects; when wintering along the coasts and in the southern states, they feed mainly on marine crustaceans and mollusks. Buf-fleheads nest in tree cavities, sometimes using old flicker holes. *Bucephala albeola* is a common spring and fall migrant in Pennsylvania.

Hooded Merganser *(Lophodytes cucullatus),* **Common Merganser** *(Mergus merganser),* and **Red-Breasted Merganser** *(Mergus serrator).* Hooded and red-breasted mergansers average 16 to 18 inches in length; the common merganser is 23 to 25 inches. Weight: around 1.5 pounds for the hooded and red-breasted species, and 2.5 to 4 pounds for the common. Mergansers are also known as fish ducks or sawbills, the latter for the sharp excrescences in these birds' bills, which help them grip fish. All three species have dis-

Mergansers have serrated bills to help them catch fish. Two species breed in Pennsylvania. The red-breasted merganser, shown here, breeds in Canada and Alaska and is commonly seen on Pennsylvania's rivers during migration.

tinctive colorful plumages, and they fly fast and low over the water. They eat mainly fish, fish eggs, and other aquatic animals.

The hooded merganser nests in Pennsylvania, in swampy areas, mainly in the northern tier of counties. The hen lays ten to twelve eggs in a tree cavity; she may use an artificial box erected for wood ducks. The common merganser (known as the goosander in Britain) nests along Pennsylvania's wilder rivers and streams, mainly in the northern half of the state; it lays eight to eleven eggs in a tree cavity, rock pile, hole in a stream bank, or on the ground concealed by jumbled logs or dense brush. The red-breasted merganser breeds in Canada and Alaska.

Hooded and red-breasted mergansers winter along the coasts and in the southern United States. The common merganser winters in Pennsylvania, on the Great Lakes, and across the continent in places where the water stays unfrozen.

Ruddy Duck *(Oxyura jamaicensis).* The ruddy duck is about 15 inches long and weighs 1 pound. Small and stubby, it has a short, thick neck and a spiky, upturned tail. The male has a reddish body; both sexes have white cheeks under a dark crown. Ruddy ducks prefer to dive, rather than fly, away from danger. In flight, ruddy ducks skim low over the water in compact flocks. They feed mainly on vegetation such as wigeon grass, pondweeds, and bulrush seeds, and on midge larvae and mollusks. Juveniles eat a larger proportion of protein-rich animal food than adults do. The species nests in the western United States and Canada and winters along both coasts. Ruddy ducks are common spring and fall migrants on Pennsylvania's major waterways; most follow a migration corridor from North Dakota across Minnesota, Wisconsin, and western Pennsylvania, leading to wintering areas on Chesapeake Bay.

Common Loon *(Gavia immer)*. Common loons (or divers, as they are also called) are heavy-bodied birds about 2 feet in length. In spring, the common loon has a black-and-white plumage in a beaded, almost a checkerboard, pattern. Look for a bird that swims low in the water like a submarine and, when diving, hops up and forward before plunging into the depths. The fortunate observer may hear a loon give its eery, high-pitched laughing call. There are old records of common loons breeding on lakes in the Poconos, but the species does not nest in Pennsylvania at this time. Loons breed in New England, the Upper Midwest, and across Canada and Alaska. The female usually lays two eggs in a heap of vegetation along the water's edge; both sexes incubate the eggs and rear the young. Loons eat fish less than 10 inches long, including minnows, perch, and suckers, as well as crustaceans, mollusks, and aquatic insects. They can dive as deep as 200 feet. Common loons sometimes winter on ice-free water in Pennsylvania, although most keep to coastal areas. The fall and winter plumage is a blend of grays.

Red-Necked Grebe *(Podiceps grisegena)*, **Horned Grebe** *(Podiceps auritus)*, and **Pied-Billed Grebe** *(Podilymbus podiceps)*. The best time to see grebes is during the spring migration, as these species move northward from wintering areas along the Atlantic coast to their breeding ranges farther north. The red-necked grebe is large and has a long neck, which is red in the male, and a heavy pointed bill. The male horned grebe wears luxuriant buff-colored ear tufts. The pied-billed grebe is a small, dark bird with a chickenlike bill; it dives quickly at the approach of a human, and when it emerges again, often many feet away from where it submerged, this shy diver may remain underwater with only its bill and eyes projecting above the surface. Grebes eat mainly aquatic insects; they also take fish, crustaceans, and amphibians.

Of the three species, only the pied-billed grebe (also called the hell-diver) breeds in Pennsylvania, in dense marshes, farm ponds, and flooded quarries. Although nowhere common in Pennsylvania, pied-billed grebes nest across the state. The nest is a sodden raft of decaying vegetation hidden among cattails, bulrushes, and other emergent plants. Eggs have been found from early May through late July. The female lays four to seven pale blue or green eggs and covers them with debris when she is away from the nest. Both parents (but principally the female) incubate the clutch for about twenty-three days. Chicks can swim and dive soon after hatching; both parents feed them, and the young may ride on their parents' backs. Pied-billed grebes eat insects, small fish, crustaceans, frogs, tadpoles, salamanders, spiders, and aquatic plants. Adults pluck and swallow their own feathers; the half-digested feathers may protect the stomach lining from sharp fish bones. In autumn, pied-billed grebes migrate to wintering areas along the coast.

Double-Crested Cormorant *(Phalacrocorax auritus)*. Black birds with snaky-looking necks, cormorants often stand on rocks and snags holding their wings outstretched. Their feathers are not as waterproof as those of loons and grebes, which helps cormorants to dive but requires frequent drying-out sessions. Double-crested cormorants dive after fish, eels, crayfish, frogs, salamanders, and snakes. They forage as individuals and in groups. They nest in colonies, often sharing habitats with herons. The nest is a platform of sticks and rubbish, usually built in a tree but sometimes placed on the ground. The three to four bluish white eggs are incubated by both parents for about a month. The young stay in the nest for three to four weeks after hatching, then leave the nest but remain close to the colony; they can fly after another two to three weeks. Until 1996, double-crested cormorants did not breed in Pennsylvania, although they did so in nearby states. During that year, and in years since, cormorants have nested on Wade Island in the Susquehanna near Harrisburg. Cormorants are being seen in increasing numbers in Pennsylvania and are now common on most of the larger river systems.

VULTURES

Vultures, sometimes called buzzards, are large, dark-colored birds with broad wingspans, often seen soaring in wide circles in the sky as they search for carrion to eat. Although graceful when flying, they are clumsy on the land. When resting, they sit perched in trees or stand on the ground, often near a carcass on which they've been feeding.

Vultures have long been grouped with the hawks, eagles, and falcons, but recent genetic research links them more closely to the storks, another family of naked-headed birds. Seven species of vultures inhabit North America, including the endangered California condor. Pennsylvania has two species: the turkey vulture and black vulture. The turkey vulture, by far the more common bird, is found statewide. The black vulture is a southern species that extended its range into Pennsylvania during the second half of the twentieth century.

Turkey Vulture *(Cathartes aura)*. *Cathartes* means "purifier," and the turkey vulture is the chief avian scavenger in North America, cleaning up huge quantities of unsanitary carrion from roads, fields, and forests. Adults are 30 inches in length, their wings span up to 6 feet, and they weigh around 4 pounds. A turkey vulture's body is covered with blackish brown plumage; seen from below, the wings appear to be two-toned, with the flight feathers (on the trailing edge) lighter in color than the rest of the feathering. Turkey vultures soar with their wings held slightly above the horizontal, forming a gentle V, and they rock and teeter as they wheel through the air.

Turkey vultures use both their eyesight and their sense of smell to find dead animals. A turkey vulture's olfactory organs are large and well supplied with nerve endings; with its sense of smell alone, a vulture can find a carcass completely screened by the forest canopy. The heavy ivory-colored bill has a sharp hook on the end for opening a carcass and removing flesh. A vulture uses its strong, curved talons to tear food apart or to hold it in place while the bird picks out flesh with its bill. To let a vulture probe deeply into carrion without getting too messy, its head and neck are unfeathered—"like the bare arms of a butcher," observed one naturalist. Adults have pink heads and necks; in young birds, these bare areas are blackish. Because vultures lack a syrinx, or voice box, the only sounds they can make are grunts, hisses, and whines.

Vultures' long, broad wings hold them aloft like kites. When the air is rising, a vulture can maintain or even increase its altitude without flapping

*The turkey vulture's head and neck are unfeathered, letting it
probe deeply into carrion without getting too messy.*

its wings. Like many hawks and falcons, vultures often travel along mountain ridges, using thermal updrafts to stay airborne. The lift from rising air
is so important to their ability to fly that vultures may roost for several days
when rainy weather suppresses updrafts. Observations made from gliders
show that the turkey vulture has a lower sinking speed than the black vulture; this heightened soaring ability may allow the turkey vulture to range
farther north than the black vulture.

Vultures eat all kinds of carrion, including winter- and highway-killed
wild mammals ranging in size from mice to deer; domestic animals; and
slaughterhouse refuse. Both wild and captive turkey vultures have been
observed killing small birds. Turkey vultures occasionally eat insects and fish
that have become stranded in shallow water. Recently on the road near our
house, I watched two vultures begin feeding on a freshly killed porcupine; I
don't know if any other animals visited the carcass, but after only forty-eight
hours all that remained of the porcupine were the backbone, skull, and a
scattering of quills.

When forming pairs, turkey vultures gather in a circle on the ground
and perform hopping movements with their wings partly spread; in the
air, one amorous bird may follow another closely, both of them flapping
their wings and diving. As common as vultures are, you'd think people
would find their nests more frequently. But vultures hide their nests well,

in caves, on steep cliffs, among tumbled boulders, in hollow logs or stumps, and in dense thickets. (Some unusual nest sites include abandoned farm buildings, duck blinds, a cavity in a dead tree with the entrance 14 feet above the nest, and 6 feet below ground in a rotted stump.) Vultures do not build a nest but deposit their eggs directly on gravel, rotted sawdust, or bare earth.

The female typically lays two eggs, although some clutches contain one egg and others have three. The eggs are white, blotched with lavender or brown. Both parents share in the incubating. When disturbed on the nest, the adults may fly off, or they may stand their ground while hissing or vomiting at the intruder; they may even feign death. After thirty to forty days' incubation, the eggs hatch into altricial young. The young birds eat carrion that their parents regurgitate to them. They can fly after nine to ten weeks. They stay close to the nest for another one to three weeks, perching and roosting nearby.

Family groups stay together until the fall migration. Vultures are gregarious, and eight to twenty-five or more adults and juveniles may roost and soar together; several hundred individuals may roost communally, sometimes with black vultures included in the mix, perched on the limbs of dead snags. Although turkey vultures nest in caves, they rarely enter them at other times of the year and do not use them for winter shelter. They molt once each year, a gradual process that lasts from late winter or early spring into early fall.

Turkey vultures favor farmland with pastures and abundant carrion near wooded areas for nesting, perching, and roosting. In Pennsylvania, *Cathartes aura* is a common northbound migrant in late February and March. In summer, it breeds throughout the state except in urbanized areas. Historically, the turkey vulture has been a common breeder in the Ridge and Valley province in south-central Pennsylvania; over the last fifty years, it has expanded its range northward, perhaps in response to a growing population of deer that provide food in the form of road-killed or winter-starved animals. Over the last several decades, more and more turkey vultures have stayed on and wintered in Pennsylvania, mainly in the southeast. Most turkey vultures spend the winter in the southern United States and in Central and South America.

Turkey vultures and their eggs and nestlings are preyed on by opossums, foxes, dogs, and owls. *Cathartes aura* has been identified as the main avian species causing damage and deaths in military aircraft accidents in the United States.

Black Vulture *(Coragyps atratus).* The black vulture's shorter wings and tail make it look smaller than the turkey vulture, although its body is about the same size. The black vulture has a black head. Because its wings form a smaller sail area, it is not as efficient at soaring as the turkey vulture: the black vulture often must punctuate its sailing with bouts of rapid wing flapping. Seen from below, the black vulture shows distinctive pale patches on the undersides of the wings near the tips. The black vulture holds its wings more horizontally than the turkey vulture does.

Behavior, feeding, and nesting habits are similar to those of the turkey vulture, except that black vultures are more aggressive and apt to prey on live animals, including the young of livestock. In confrontations over carrion, black vultures usually drive off turkey vultures. In addition to carrion, black vultures eat rotting vegetation and garbage. They nest in thickets, hollow logs and stumps, caves, and cavities beneath boulders. The eggs, usually two per clutch, are pale greenish gray blotched with brown. Both sexes incubate, and the young hatch after thirty-seven to forty-one days. They stay in the nest for about two months and can fly capably when seventy-five to eighty days old. The black vulture breeds mainly in south-central and southeastern Pennsylvania and is expanding its range into the northeastern part of the state.

OSPREY AND EAGLES

Large, striking birds of prey, the osprey, bald eagle, and golden eagle seem to embody power and majesty. All can be seen in the Keystone State, although none are common here. In North America, environmental contamination and human encroachment on natural habitats have reduced these birds' numbers and lowered their breeding success, although in the last several decades, those trends have been reversed. Taxonomists place the bald and golden eagles with the buteos: soaring hawks with broad wings and broad, rounded tails.

Osprey *(Pandion haliaetus).* The osprey is an eaglelike hawk found throughout North America and in the Eastern Hemisphere. The species inhabits seacoasts and the shores of large rivers and lakes. In the East, many ospreys live on Chesapeake Bay. In Pennsylvania, ospreys are found along the Susquehanna and Delaware rivers and near creeks, ponds, lakes, and reservoirs throughout the state, depending on the season.

Osprey.

The plumage is dark above and white below; the head is largely white, with a black patch across each cheek. A conspicuous crook to the wings and black "wrist" markings are good field identifiers. Adults are 21 to 24 inches long, and their wings span 5 feet; standing, they are 1.5 feet tall. The call is a series of piercing whistles: *cheep, cheep, cheep.* Except when migrating, ospreys flap more than they sail; their wingbeats are slow and deep. They hover 50 to 150 feet up and plunge into the water for their prey, sometimes going all the way under. Their dense, oily plumage keeps them from getting soaked. The diet is almost entirely fish: suckers, bullheads, catfish, carp, sunfish, perch, or whatever is locally common.

Ospreys build bulky stick nests in trees, on telephone poles and billboards, and on platforms put up on their behalf. Females lay two to four (usually three) eggs, which hatch after around thirty-eight days of incubation. In the 1980s, biologists from East Stroudsburg State University began an innovative hacking program to return the osprey to Pennsylvania as a breeding species. Other releases have been made in Tioga and Butler counties, and birds reintroduced in West Virginia have moved into Somerset, Fayette, and Westmoreland counties. Currently forty to fifty pairs nest in Pennsylvania, and the population is rising. Probably more ospreys inhabit Pennsylvania today than were here during the settlement era, and we now have more suitable habitat for them, in the form of reservoirs and man-made lakes.

In spring, ospreys migrate through Pennsylvania in April and May. Fall finds these fish hawks heading south along the state's mountain ridges in August, September, and October. Ospreys rarely winter in Pennsylvania; most fly to the southern states, Central America, and northern South America.

Bald Eagle *(Haliaeetus leucocephalus).* The Latin binomial means "sea eagle with a white head." Settlers in North America gave the bird its common name at a time when "bald" meant white rather than hairless. The head of a mature eagle is covered with gleaming white feathers. Its body is dark brown, its tail white. Immature birds are dusky brown marked with white on their wings and body; full adult plumage and the ability to breed are not attained until the fourth or fifth year. In both adults and young, the massive bill is yellow. The feet are also yellow, with the legs feathered halfway down.

Adults are 30 to 40 inches long and weigh 8 to 14 pounds. Their wings span 6 to 8 feet; the standing height is about 2 feet. As with other birds of prey, the female is slightly larger than the male. Bald eagles fly with strong, deep wing strokes or soar with their broad wings held flattened and steady. Their eyesight is among the keenest in the animal world, five or six times sharper than a human's. An eagle's call is a rapid harsh cackle, *kweek-kik-ik-ik-ik-ik,* or a lower *kak-kak-kak.*

Bald eagles feed mainly on fish (60 to 90 percent of the diet) taken alive or as carrion. They also eat birds, including waterfowl and gulls, and small mammals. Eagles soar above the water or sit on a convenient perch; when they spot a fish near the water's surface, they swoop down and snag it with their talons. They use their talons for killing prey and their heavy bills for tearing it apart. An eagle may fly at an osprey and force it to drop a captured fish, which the eagle then grabs in midair.

Eagles mate for life, although when one partner dies, the other quickly finds a mate if one is available. Nesting is preceded by a spectacular aerial courtship in which the birds may lock talons, scream, dive, and somersault in midair. An eagle's nest, known as an aerie, is a bulky affair usually built in a large tree 40 to 100 feet above the ground—or, less commonly, on a cliff. Most nests look out onto lakes, rivers, reservoirs, or seashores. A new nest is about 5 feet wide and 2 feet high, with an inside depression 4 or 5 inches deep and 20 inches in diameter. Often a pair returns to the same nest year after year, adding a new layer of sticks and branches, plus a lining of grass, moss, twigs, and weeds; enlarged annually, some treetop nests grow so heavy that they break the branches supporting them.

The female lays two eggs (sometimes only one and rarely three) in March or April. The eggs are about 2.5 by 2.75 inches, dull white and unmarked. Both parents incubate them, and they hatch after about thirty-five days. The young eaglets are fed by their parents. Occasionally a large, healthy hatchling will kill a smaller, weaker one. Eaglets develop most of their feathering by three to four weeks; they walk about in the nest at six to seven weeks and begin to fly when twelve weeks old. The young separate from their parents in autumn. An estimated 55 percent of eaglets survive their first year, and only one in five reaches breeding age. Adults can live thirty years or longer in the wild.

The bald eagle was chosen as the United States' national symbol in 1782. At that time, an estimated twenty-five thousand eagles lived in what is now the lower forty-eight states. Today the same area supports many fewer breeding pairs, most of them in the South, West, and Pacific Northwest. However, the population has been rebounding since the mid-twentieth century, thanks to limits on the use of organochlorine insecticides, protection of eagle habitat, restoration programs undertaken by wildlife agencies, and a greater public acceptance of eagles and other predators as a natural, necessary part of a healthy ecosystem. In 1999, the U.S. Fish and Wildlife Service recommended that the bald eagle be removed from the federal endangered species list; it currently remains listed federally as a threatened species, and in Pennsylvania, it is a state endangered species.

At one time, many bald eagles lived along Pennsylvania's rivers and the shores of Lake Erie. Their numbers plummeted until only three pairs were known to be breeding in the state between 1963 and 1980. Since then, the population has risen steadily. In 1983, the Pennsylvania Game Commission began taking eagle nestlings from Canada and "hacking" them, placing them on artificial nest platforms and hand-rearing them to fledging. Biologists also introduced eaglets into nests where wild parents raised them. These young birds have boosted the state's population, and many have themselves begun to breed. In 1999, forty-one active nests produced more than forty eaglets. Nearly half of the nests were in northwestern Pennsylvania, on Pymatuning Reservoir and Geneva Marsh. Eagles also nested in

Pine Creek Gorge in Tioga County; along the Susquehanna River in Dauphin, Lancaster, and York counties; in the Pocono region; on Raystown Lake in Huntingdon County; and along the Allegheny River from the New York border south to Venango County.

In November, bald eagles migrate in their greatest numbers along the south-trending ridgetops of Pennsylvania's Ridge and Valley region. Eagles sometimes winter on larger lakes and rivers that remain ice-free, including Pymatuning Lake, Geneva Marsh, Lake Raystown, the lower Susquehanna, Lackawaxen River in Pike County, Delaware River, and Octoraro Creek in Lancaster County.

Golden Eagle *(Aquila chrysaetos).* The golden eagle is a denizen of remote mountains, deserts, and tundra lands. The species occurs in Eurasia and North Africa as well as in North America, where it is most common in the western states, Canada, and Alaska. The golden eagle is rare in the Northeast. Most of the golden eagles that migrate through Pennsylvania breed in northern Quebec and Labrador, where they hunt over marshes and bogs and along rivers. Some of the birds winter in the Appalachians, mainly south of Pennsylvania.

Adults and immatures have a rich, dark brown body with gold-tipped feathers on the head and neck. Unlike the bald eagle, the golden eagle has trouserlike feathering on its legs that reaches all the way to the toes. Adult

Golden eagles are powerful predators that inhabit wilderness areas.

golden eagles resemble young bald eagles, but the golden eagles are darker. Immature golden eagles have white wing patches and, for their first several years, a white band at the base of the tail.

Golden eagles are about the size of bald eagles. Classic buteos, they have long, rounded wings. They flap less and soar more than bald eagles do. The call is a series of rapid, sharp chirps. Golden eagles are more predatory and aggressive than bald eagles. They take small rodents, rabbits, hares, and other mammals up to the size of foxes; they also eat birds, reptiles, and fish. They crush prey in their sharp talons and use their hooked beaks to rip flesh free for eating. In the West, these powerful predators have been known to kill young deer and knock young mountain sheep off high ledges, then feast on the gore below.

Golden eagles pass through Pennsylvania in February and March, with stragglers in April and May. They do not breed in our state, although they were reported to have nested along the Susquehanna before 1850. They also migrate through our state in autumn, generally following along the higher ridges in November and December. Occasionally individuals winter here in rugged terrain. Golden eagles do not breed until they are five years old. Breeding habits are similar to those of bald eagles, except that golden eagles often nest on cliffs. Although the North American population has declined since the settlement era, the current population is thought to be stable.

HAWKS

Hawks are also known as raptors, from a Latin word meaning "plunderer." They are quick, efficient predators with sharp talons and hooked beaks; in the different species, beaks and feet vary in size and shape according to prey preferences. The eyesight of some hawks is thought to be as sharp as that of a human looking through eight-power binoculars. A hawk's eyes are located in the front of the head for binocular vision, letting the bird judge distances and strike its prey accurately. Hawks have sharp hearing.

A hunting hawk may soar on high, sit and watch from a perch, or fly through cover and flush out its prey. At the moment of impact, a special bone-and-tendon arrangement causes the toes to clench, driving the talons deep into the prey's vitals. A snap from the stout, hooked bill can crush a prey animal's skull or break its spine. Hawks often "mantle" their prey, crouching and spreading their wings to hide the kill from other predators. The bird may eat its meal on the ground or carry it to a feeding station, often a fence post or tree limb, where, with its beak, it plucks or opens the skin and tears out pieces of meat. Hours after feeding, the hawk regurgitates a pellet containing feathers, fur, or small bones swallowed accidentally.

In addition to the familiar red-tailed hawk (described in a separate chapter), seven other hawks breed or are seen frequently in Pennsylvania. Identifying hawks can be difficult. Although males and females of a given species are generally of similar colors, there can be much individual variation in plumage. Juveniles are especially hard to identify. Within the same species, adult females tend to be larger than adult males—in some cases, nearly twice as heavy.

Hawks nest high above the ground on tree limbs and rock ledges. Nests are loosely built of sticks and twigs; some species line their nests with feathers and down. A mated pair will either remodel an old nest or build a new one, sometimes starting on top of a squirrel or crow nest. The female usually begins incubating her clutch before the last egg is laid, so that some young hatch earlier than others, resulting in hatchlings of different sizes in the same nest. The female does most of the incubating and is supplied with food by her mate.

Newly hatched hawks are altricial: blind (or nearly so), unfeathered, and covered with down. They grow rapidly. After about two weeks, when the young no longer require constant brooding, the female joins the male in hunting to feed them. In the nest, the young learn to tear meat apart and feed themselves. After five or six weeks, when their flight feathers have grown in, they begin taking short flights; several weeks later the fledglings start hunting on their own.

Hawks help control insect, rodent, and small bird populations. They improve a prey species by forcing animals to develop alertness, speed, and other survival attributes, and by weeding out unfit individuals. Hawks and other raptors are environmental indicators. If toxins accumulate in natural food chains, the predators—which eat many prey animals and thus concentrate toxins in their own bodies—are usually the first to show effects, which can include failure to reproduce, eggs whose shells are so thin that they break, eggs that do not hatch, or outright death of the adults from poisoning. Heavy metals and chlorine-based pesticides such as DDT have severely depleted hawk numbers in the past; laws enacted during the late twentieth century have limited the use of pesticides in North America, and hawk populations have increased.

Until recently, birds of prey were often labeled "chicken hawks" and shot on sight. Research has shown that while hawks do kill some poultry

and game, in most cases they do not cause great economic damage or drastically deplete game populations. Today it is illegal to shoot hawks in the
United States. Although hawk populations have rebounded here, the species
are not secure. Many hawks winter in Central and South America, where
deforestation, hunting, and the use of pesticides continue to endanger them.

The seven hawks described in this chapter include three accipiters,
three buteos, and one harrier.

Accipiters (sharp-shinned hawk, Cooper's hawk, and northern goshawk)
have small heads, long tails, and short wings that are rounded at the tips.
Accipiters fly with rapid wingbeats followed by a long glide. Extremely
maneuverable, they can flash through forested and brushy areas, surprising
smaller birds and overtaking them with a burst of speed.

Buteos (red-shouldered hawk, broad-winged hawk, and rough-legged
hawk) possess stocky bodies, broad wings, and short tails. Buteos perch in
wooded or open country or soar in wide circles when hunting. Small
mammals form the bulk of their prey.

The northern harrier is the only harrier in North America. Harriers
are long-legged hawks with long, narrow wings and elongated tails. The
northern harrier glides with its wingtips held perceptibly above the horizontal (much like a turkey vulture), quartering above open country in
search of prey.

Sharp-Shinned Hawk *(Accipiter striatus)*. Length, 10 to 14 inches; wingspread, 20 to 27 inches; weight, 5 to 9 ounces. It can be hard to tell this
species from the similarly marked Cooper's hawk, since large female sharp-
shinned hawks are nearly the size of small male Cooper's hawks. Adults have
red eyes and are blue-gray above, with rufous barring on the breast. Immatures are brown above and heavily streaked below.

Sharp-shinned hawks feed mainly on small birds such as sparrows, warblers, and vireos, and on prey as large as robins. They also take rodents, bats,
squirrels, lizards, frogs, snakes, and large insects. These accipiters inhabit
woodlands, thickets, and woods edges. They breed throughout the United
States and Canada, and they winter south to Panama. In recent years, backyard bird feeders have kept purple finches, pine siskins, and other seedeaters
farther north than in the past; sharp-shinned and Cooper's hawks have
stayed, too, to take advantage of the small birds clustered near the feeders.

Sharp-shinned hawks nest in May in Pennsylvania. They often hide
their nests in dense conifers or in the thick foliage of deciduous trees 20 to
60 feet above the ground. The four or five eggs are white or bluish and

*Accipiters such as the sharp-shinned hawk have learned
to hunt for prey around bird feeders.*

blotched with brown. Incubation is mostly by the female and takes thirty to
thirty-five days. Around the nest, adults voice a *kek kek kek* call; in flight
they utter a shrill scream. Since the 1980s, counts of migrating sharp-
shinned hawks in the East have declined significantly, perhaps caused by the
loss of wintering habitat farther to the south or by increased residential
development in the species' northern breeding range.

Cooper's Hawk *(Accipiter cooperii).* Length, 14 to 20 inches; wingspread, 27
to 36 inches; weight, 10 to 20 ounces (slightly smaller than a crow). Adults
look like large sharp-shinned hawks, with red eyes, blue-gray backs, and
rusty breasts, except that the Cooper's hawk has a rounded tail, in contrast
to the square-tipped tail of the sharp-shin. Named in 1828 after William
Cooper, a New York naturalist, Cooper's hawks prey mainly on birds the
size of robins, jays, and flickers. They eat small mammals, including squirrels, chipmunks, mice, and bats. They hunt by moving from perch to perch
and by flying stealthily through dense cover, then darting in on prey. Over
the years, I've found several Cooper's hawks dead in the woods, apparently
killed by crashing into tree limbs.

Cooper's hawks breed throughout North America. In Pennsylvania and
New Jersey, the population is currently expanding. Cooper's hawks inhabit

open woods, woods edges, bottomland forests, and, increasingly, suburbs where there are mature trees for nesting. Cooper's hawks often place their nests on older structures such as abandoned crows' nests. The nest, 25 to 50 feet above the ground, is a bulky mass of sticks sometimes lined with softer tree bark. The female lays three to five bluish white eggs; incubation takes thirty-four to thirty-six days. The male brings food to his mate and incubates the eggs while she feeds. Cooper's hawks shift southward in the winter.

Northern Goshawk *(Accipiter gentilis)*. Length, 20 to 26 inches; wingspread, 40 to 47 inches; weight, 1.5 to 3.5 pounds. Adults are blue-gray above and white below, with light barring on the breast. Immatures are brown above and creamy white below, with heavily streaked undersides. Immatures and adults have a prominent white line over each eye, and the eyes of the adults are bright red. Goshawks are also called blue darters.

The largest of our accipiters, goshawks prey on medium-size birds such as grouse and crows, and on smaller birds. They take squirrels, rabbits, snowshoe hares, small rodents, snakes, and insects. Swift-flying, maneuverable, and relentless, they may land and pursue their prey on the ground, running through thick underbrush with their wings trailing behind. Goshawks breed in open woods and on woodland edges, with a preference for remote areas. They nest up to 75 feet above the ground in trees, often on a north-facing slope. The bulky, stick-built structure is 3 to 4 feet in diameter. A pair may use the same nest year after year. The female lays three or four bluish white eggs and incubates them for thirty-two to thirty-eight days. Goshawks defend their nests fiercely, diving at intruders, including humans, and sometimes drawing blood. Pennsylvania is on the southern edge of the goshawk's breeding range; nests have been found mainly in the northern half of the state. In winters when prey is scarce in the northern latitudes, many goshawks invade the Keystone State.

Red-Shouldered Hawk *(Buteo lineatus)*. Length, 18 to 24 inches; wingspread, 33 to 50 inches; weight, 2 to 3 pounds. Adults are dark brown above, with chestnut-red shoulders (the shoulder markings are not always visible); the breast and belly are reddish brown. The tail is strongly barred with black and white. Many individuals have a translucent "window" area near the wingtips, which can be seen from below.

Red-shouldered hawks inhabit dense lowland woods interspersed with small clearings and marshes. They prey on voles, mice, chipmunks, frogs, toads, snakes, and insects. Red-shouldered hawks hunt from an exposed perch offering a wide field of view or by flying across open areas and taking creatures by surprise. Their voice, a piercing *kee-yer,* is often mimicked by blue jays. They build their nests in trees, 35 to 65 feet above the ground, laying three or four bluish white, brown-blotched eggs. Incubation takes a little over a month. Young leave the nest after six weeks, and both parents feed them for another eight to ten weeks. In Pennsylvania, the population has declined over the last century as forested lowlands have been cleared for agriculture and development. On the plus side, bird watchers are reporting more red-shouldered hawks nesting in suburban areas with stands of mature trees. To protect *Buteo lineatus,* we should preserve large tracts of mature woods near wetlands and waterways.

Broad-Winged Hawk *(Buteo platypterus).* Length, 13 to 19 inches; wingspread, 32 to 39 inches; weight, 13 to 20 ounces. This chunky, crow-size buteo is easily recognized by its heavily banded tail with two dark and two light bands. The upper plumage is a dark gray-brown; the underparts are white, heavily streaked with brown. The broad-winged hawk inhabits forests and preys on mice, voles, squirrels, small birds, snakes, amphibians, and large insects. Common and abundant, it is rather unwary and approachable. The voice is a high whistled *p-we-e-e-e.*

Broad-winged hawks breed in deciduous and mixed woods, often near clearings or edges. They build their nests in the lower branches of large trees 24 to 40 feet up. One year a pair nested in an oak about 100 feet from our house; I set up a spotting scope in my office and watched as they finished building the nest (formerly a squirrel's nest) and as the female laid eggs and began incubating them. The male brought food to the female and incubated while she ate. After a few weeks, I could see three white-feathered nestlings poking their heads up. The parents brought them small rodents, birds, and snakes. The young fledged and departed after about six weeks.

Broad-winged hawks nest across eastern North America; they arrive in Pennsylvania in late April and early May. The birds head south again mainly in the second and third weeks of September, in loose flocks that often congregate in "kettles" of rising air, which they use to gain altitude. The flocks are particularly visible along Pennsylvania's coastal plain; residents of Lancaster, Philadelphia, Chester, Delaware, Bucks, and Montgomery counties can often look up from their backyards and see groups of broad-winged hawks circling and climbing. *Buteo platypterus* winters in the tropical rain forest, with some individuals going as far as Brazil and Peru, up to 4,800 miles one way.

Rough-Legged Hawk *(Buteo lagopus)*. Length, 19 to 24 inches; wingspread, 50 to 56 inches; weight, 2 pounds. This species comes in light and dark color phases, with much individual gradation in between. The feet are feathered to the toes, givings the legs a ragged appearance: its feathery pantaloons help keep this northern bird warm in winter. The rough-legged hawk nests in the Arctic, mainly on rock ledges north of the boreal forest. It feeds on lemmings, voles, and birds. We see rough-legged hawks in the winter, when prey scarcities in the north may drive them as far south as Virginia. Rough-legged hawks seek out open land that resembles their tundra home. They hover above fields, beating their broad wings in short, rapid strokes while scanning the ground for prey.

Northern Harrier *(Circus cyaneus)*. Length, 18 to 24 inches; wingspread, 40 to 54 inches; weight, 12 to 16 ounces. Harriers—also called marsh hawks—are slender raptors with long wings and tails and a ruff of feathers around the face much like the facial disk of an owl. Males are bluish gray above and white below; the tail is gray with dark bands. Females are brown above, light brown with dark streaks below, the tail barred black and buff. Both sexes show a white rump patch. The voice is a nasal *pee, pee, pee*.

Harriers inhabit freshwater and saltwater marshes, wet meadows, bogs, flat open farmland, and revegetated strip mines. Harriers hunt by quartering low across the land or hovering on rapidly beating wings. They take mice, voles, rabbits, birds, small ducks, snakes, frogs, toads, and insects, especially grasshoppers; like owls, they may rely on their hearing to locate prey concealed by grass. Harriers nest on or near the ground in brush, sometimes on a branch above water. The female lays four to six eggs, which hatch after about a month. Harriers sometimes nest in loose colonies, with one male having two or more mates. The population in North America has been declining for more than thirty years as the loss of wetlands, urban and suburban sprawl, reforestation, conversion of hayfields to row crops, and intensive farming have all reduced harrier habitat. In Pennsylvania in recent years, harriers have nested on reclaimed strip-mined lands, especially in Clarion County.

Weidensaul, S. *Raptors: The Birds of Prey.* New York: Lyons & Burford, 1996.

RED-TAILED HAWK

In the 1960s, seeing a hawk—any hawk—was a rare and stirring event. Today, with laws in place to protect avian predators and an ongoing effort to cleanse the environment of toxins that once harmed their breeding, we see hawks much more frequently, and the red-tailed hawk is one of the most common species. *Buteo jamaicensis* belongs to the buteos, a group of hawks known for their soaring abilities. Found from Alaska to Atlantic Canada and south to Panama and the Virgin Islands, the red-tailed hawk is the most widespread large hawk in North America. It is a year-round resident in the Northeast.

Biology. This hawk has a chunky body, broad wings with rounded tips, and a broad, rounded tail. Its color varies from pale tan to dark brown. In light-phase birds, dark markings are often visible on the leading edges of the wings. In adults, the tail is an eye-catching brick red when seen from above, and a paler buff when viewed from below. Look for red-tailed hawks hanging above fields on motionless wings, a behavior known as "kiting." Another way to locate them is by hearing crows "mob" a hawk by flying around the perched or fleeing raptor while cawing angrily.

Adults are about 2 feet long and have a 4-foot wingspan. Males weigh around 36 ounces and females 42 ounces. The call, given most often in flight, is a hoarse, high-pitched scream: *kee-eeearr.*

When hunting, red-tailed hawks usually sit on perches such as dead snags or phone poles in open areas where they can see for long distances. Spotting their prey, they swoop down and seize it in their talons. Red-tailed hawks can reach speeds of up to 120 miles per hour when diving. They hunt while kiting, soaring, and quartering low over fields. They take a variety of small to medium-size mammals, including voles, mice, rats, squirrels, muskrats, and cottontails. They catch and eat starlings, sparrows, flickers, red-winged black-birds, pigeons, waterfowl, and ring-necked pheasants. Small mammals are swallowed whole; birds are usually beheaded, plucked, and then eaten. Red-tailed hawks occasionally feed on insects, toads, snakes, and fresh carrion. They eat 2 to 5 ounces of food per day—less in summer, more in winter.

One study found that during summer, adults spent 3 percent of their active hours soaring, 1 percent in flapping flight, and 96 percent of their time perching. Another survey indicated that 25 percent of time was spent in soaring. At night, red-tailed hawks roost in trees with dense foliage. In cold weather, they fluff up their plumage and draw one leg back into their feathers.

The species breeds in moist woodlands, either hardwood or mixed conifers and hardwoods. Breeding territories are 1.5 to 2 square miles. Red-tailed hawks are monogamous and probably mate for life, although an individual whose mate has died will quickly choose a replacement. Courting begins in late winter. Birds soar in wide circles high in the sky, crying shrilly; the male may pass food to his mate, and sometimes the pair engages in mock combat, coming together and briefly locking talons.

Red-tailed hawks nest high up in mature trees and occasionally on cliffs. Pairs may add material to a crow or squirrel nest, remodel their last year's nest, or build anew. Nests are about 30 inches in diameter, constructed of sticks, with a 14-inch central cup that may be lined with strips of bark, fresh conifer sprigs, corn husks, or aspen catkins. The female may start laying as early as mid-March, although most eggs are laid from early April to mid-May. Clutches number two to four eggs, which are whitish and blotched with brown. In the early stages of breeding, red-tailed hawks are nervous and sensitive and may abandon their nest if disturbed.

The female does most of the incubating, and the male brings her food. The eggs hatch after thirty to thirty-five days. The male continues to supply prey, while his mate broods the nestlings and tears the prey apart for the

Red-tailed hawks perch on telephone poles and in trees, from which they swoop down to capture smaller birds and mammals.

young birds. If the male cannot find enough food—ten to fifteen deliveries per day are needed—the female joins him in hunting. Left exposed, nestlings and unhatched eggs may fall to crows or ravens; great horned owls also take young red-tails.

Nestlings start stretching their wings and exercising when a month old. They leave the nest about six weeks after hatching, and their parents continue to feed them as they hone instinctive hunting skills. Juveniles begin catching their own prey six to seven weeks after fledging. They may stay with their parents for up to six months.

Red-tailed hawks migrate singly or in small groups. Individuals from northern areas begin shifting south during autumn, although they may delay or shorten their migration if temperatures stay warm and snow does not cover up rodents' runways. Migrating hawks travel at an average of 30 miles per hour. At Hawk Mountain, the well-known sanctuary on Kittatinny Ridge near Reading, 90 percent of all red-tailed hawks seen by observers pass between October 9 and December 1, with the peak during the first week in November.

Hawks glide along above Kittatinny and other south-trending ridges of the Appalachians, "surfing" on currents of air pushed up when the mountains deflect autumn's fresh northwest winds. The late Ned Smith, a noted Pennsylvania naturalist and artist, told me of a day when, around midmorning, he noticed that every red-tailed hawk passing by his ridgetop observation post had a bulging crop. He wondered what they were feeding on until he noticed the repetitive "knocking" calls of gray squirrels all over the mountain. On days of light or no wind, red-tailed hawks and other raptors may not fly at all, or they may climb high on thermals (rising columns of air caused by the sun's heating the ground) and then glide southward, slowly losing altitude until they encounter another thermal.

Based on data from returns on banded individuals, biologists believe that about 20 percent of adults perish in any given year. About half of all juveniles reach one year of age. The maximum lifespan is estimated at twenty-five years.

Habitat. Red-tailed hawks favor open areas interspersed with patches of trees. They inhabit pastures, urban parks, and broken woodland. In general, they prefer more open situations than do their relatives the broad-winged and red-shouldered hawks. The red-tailed hawk has replaced the red-shouldered hawk in partially cleared bottomland forest in much of the eastern and midwestern United States. Several studies imply that the most critical habitat requirement for the red-tailed hawk is an adequate number of perches from which to hunt.

Population. *Buteo jamaicensis* is Pennsylvania's most widespread breeding hawk. It nests in every county in the state, including heavily urbanized Philadelphia County. The population mushroomed in the late 1800s; declined in the early 1900s, because of persecution by humans and the clear-cutting of vast tracts of land; continued in its free-fall while pesticides were widely and carelessly used; and has risen again over the last forty years. Scientists believe that in much of North America, the number of red-tailed hawks is increasing.

FALCONS

A peregrine flying down the dunes at Assateague Island: tapered wings rowing, creamy bib catching the light, its back the color of wet slate. It whipped past me and veered toward a band of teal in flight. The falcon overtook the teal and knifed through them. The peregrine must have fed not long before, because it only yawed slightly in its course and struck at one of the teal in a casual, almost playful, manner; the flock scattered, ducks splashing into a pond and swimming in panic for the grass-choked edge. The falcon leveled off, bore away into the haze, and was gone.

Falcons are streamlined raptors superbly adapted to taking prey in midair. They have large heads, broad shoulders, long pointed wings, and long tails that narrow at the tip. Three species occur regularly in Pennsylvania. The powerful crow-size peregrine—the fastest of all birds—is a rare breeder. The merlin, a shade larger than a grackle, does not breed in Pennsylvania but passes through the state in spring and fall. The American kestrel, a small, colorful falcon, breeds widely across the region. (The kestrel is covered in a separate chapter.)

Peregrine Falcon *(Falco peregrinus)*. The Latin name comes from the species' peregrinations, or wanderings: this falcon inhabits every continent but Antarctica, and some individuals make long transcontinental migrations. Length is 15 to 20 inches; wingspread, 43 to 46 inches; weight, around 20 ounces in the male and 29 ounces in the female. The peregrine has a dark crown and bold, dark "mustache" markings on the cheeks. The back is gray, the belly pale with dark barring. The tail is banded dark and light, and ends in a white tip. Juveniles appear more brownish, with heavily streaked breasts. The large, knouted feet are pale yellow. The male is sometimes called a tiercel, denoting the fact that he is about a third smaller than the female. Flying peregrines show a distinctive crossbow shape.

Peregrines favor open country and often live near seacoasts. Some have moved into cities, where they feed mainly on pigeons. *Falco peregrinus* was once called the duck hawk, reflecting its penchant for taking waterfowl. Peregrines prey on many other birds as well: blue jays, mourning doves, nighthawks, nuthatches, killdeers, herons, robins, gulls, grouse, grackles. Much less frequently, they feed on mammals, insects, and carrion. The normal flight speed is 60 miles per hour. Flying with deep, pumping wingbeats, a peregrine will skim low over plain or water, surprise its prey, and chase it down. Sometimes it launches from a perch. Or it flies high in the air, spots its target below, and stoops, or dives, at a speed in excess of 200 miles per hour. The falcon strikes with its heavy feet, often killing its victim outright and sending it plummeting earthward, streaming feathers. The raptor eats its prey on the ground or carries it to a perch.

Peregrines may mate for life. Pairs perform courtship flights with circling and diving maneuvers and chases by both sexes; the male presents food to the female. Nesting may begin as early as March. Peregrines nest mainly on cliffs, rarely in tree cavities and the abandoned nests of other birds, and sometimes on bridge girders and the roofs and ledges of tall buildings. No actual nest is built; the female lays two to six eggs (usually three or four) in a rude scrape. The eggs are whitish to pale brown, heavily fretted with brown. The female does most of the incubating and is fed by her mate; the smaller size of the male may make him a more successful predator on smaller birds, capable of bringing back greater numbers of prey. The eggs hatch after thirty-two to thirty-five days. The female stays with the young for a week or longer, with the male continuing to supply food. Later, she leaves the nest and joins him in hunting to feed the brood. The young begin taking their first flights when they are about six weeks old and fledge in late May or June.

From the 1940s to the 1960s, peregrine falcons were driven to the edge of extinction in North America by reproductive failure caused by DDT and other persistent pesticides (the population had already been knocked back by human persecution and egg-collecting activities). Starting in 1974, biologists reintroduced the peregrine in many states, and populations began to recover. A private organization, the Peregrine Fund, originally based at Cornell University, used captive breeding to rear thousands of chicks for

*When diving on prey, a peregrine falcon can reach
a speed of more than 200 miles per hour.*

hacking, or releasing into the wild. Chicks that survived and matured to
adulthood have returned to reproduce in areas where they were hacked. In
Philadelphia peregrines have nested on five interstate bridges and on City
Hall, where they formerly bred until the early 1950s. In 1998, peregrines
nested in Wilkes-Barre, Pittsburgh, and Harrisburg, but there is no evidence that the falcons have returned to the cliffs along Pennsylvania's rivers
where peregrines once reared young.

Probably the best way to see a peregrine falcon is by "hawk watching"
on ridges during the autumn migration, when peregrines that bred in
Canada travel to wintering grounds along the coast in the southern states
and as far south as South America. At Hawk Mountain Sanctuary, sightings
of peregrines peak in early October.

Merlin *(Falco columbarius).* This compact falcon breeds across northern
North America and Europe. Length, 10 to 13 inches; wingspread, 23
inches; average weight, 7 ounces for the female and 5 ounces for the male.
In males, the upper parts are blue-gray; in females, brownish. The tail is

banded and the breast streaked brown in both sexes. Merlins seen in Pennsylvania are migrating between breeding grounds in Canada and marshy and open-country wintering areas in the coastal South.

Merlins are also called pigeon hawks; the noted ornithologist Alexander Sprunt contended that the name came from the merlin's resemblance to a pigeon, both in flight and at rest, rather than from any prey preference. Merlins prey mainly on birds, overtaking them and snatching them out of the air: sandpipers, swallows, sparrows, larks, robins, and a host of others, and occasionally birds as large as pigeons and ptarmigan. Merlins take a few rodents and bats and also eat large insects, including dragonflies caught on the wing.

In their northern breeding grounds, merlins lay their eggs in abandoned nests of ravens, crows, hawks, and magpies, and sometimes in tree cavities, on cliff ledges, or on the ground. Four or five eggs are the norm. In North America, *Falco columbarius* is thought to be increasing on the northern prairies while remaining stable elsewhere. In Pennsylvania, merlins are occasional spring migrants in March, April, and early May; in fall, they pass through in September, October, and early November. Biologists have noted that more merlins are wintering in cities and suburbs in the eastern United States; they credit their presence to an increased food supply in the form of songbirds attracted to bird feeders, as well as starling flocks.

KESTREL

The American kestrel, *Falco sparverius,* was formerly called the sparrow hawk because of its small size. It is our smallest falcon and one of the most abundant raptors in the Northeast. Taxonomists currently recognize ten subspecies, or races, ranging from Alaska and southern Quebec south to Florida, the Caribbean islands, and Central America. Because kestrels inhabit open terrain where they can be easily observed, they make interesting subjects for nature study. Persons wanting to attract kestrels can put up nesting boxes.

Biology. I was watching for birds one day, standing by an old stone wall, when I turned to look in a different direction and saw a bird at face level, 10 feet away and closing fast. Adrenaline rushed through my body, and every nerve jangled as I ducked. I realized I'd been buzzed by a kestrel when I turned and watched the little smart aleck flying away. I retained an image of black and white vertical bars marking a fierce face—and a sense of what it must be like to become a falcon's prey.

The male kestrel has slate-colored wings, a buffy breast speckled with black, and a long, rust-red tail. The female kestrel differs somewhat: her wings are the same rusty color as her tail, and her breast is streaked rather than spotted. Dark markings toward the back of the head may act as "eye spots" to deter attacks by other predators. A kestrel's body length is 9 to 12 inches, wingspread is about 21 inches, and weight averages 4 ounces in the male and 5 ounces in the female. The call is a high, strident *killy-killy-killy*.

Kestrels often perch on utility wires or snags while scanning fields for prey. When sitting, they often move their tails up and down. They have a nimble, buoyant flight and may hang in the air above a grassy area, facing into the breeze, their long, tapered wings fluttering as they hover while scanning the ground. (Another name for the species is windhover.) Flight speeds range from 16 to 36 miles per hour; migrating individuals average 24 miles per hour. In summer, kestrels eat many insects: grasshoppers are a favorite, along with crickets, moths, beetles, dragonflies, caterpillars, and others; kestrels also take mice, voles, bats, small birds, frogs, earthworms, spiders, and crayfish. When insects are unavailable in winter, kestrels concentrate on small mammals and birds. An individual requires food totaling 20 to 25 percent of its body weight daily.

In the Northeast, kestrels breed in April and May. During courtship, the male may fly high into the air, calling, and then dive conspicuously toward the ground. He brings food to the female near the nest—a practice that continues until one or two weeks after the young have hatched. Pairs defend a breeding territory measured at 0.43 to 0.7 square mile in a Wyoming study and 0.8 square mile in a Wisconsin investigation. Kestrels are cavity nesters; they favor the abandoned holes of flickers and other woodpeckers but will also nest in natural tree cavities, crevices in barns or other buildings, and nest boxes. Most nests are 10 to 30 feet in the air. The female lays four to six eggs, which are white to pale brown and spotted with gray or brown. The eggs hatch after twenty-seven to thirty-one days of incubating. The young remain in the nest for around a month, with both parents bringing food to them. After they fledge from the nest, their parents feed them for another week or so. Juveniles molt their body plumage in September or October. In the Northeast, kestrels raise one brood a year; birds in the Southeast, or in areas where rodent populations are high, may produce a second brood.

In winter, kestrels from northern areas migrate to the south; young birds seem to go farther than adults. Birds in southern and middle latitudes

Dark markings on the back of a kestrel's head may act as "eye spots" and ward off attacks by other predators.

are often permanent residents. The species winters in the United States, Mexico, and Central America.

The maximum recorded age for a kestrel is seventeen years. The average lifespan is a little over one year, and in one survey, ornithologists found that only fifteen out of more than five hundred birds survived beyond age four.

Habitat. Kestrels live in open and semiopen terrain, including grasslands, farming country, and forest clearings and edges. They require cavities for nesting and perches from which to hunt. Kestrels are fairly common in towns and cities and sometimes nest in bird boxes in suburban yards; biologist Daniel Brauning, editor of the *Atlas of Breeding Birds in Pennsylvania,* observed a pair of kestrels nesting on Philadelphia's City Hall. Studies of wintering kestrels have revealed that females are more likely to hunt in open grasslands, while males seek prey in brushy areas and fields with taller vegetation.

Population. Kestrels nest in every county in Pennsylvania. They are most common in the Piedmont Province in the southeast and least common on the heavily wooded Allegheny High Plateau. Apparently the American kestrel was not affected as seriously by DDT from the 1950s through the 1970s as were the larger falcons and hawks. Today, kestrel numbers seem to be stable in Pennsylvania. However, counts of migrating kestrels suggest that populations have declined in other parts of the Northeast.

RUFFED GROUSE

Designated Pennsylvania's state bird in 1931, the ruffed grouse *(Bonasa umbellus)* is found in most of the Keystone State except for urban areas and extensive farmland. The ruffed grouse is a member of the pheasant family, which in North America includes forest and prairie grouse, pheasants, turkeys, quail, and ptarmigan. These chickenlike birds feed mainly on plant matter, nest on the ground, and often run rather than fly to avoid danger. The ruffed grouse inhabits brushy and wooded terrain from Alaska to northern California and from Labrador to Georgia.

Biology. An adult grouse weighs about 1.5 pounds; body length is 15 to 19 inches; and the wings span 22 to 25 inches. Two color variations, or morphs, occur: gray (found with greater frequency to the north of Pennsylvania) and brown (the color of the vast majority of grouse in Pennsylvania). I hunt grouse and have occasionally bagged a third variant, which I would describe as a metallic copper tone. In the common brown morph, the plumage is a variegated brown sprinkled with white and black above; the breast is pale, with horizontal gold barring. The tail is chestnut brown tipped with a wide black band bordered by two narrower grayish bands. In the male, or cock, the black band is continuous; in the female, or hen, the band is broken or indistinct in its central part. The word *ruffed* refers to a ruff of iridescent black feathers encircling the neck. Cocks are bigger and heavier than hens and have larger and more prominent ruffs.

Like most birds, grouse have keen eyesight and hearing. At one time they were not nearly as wary as they are today: early settlers killed them with sticks and stones. Although a grouse takes off, or flushes, with great power, it cannot fly for long distances because its pectoral muscles are poorly supplied with blood vessels. (For this reason, the breast meat of grouse and most other pheasantlike birds is white rather than red.) Maximum flight speed is about 20 miles per hour. After taking off, a grouse dodges between trees and branches, then sets its wings and glides, usually traveling less than 100 yards.

Grouse are year-round residents and do not migrate, although they may make short movements to areas having dense escape cover or abundant food. In spring, grouse feed on new leaves of trees and shrubs, and shoots of dandelions, clover, wild strawberry, ferns, and other plants. In summer, they eat blackberries, blueberries, and other fruits, the leaves of various plants, and insects, spiders, and snails. In autumn, they turn to small acorns, beechnuts, cherries, barberries, wild grapes, apples, hawthorn and dogwood fruits, various buds and leaves, and flowers of witch hazel. Buds are a winter staple: aspen, birch, beech, maple, cherry, and apple are favorites. Late on a December day, I watched four grouse fly up into an aspen tree, clamber about on the branches snapping up buds for about ten minutes, and then fly off to roost. By feeding quickly on high-energy buds, grouse expose themselves to aerial predators for a minimum amount of time while maximizing their caloric intake.

Ruffed grouse feed on a variety of leaves, shoots,
flowers, nuts, buds, berries, and insects.

Grouse roost in the dense foliage of shrubs and trees, often conifers. In winter, when the snow is a foot deep or deeper, they may fly or burrow into it and "snow roost," letting the insulating properties of the snow keep them warm. (The subnivean environment can be up to 50 degrees warmer than the air temperature.) During winter, a grouse has a fringe of horny skin around each toe, increasing the surface area to provide traction on ice and to let the bird walk on powdery snow without sinking in deeply.

Grouse mate in April and May. Cocks "drum" to proclaim their territories and to attract hens. With tail fanned, the male stands on a prominent log or rock and beats the air with his wings. The rush of air created by his wingbeats sounds like a drum being thumped. The drumming starts slowly and increases in speed until the individual beats merge into a steady whir lasting several seconds. The sound of drumming may carry up to a quarter mile. It probably attracts some predators, but owls apparently cannot hear the low-frequency sound. In prime habitat, drumming grouse will be spaced about 500 feet apart. When a female arrives, the cock fans his tail, erects his ruff, and struts while hissing and dragging his wingtips along the ground. A male may mate with several hens. Males play no role in nesting or rearing the young. Cock grouse drum at other times than in the spring: I often notice them in the fall, and once, inexplicably, I heard a drummer in the moonlight on New Year's Eve.

The hen picks a nesting site on the ground at the base of a tree, in thick brush, or tucked against a log, rock, or root. She lines a shallow depression with leaves, pine needles, and a few feathers. In it she lays nine to twelve eggs. The eggs are buff-colored, sometimes speckled with brown. The female incubates her clutch for twenty-one to twenty-four days until the eggs hatch, all within a few hours of each other. The incubating hen sits tight on the nest, relying on her protective coloration to remain undetected. If frightened off, the hen may not return to her nest but may renest later.

Newly hatched grouse are precocial: they can leave the nest as soon as their downy feathers dry. The chicks follow their mother. She does not feed them, but they immediately begin picking up insects, spiders, and vegetation on their own. At first the hen broods her chicks at night and in bad weather. She can be fearless in their defense. Several times I've been startled by a hen grouse making agitated perking sounds and headed straight toward me; the standard ploy is to veer aside and run off cheeping, dragging a wing behind to draw my attention while her chicks hide. The chicks are marked with dark brown stripes against a tan background, a scheme that blends them in with the leaf duff and ground cover. The home range of a hen and brood is 0.25 to 0.5 mile in diameter.

Young grouse rely on high-protein insects to grow quickly. They can make short flights when two to three weeks old. In years when cold, drenching rains fall in May and June, broods may perish because they cannot find enough insects. Opossums, raccoons, and black rat snakes are major predators on eggs; foxes, hawks, and owls take young birds. Adults fall to raptors, particularly Cooper's hawks, goshawks, and great horned owls, as well as foxes, bobcats, and coyotes. Most grouse die before they are a year old, and few live beyond two years of age.

By autumn, the young are virtually identical to the adults, and in September, they begin dispersing from the areas where they were hatched. Young grouse may take off in apparently undirected flight, and some are killed when they crash into trees, fences, windows, and the sides of buildings; this season of restless movement is known as the "fall shuffle" or "crazy flight." Most young birds travel about 1 mile, then settle into a solitary existence, although loose coveys may form in places where food is concentrated.

Habitat. Prime ruffed grouse habitats include abandoned upland farms with pastures overgrown in shrubs; areas that were logged within the last five to fifteen years, where briers and saplings are pushing up; woods edges; wooded hillsides and ravines with scattered clearings; and wetland fringes. Grouse are less plentiful in uniform mature woods. Both adults and broods need a mix of feeding and sheltering areas. In winter, grouse seek out dense brush in which to hide from predators; they fly or walk to food sources, such as aspen stands and spring seeps where green vegetation remains. Grouse sometimes shelter from inclement weather in conifers and mountain laurel thickets.

Population. Probably grouse were not common in Pennsylvania before European settlers began opening up the old-growth forests. Today *Bonasa umbellus* is found in wooded and brushy habitats throughout the state. The grouse population seems to be cyclical, fluctuating between high and low numbers of birds. The cycles span periods of five to ten years; they may differ from region to region and may be modified by local cover, food, and weather conditions. Some researchers suggest that peak grouse years often end in the number two (1992, 2002, and so on) and that ebb years end in the number seven. In peak years in excellent habitat, two hundred or more birds may inhabit a square mile.

In Pennsylvania, ruffed grouse are not found in urban areas, and they are scarce or absent from the heavily cultivated, highly developed southeast. They are plentiful in the hills and mountains of the Ridge and Valley region, on the Allegheny High Plateau, and in the Poconos. Timber management practices on public lands can be used to maintain good populations.

RING-NECKED PHEASANT

The ring-necked pheasant *(Phasianus colchicus)* is a native of Asia. Wild pheasants found in Pennsylvania are hybrids descending from populations established in Britain as early as the 1500s, combined with Chinese pheasants stocked in North America in the 1890s. Pheasants inhabit semi-open land in farming areas. Plentiful in Pennsylvania in the 1950s and 1960s, pheasants have today disappeared from much of their former range; some observers fear that the species may become extirpated as a breeding bird in the Keystone State. Pheasants remain abundant in the Midwest, the northern prairies, and parts of Canada and the Rocky Mountain West.

Biology. Male pheasants are called roosters, cockbirds, or cocks; females are known as hens. An adult rooster weighs 2.5 to 3.5 pounds; an adult hen, about 2 pounds. The standing height of a rooster is about 1 foot, and its length from beak to tail tip averages 36 inches. A pheasant is long-legged and rangy, with a 20- to 23-inch tail and short, rounded wings.

The hen's plumage is a camouflaging mix of brown, black, and gray. In contrast, the cock's feathers are a gaudy play of reds, browns, golds, and black. A rooster has scarlet cheek patches, a white neck ring, and iridescent plumage that is greenish black on the head, golden brown on the breast, and greenish gray or bluish on the rump and lower back. The tail feathers of both sexes are golden brown with black bars. The rooster's call is a hoarse double squawk, followed by rapid muffled wingbeats that may or may not be audible depending on distance. Roosters crow most often during the spring and summer mating season, especially at sunrise and sunset; they also emit a loud cackle when flushed into flight.

Pheasants eat weed seeds (ragweed, smartweed, foxtail, and others), farm crops (corn, wheat, barley, oats, beans, and buckwheat), fruits and berries (raspberries, dewberries, strawberries, elderberries, barberries, thornapples, rose hips, apples, wild grapes), green shoots, leaves, and grasses. In summer, they eat insects, spiders, earthworms, snails, and the occasional rodent, snake, lizard, or frog. They feed mainly on the ground but may climb into trees for fruit. When pursued, pheasants would rather run than fly, dodging through weeds and brush. Cornered or surprised, they take to the air. Strong fliers over short distances, they can reach a maximum speed of 45 miles per hour. At night, pheasants roost in trees and on the ground among tall weeds. Individuals generally remain within 1 square mile.

When courting a hen, a rooster spreads his tail and wings, and struts; his red cheek patches swell, his head is held low, and his neck feathers are ruffled. One rooster may collect a harem of four to a dozen hens. Pheasants begin breeding in late March or early April, and nesting runs from April to August. The male does not help the hen incubate eggs or raise young.

The hen picks a site on the ground in a hayfield, weedy field, overgrown pasture, or brushy fencerow. A natural hollow, or one scraped out by the hen, is lined with weeds, grasses, and leaves. Surrounding vegetation helps to conceal the nest. The hen lays six to fifteen eggs (usually ten to twelve) over a two-week period; the eggs are plain olive-buff in color. She does not start incubating until all the eggs are laid, so that all hatch at about the same time. Nests are destroyed by farm operations such as hay mowing; by crows, opossums, raccoons, and other predators; and by fires and floods. If she loses her first clutch, a hen will usually renest, with some individuals making three attempts.

After twenty-three to twenty-eight days, the eggs hatch. The young are covered with down, their eyes are open, and they are able to walk about and eat soon after hatching. The hen leads them away from the nest. She sets or "broods" the chicks at night and during cold or rain. Instinctively, chicks will squat and remain motionless at a soft call given by their mother; their coloration, tan with darker streaking, conceals them. Foxes, raccoons, weasels, house cats, dogs, crows, and hawks prey on the young. Foxes, great horned owls, and red–tailed hawks take adult pheasants.

The tail of a wild ring-necked pheasant male,
or cock, can be more than 20 inches long.

The hen guides her brood to food: the farther that feeding areas lie from nesting grounds, the longer the chicks are exposed to predation. Young pheasants concentrate on protein-rich insects for about a month before switching to weed seeds and cultivated grains. By two weeks of age, chicks can fly short distances; after six weeks, their adult plumage starts to come in; and by autumn, birds of the year look like adults. Young roosters can be told from older males by the length and hardness of their spurs: hard, pointed growths that protrude from the backs of their legs. In young birds, the spurs are relatively soft, blunt, and short (0.25 inch or less); older roosters have hard, sharp spurs up to 1 inch in length from the spur tip to the front of the leg. The spurs are defensive weapons used against predators and rival cockbirds.

Except during the breeding season, when roosters stake out territories and hens remain with their broods, pheasants are fairly gregarious, feeding and roosting in groups. Winter flocks are often segregated by sex: small bands of males, larger groupings of females. In the winter, pheasants shelter in conifers, brushy hollows, fencerows, marshes, and forests with a dense understory. Winter cover may lie several miles from the summertime breeding and brood-rearing areas.

Habitat. Pheasants are restricted almost entirely to farmland. Ideal habitat consists of at least 70 percent cropland in an even mixture of hay, small grains, and corn, with adequate winter cover nearby. In Pennsylvania, the agricultural areas southeast of the Appalachians constitute the major pheasant range, but in those areas much land has been lost to development and urban sprawl.

As agricultural practices have changed, good farm habitat has dwindled. The following factors have contributed to the decline in the population of pheasants, as well as bobwhite quail, cottontails, and other farmland species: removing fencerows to accommodate larger farm equipment; a switch to corn and away from small grains; increased production of alfalfa hay, which is mowed early in the year, wrecking nests and killing the incubating hens; putting all possible acreage into production rather than leaving some of it fallow; spraying herbicides to kill seed-producing weeds; and fall tillage that buries waste grain and crop residues below the soil. Pheasants seem to need large blocks of habitat for the dispersal of broods; when development fragments farmland, pheasants cannot repopulate otherwise suitable areas. Between 1950 and 1990, Pennsylvania's farm acreage fell nearly 50 percent, from 14.1 million to 7.8 million acres, and much of the remaining farmland is highly fragmented and intensively farmed.

Population. Pheasants were first successfully introduced into Pennsylvania in 1892, and the state Game Commission began releasing pheasants in 1915. By the 1920s the pheasant was common in agricultural areas in the commonwealth. The population peaked during the 1960s and has gone down since then. The Pennsylvania Game Commission continues to stock pheasants for the fall hunting season, and it is hard to tell how many birds seen during winter are stocked birds and how many are wild-reared pheasants. The southeast has the greatest densities of breeding birds in the state.

Pheasants perish from collisions with cars, hunting, predation, and disease. Biologists estimate that in areas where pheasants are abundant, ninety-five out of one hundred cocks can die each year without affecting long-term population levels, and 65 to 70 percent of hens can perish. However, if 80 percent or more of the hens die in a year, the population will decline.

WILD TURKEY

The wild turkey is a permanent nonmigratory resident of Pennsylvania's woods and mountains. All North American turkeys—including the domesticated bird—belong to the species *Meleagris gallopavo*. Taxonomists recognize at least six subspecies; the race found in Pennsylvania and in adjoining states is known as the eastern wild turkey. Turkeys range from Pennsylvania to Florida, and from the Atlantic lowlands west to Mexico.

Turkeys are related to quail, pheasants, and grouse. Native Americans hunted them for food, and some cultures domesticated the birds. Later, wild turkeys became a reliable food source for white settlers. Benjamin Franklin so admired the big bronze bird that he lobbied to have it designated our national emblem. Comparing it to the bald eagle, he said: "The Turkey is a much more respectable Bird, and withal a true original Native of America."

Biology. Adult males, also called gobblers or toms, are 3 to 4 feet long and stand 2.5 to 3 feet tall. Females, or hens, are shorter by a third. Gobblers average 16 pounds, although some weigh up to 25 pounds. Hens weigh 9 to 10 pounds.

Compared to the domesticated variety, the wild turkey is much slimmer and has a smaller head, a longer neck, and longer legs. Plumage is an overall rich brown. In shade, a wild turkey appears black; in bright sunlight, its feathers gleam with copper, blue, green, and mahogany highlights. The hen's plumage is duller than the tom's, and her breast feathers end in a brown or buff band, while those of the gobbler are tipped black. A gobbler has spurs—bony spikes on the back of each leg, used in fighting—and a "beard," a growth of black hairlike feathers protruding from the breast. The beard, which is not shed during molting, grows quickly for four or five years, and then grows more slowly; it may reach 12 inches in length. A fleshy appendage called a caruncle or a snood dangles from between the gobbler's eyes. In both sexes, the head is bluish gray. In the mating season, a gobbler's head and neck turn deep red; during courtship display, the colors of the head and neck change quickly to blue, purple, and white.

Turkeys make many sounds. Best known is the tom's loud, ringing gobble, *il-obble-obble-obble,* used in the spring to attract females. Other calls include yelps *(keouk keouk keouk),* made by both sexes; the cluck, *kut,* an assembly note; the whistle, or "kee-kee run," of a young bird, *kee kee kee;* and the alarm note, an explosive *putt.* Turkeys have keen eyesight and hearing. They fly at speeds of 40 to 55 miles per hour and can glide for over a mile. Usually they run to escape danger; the strides of fleeing gobblers have been measured at 4 feet, and the top speed is an estimated 15 miles per hour. In the evening, turkeys flap up into trees to spend the night; a flock of six to forty birds may roost in the same tree or in nearby trees. In the morning, the birds glide to the ground, call, and reassemble.

In spring, turkeys eat tender greens, tubers, and nuts left from the previous fall. As the weather warms, they turn more to insects, including grasshoppers, beetles, dragonflies, and various larvae, as well as spiders, centipedes, snails, and slugs. But even in summer, up to 90 percent of the diet is vegetable: fruits, seeds, roots, bulbs, stems, leaves, flowers, and buds of many plants. In the fall, turkeys eat mast (beechnuts, chestnuts, acorns, and other nuts); fruits (dogwood, grape, cherry, gum, and hawthorn); seeds (grasses, sedges, and weeds); and crops (corn, oats, and wheat). In winter, they kick down through the snow to find nuts, seeds, and fruits; they also eat green plants, crustaceans, and insect larvae found in spring seeps kept ice-free by continually emerging groundwater. Turkeys scratch for food by raking backward with one foot, raking back twice with the opposite foot, and then repeating with a single stroke of the first foot. The forest floor may be scraped almost free of leaves where turkeys have fed. Turkey droppings are diagnostic: J-shaped scats are left by toms, and compact spiral-shaped piles are deposited by the hens.

In March, toms begin gobbling, mostly early in the morning. Blowing a car horn, beating on a tin pan, or making almost any loud noise may provoke a tom's lusty gobbles. Hens are tolled in by the calling. In the presence of a hen, the gobbler fans his tail, erects his feathers, and tucks his swollen head back against his body. He struts back and forth, hissing and dragging his wingtips on the ground. Rival males fight: each grasps the other's head or neck in his beak and tries to shove or pull his foe off balance. A dominant male may collect a harem of eight to twelve or more hens. Males play no part in nesting, incubating eggs, or rearing young.

In late April, after mating, the hens slip away from the harem. They nest on the ground in wooded or brushy areas, near clearings, old roads, or weedy fields. The nest is a leaf-lined depression under the curve of a fallen log or at the base of a tree. The gobbler's sperm is stored in the hen's oviduct, and fertilized eggs can be laid four weeks after mating. The hen lays an egg nearly every day until her nest contains eight to fifteen (the average is twelve); smaller clutches are produced by younger birds. The hen may cover the eggs with leaves when she departs the nest. She begins incubating with the last egg laid. If disturbed during laying or incubating, she may abandon her clutch. Eggs are pale buff, spotted with reddish brown. Coyotes, foxes, bobcats, and horned owls prey on nesting hens; these predators also eat eggs, as do raccoons, opossums, skunks, minks, red squirrels, crows, and snakes.

Incubation lasts about twenty-eight days. Within a few hours of hatching, the downy poults leave the nest. The hen leads them to open feeding areas. Poults concentrate on insects—leafhoppers, crickets, grasshoppers, and others—and also eat tender greens and fruits. The hen broods her young nightly for at least two weeks, until their wings develop and they can roost in trees. When poults are about three weeks old, several families may merge into a large flock that forages over several hundred acres daily. By six weeks, poults are fairly strong fliers, and by autumn, they are practically self-sufficient.

Autumn flocks often contain several hens and their young. Old toms usually remain apart, in pairs or trios. During early winter, the family groups disperse, and new flocks form: hens, with adults and juveniles mixed; young toms, called jakes; and mature toms. In cold or stormy weather, turkeys may roost for a week or more without feeding. They often forage in areas where white-tailed deer have pawed through the snow to uncover ground-level food sources. In harsh winters, turkeys starve to death if they can't reach food beneath the snow. Snow depth appears to limit the northern extent of the species' range.

Most turkeys die before they reach two years of age. Only a few predators dare to tackle adult birds; turkeys are more likely to perish from disease, starvation, and hunting. Some individuals reach age five, and a few survive for ten years or longer.

Habitat. Turkeys do best in mature deciduous and mixed woods with water sources and grassy openings. Shy and secretive, they shun areas of high human activity. Hens and poults use forest clearings, where the poults hunt for insects. Spring seeps provide insects and green vegetation in winter; at this time, turkeys feed and loaf in wooded gullies and creek bottoms and on sunny south-facing slopes.

Pennsylvania's prime turkey range is in the mountainous north-central plateau in the area bordered by the Allegheny River on the west and the North Branch of the Susquehanna on the east. In the south-central part of the state, hardwood forests on the Appalachian ridges provide another excellent habitat. *Meleagris gallopavo* is found throughout Pennsylvania, except in highly urbanized zones and the area around Philadelphia.

Population. By the late 1800s, few wild turkeys were left in the eastern United States, because widespread logging had destroyed their woodland habitat and unrestricted hunting had further slashed their numbers. The only part of Pennsylvania where turkeys still lived was the south-central mountains, particularly in Huntingdon County. A restocking program may have helped to boost the population starting in 1915, but what really brought the turkey back was a trap-and-transfer effort begun by the Pennsylvania Game Commission in 1956, combined with natural range expansion. Today, following summers of good poult production, statewide populations exceed 150,000. Other states have also worked to restore the wild turkey and to establish new populations, and biologists believe the species now has a larger range than it did before European settlement.

BOBWHITE QUAIL

The northern bobwhite quail, *Colinus virginianus,* nears the northern edge of its range in Pennsylvania. Quail are heard more often than they are seen, and the bobwhite is named for its call, a whistled *bob-bob-white* whose final note is loud and ringing. The species lives throughout the South, and west to Texas, Mexico, and Guatemala. Quail are year-round residents and do not migrate. In Pennsylvania, the quail population is not large and has been declining over the last century.

Biology. An adult bobwhite is 8 inches long, stands about 6 inches high, and weighs 7 ounces. Stout and chunky through the body, a bobwhite has a small head, short wings, and a short, rounded tail. Plumage is chestnut brown, white, and black, with the brown graying toward the tail. The sides are streaked with orange-brown, and the underparts are white or creamy, barred lightly with black. The plumage of the male, or cock, differs from that of the female, or hen: the cock has a white throat and eye line separated by a dark brown band, while the hen has a buffy throat and eye line with a light brown dividing band. In addition to the *bob-white* call, quail sound subdued clucks, and individuals separated from their fellows give an assembly call, a shrill *ka-loi-kee.*

Social birds, bobwhites gather in groups called coveys. Depending on the time of year, a covey will contain ten to thirty members. The birds range up to a quarter mile daily, covering 10 to more than 100 acres. Covey members forage together, rest together, and sleep as a unit at night. Sleeping bobwhites crouch on the ground, forming a circle with their tails together and their heads pointing outward like the spokes of a wheel. Such group roosting helps each individual to maintain body heat. If disturbed, the quail all flush in different directions.

Bobwhites eat seeds of ragweed, poke, beggarweed, foxtail, switch grass, pigweed, and other plants. They also consume insects, including Japanese beetles, June beetles, potato beetles, grasshoppers, crickets, and aphids. They glean fields for waste grains such as corn, wheat, and sorghum. They eat young greens in spring; insects in summer; nuts, berries, wild fruits, and green plants in fall; and weed seeds in winter. They find most of their food by scratching through the litter covering the ground.

When a cock courts a hen in the spring, he turns his head to one side, showing off the black and white facial pattern. He bows low, elevates his tail, droops his wings, and puffs out his body feathers. He makes short rushes at the hen, then walks around her with his tail fanned. Both members of the pair may choose a nesting site: a shallow depression in high grass or weeds along a fencerow, roadside, or stream bank, or in a field of timothy, alfalfa, or clover. In a study of over six hundred nests, 56 percent were in brown-sedge fields, 16 percent were in woodlands, 15 percent were in fallow fields, and 13 percent were in cultivated fields.

Both sexes build the nest, with the hen doing most of the work. The nest is lined with grass and leaves; weeds may be woven into an arch overhead, completely hiding the nest. The hen lays twelve to twenty eggs (typically fourteen to sixteen), which are dull or creamy white and pyriform—pointed at one end and shaped like a top. Both sexes incubate the eggs, with the

In spring, the male bobwhite calls to attract, court, and keep in contact with his mate.

male setting about a quarter of the time. The eggs hatch after twenty-three or twenty-four days, usually around the middle of June in Pennsylvania. In the past, ornithologists believed bobwhites were strictly monogamous, and the pair raised their brood cooperatively. Recent research shows that some females lay a clutch of eggs, then go off and mate with another male while the first male takes over incubating the eggs and rearing the young. In much of its range, the bobwhite produces two broods each year. Nests are destroyed by farm operations, such as early hay mowing, and by predators, including skunks, opossums, raccoons, crows, and snakes.

The eggs hatch on the same day, and the precocial young leave the nest shortly thereafter. They feed themselves, at first concentrating on protein-rich insects. The chicks are fuzzy, buff beneath and mottled chestnut brown above, with a dark streak extending back from each eye. The parent or parents brood the young, settling on top of them at night and during rain. Foxes, weasels, hawks, and house cats prey on quail, both chicks and adults. Extended periods of cold or rain can kill the chicks.

When two weeks old, young quail can fly short distances, and after ten weeks, they have achieved most of the speed and agility of their parents. After their feathers grow in, four-month-old birds are nearly identical to adults, except that birds of the year have pale tips on their outer primary coverts, or wing feathers, while the same feathers are uniformly gray on the adults. In summer, quail often take dust baths and pick up grit for their gizzards along dirt roads and field edges. In autumn, the family groups break up; later, bobwhites regroup into winter coveys that average twelve birds but can be as large as thirty. Bobwhites are vulnerable to hard winters. When deep or hard-crusted snow covers food sources, many birds starve.

Habitat. The range of the northern bobwhite contrasts with that of the ruffed grouse, with the grouse dwelling in wooded uplands and the quail inhabiting agricultural lowlands. Northern bobwhites thrive in brushy abandoned fields, open pinelands, and agricultural lands; *Colinus virginianus* has been characterized as a bird of "farmland and early successional stages." Ideal habitat consists of a mix of grassland, cropland, brushy cover, and woods. To protect them from predators and for shelter during heavy weather, bobwhites need dense cover such as blackberry thickets, fencerows, wild grape tangles, and fields overgrown with greenbrier or scrubby pines. Quail also require grassy or weedy areas for nesting and feeding.

Throughout the bobwhite's range in Pennsylvania, habitats have dwindled, both in quantity and quality, as farming has changed over the last fifty years. The clearing out of fencerows has destroyed shelter areas, and the increasing use of herbicides and pesticides has suppressed weeds and insects, both important food sources for bobwhites. At the same time, many farms have reverted to woodland or become covered with houses.

Population. Pockets of bobwhites exist in open country in the lower third of Pennsylvania. The southeastern and south-central regions once supported the largest populations in the state; now they have only remnant populations. As the twenty-first century opens, there may be fewer than two thousand wild bobwhites left in Pennsylvania. In contrast, during the 1930s an estimated one hundred thousand to two hundred thousand bobwhites were bagged by hunters each year.

When harsh winters wipe out local coveys, quail may not be able to repopulate Pennsylvania from the south, because corridors of undeveloped land no longer exist to let birds move northward out of Maryland. Although quail were once common in Chester, Lancaster, York, Cumberland, Adams, Franklin, and Fulton counties, now they are rare. Small local populations may arise from pen-reared birds, released mainly for training hunting dogs, but these populations probably are not self-supporting. In fact, some observers question whether there are any wild, self-supporting quail populations left in the state today.

The northern bobwhite is listed as a candidate for inclusion on Pennsylvania's state endangered or threatened species list.

RAILS, MOORHEN, COOT, AND SANDHILL CRANE

Ease a canoe down a twisting marsh channel, and if you're lucky, you may glimpse one of the five kinds of rails that migrate through Pennsylvania or breed here. In areas of deeper water, look for common moorhens and American coots. The best times to see these wetlands birds are during their spring and fall migrations. The sandhill crane, a midcontinental marsh species, has been migrating through Pennsylvania with increasing frequency and has begun nesting in the northwestern corner of the state.

Rails. Highly secretive, rails creep about through thick marsh vegetation. Many are more active at night than during the day and often are heard rather than seen; when a hiker or duck hunter does stumble upon a rail, it will usually run away through the grass rather than take to the air. Rails' narrow bodies let them slip between dense cattails and sedges. They search for food by walking about on their long-toed feet, clambering over lily pads and other emergent or submerged vegetation. Some of the rails swim readily; flanges of skin on each toe push against the water to provide propulsion, then fold backward on the return stroke to reduce resistance.

Most rails are omnivorous feeders. Some species concentrate on plants (mainly seeds, leaves, and roots of aquatic plants), and others dine on insects, spiders, snails, crayfish, and other invertebrates. The shapes of rails' bills varies with their feeding habits: the king rail has a sharp bill like a heron's, for snatching prey, while the black rail uses its shorter, more generalized bill for picking up seeds of bulrushes and other marsh plants, as well as for catching insects and crustaceans.

Rails nest among dense vegetation, either on the ground or in reeds or shrubs above shallow water. Most females lay five to twelve cryptically colored eggs and incubate them for three to four weeks. The downy chicks leave the nest soon after hatching. Both parents are thought to feed the young.

Yellow Rail *(Coturnicops noveboracensis)*. Ornithologists frequently describe the yellow rail as one of the most secretive birds in North America. This yellowish swamp dweller breeds mainly in southern and central Canada and winters along the Atlantic and Gulf coasts. In Pennsylvania, the yellow rail is a rare migrant, passing through the state in April and May and again in September and early October. The species eats a variety of insects, as well as many seeds.

Black Rail *(Laterallus jamaicensis)*. This slate-colored, sparrow-size bird breeds in tidal marshes from New Jersey to Florida and in inland marshes south of the Great Lakes; most authorities doubt that it nests in Pennsylvania. Marsh-visiting bird watchers sometimes glimpse black rails during the spring and fall migrations. The species winters in the southern United States, Central America, and the Caribbean islands.

King Rail *(Rallus elegans)*. The king rail is one of Pennsylvania's rarest breeding birds and has been designated a state endangered species. Its breeding range includes the southeastern and midwestern states. Three Pennsylvania breedings documented during the 1980s were in Philadelphia, Tioga, and Mercer counties. The king rail is reddish in color and about the size of a chicken. The largest of the North American rails, it preys on frogs and small fish as well as aquatic insects.

Virginia Rail *(Rallus limicola)*. This rail breeds in wetlands with sedges and cattails in scattered locations across Pennsylvania. Mainly nocturnal, it eats insects and their larvae, including beetles, flies, and dragonflies. Virginia rails build a nest on a platform of cattails, grasses, and reeds, in a dry zone of the marsh where living vegetation may form a canopy overhead.

Sora *(Porzana carolina)*. The origin of the sora's name is unknown. The most widespread and abundant of the North American rails, the sora breeds across Canada and the northern United States, including Pennsylvania, and winters in the southern states and Central and South America. With its short bill, the sora eats primarily seeds. I have seen this bird on several occasions in the Bog Natural Area at Black Moshannon State Park; once, when I was snooping around in a pair of hip boots, a sora lifted from the grass a

*The sora is one of five species of rails, marsh birds that
breed or migrate through the Northeast.*

few feet ahead of me and fluttered off weakly into the reeds. I hope some-
day to hear the species' song, which has been described as "an explosive,
descending musical whinny."

Common Moorhen *(Gallinula chloropus).* This bird was formerly called
the common gallinule. Its assorted cackles, clucks, croaks, and squawks help
make the marsh a magical, spooky place at night. Moorhens are mainly
dark gray, with red bills; while swimming, they make pumping motions
with their heads. They favor deeper water than the rails and often swim
among water lilies and pondweeds. Moorhens feed on buds, leaves, and
seeds of water plants, fruits and berries of land plants, and various insects
and other invertebrates. They nest mainly in thick cattails. In Pennsylvania,
moorhens nest around Pymatuning Reservoir and Conneaut Marsh and in
scattered wetlands elsewhere in the state. Moorhens migrate in spring and
fall to wintering areas from coastal North Carolina southward.

American Coot *(Fulica americana).* The coot is an uncommon breeder in
Pennsylvania but a common to abundant migrant. Hunters lump it in with
the moorhen by consigning to both the ignominious name "mud hen."
The American coot is dark gray with a bone white bill. Noisy and gregarious, coots often form flocks. They eat plant foods and also prey on insects,
fish, tadpoles, snails, crayfish, and the eggs of other birds. They feed like

ducks, upending in shallow water; they also dive to get at plants and graze on land. To take off from the water, they must first run along on the surface to build up speed. Coots need extensive marshlands for breeding. In Pennsylvania, they nest mainly in the northwest around Pymatuning Reservoir and in other wetlands areas. During mild winters when lakes and rivers don't freeze over, many coots may winter in Pennsylvania.

Sandhill Crane *(Grus canadensis)*. I was amazed, on the morning of March 22, 1998, to see what appeared to be a great blue heron feeding in the muddy clearing destined to be a horse pasture to the south of our house. I ran for the binoculars and was even more astonished to discover that my "heron" was a sandhill crane—instantly recognizable by its red forehead and tufted rear end. Sandhill cranes breed on the midwestern prairies and winter in the Gulf states. The population is burgeoning in a natural expansion that has now reached the Northeast. Recently, newly fledged sandhill cranes have turned up in several counties in northwestern Pennsylvania, and more and more migrating cranes are spotted in the state during spring and fall, mainly when they are feeding in farm fields. Sandhill cranes nest in grassy areas around marshes. They eat insects, plants, rodents, frogs, snakes, berries, seeds, and grains. Like whooping cranes, sandhill cranes have a spectacular courtship dance, in which pairs of birds leap into the air, flapping their wings and calling.

Taylor, B. *Rails, A Guide to the Rails, Crakes, Gallinules and Coots of the World.* New Haven, CT: Yale University Press, 1998.

KILLDEER, SHOREBIRDS, GULLS, AND TERNS

Unlike some neighboring states, Pennsylvania does not have a wide variety of breeding shorebirds, since the Keystone State lacks a seacoast and tidal marshes. Three species breed in Pennsylvania: the killdeer, a type of plover, nests in dry situations statewide; the spotted sandpiper breeds along rivers, lakes, and ponds; and the upland sandpiper nests in open grasslands, including reclaimed strip mines. (The American woodcock and common snipe, both of which breed in Pennsylvania, are also classified as shorebirds. They are treated in the next chapter.)

In addition to these breeding species, more than twenty kinds of plovers and sandpipers pass through Pennsylvania in spring and again in late summer. In April and May, shorebirds in their distinctive nuptial plumages are on their way to northern breeding grounds; a good time to search for them is after a strong storm, which may drive migrants in from coastal areas. In August, shorebirds begin returning to wintering grounds along the coasts of North America, Central America, and South America. In spring and fall, look for these wayfarers on sandy beaches, mudflats, marshes, wetlands, stream margins, flooded fields, plowed fields, and expanses of short grass, where they rest and feed upon insects, crustaceans, mollusks, and other invertebrates.

A number of gulls and terns can be seen along Pennsylvania's waterways (and in places far from the water, such as farmers' fields and parking lots), especially during winter.

Killdeer *(Charadrius vociferus)*. The killdeer breeds across much of North America, from Hudson Bay south to central Mexico and the Gulf coast. It winters south to Central America and northern South America. This trim-looking plover has long, narrow, pointed wings and a round head with a short, straight bill. Both the male and the female have a white forehead and a white line above the eye, two black bands across the white throat and breast, an orange-tan rump, and a white-tipped tail. Body length is about 10 inches, wingspread is 20 inches, and weight is 3 to 3.5 ounces.

Killdeers eat beetles, caterpillars, grasshoppers, crickets, ants, weevils, and fly larvae, as well as spiders, earthworms, centipedes, snails, and a few seeds and berries. They forage in typical plover fashion: quickly running ahead, stopping, and suddenly darting the bill forward to pick up food. Killdeers rely on their vision to catch prey, rather than on probing with the bill, the technique employed by most sandpipers. Sometimes they follow tractors and pick up grubs exposed by plowing.

In March and April, male killdeers arrive on the breeding range ahead of the females; most males return to the last year's breeding or natal territory. The birds seek out thinly vegetated pastures, golf courses, airports, roadsides, railroad beds, playing fields, gravel pits, unpaved parking lots, and gravel rooftops, often miles away from any open water. The observer will soon understand that the species name, *vociferus,* is apt. Males circle their territories in the air, flying with slow, deep wingbeats while noisily repeating their namesake call: *kill-dee, kill-dee, kill-dee.* They also call from the ground, uttering sixty to eighty repetitions per minute. Where territories abut, two birds

Killdeers nest and feed in open areas, including pastures,
golf courses, gravel parking areas, and railroad beds.

may run along parallel to each other, calling; killdeers are often attracted by disputes between other killdeers, and it's not uncommon to see four or five individuals calling and displaying at the same time.

Pairs are monogamous, and both the male and the female defend their territory from other killdeers. The male makes a series of scrapes on bare ground, one of which is chosen by the female for the nest; sometimes she adds small, flat pale stones to the depression. She lays four eggs directly on the ground. The eggs are gray-buff, spotted and blotched with dark brown. Most clutches are completed by the second half of April. Once when I found a clutch of killdeer eggs in an area of gravel and scanty grass, I looked away briefly at the scolding, displaying adults, then glanced back at the nest, right at my feet: several moments passed before I could bring the superbly camouflaged eggs back into focus.

After about twenty-five days of incubation, the eggs hatch over a period of four to sixteen hours. The chicks, balls of fluff on matchstick legs, can run about and feed themselves almost immediately. On the first day, their parents lead them to food sources up to 120 feet from the nest; the adults sound a *pup-pup* call to gather the chicks. They brood their young during bad weather and at night, with the male doing most of the nocturnal brooding. To lure predators—including the two-legged variety—away from their young, parent birds cry piteously, display their eye-catching rufous rumps, and drag one or both wings behind them as they lurch across the ground.

Nests may fail because of a late snowfall or predation by snakes, opossums, raccoons, foxes, and other creatures. Young birds make their first flights about a month after hatching. Most pairs raise a second brood, with the latter young fledging in mid- to late June. Especially if it is late in the season, the female may depart shortly after the second clutch hatches, leaving brood-rearing chores to the male. The female is biologically more valuable than the male, and it is in the species' interest that she turn to feeding and building up fat before migration.

In July and August, killdeers gather on mudflats and feed heavily. Sometimes they mix with other shorebirds, including yellowlegs, plovers, and sandpipers. In autumn, they travel by day in small groups, flying high in the air, their calling reaching the observer on the ground.

Killdeers breed throughout Pennsylvania except in dense forest. Some pairs nest in small, isolated patches of habitat, even in heavily wooded areas, such as a few acres of pasture or an expansive lawn. Because they find ample habitats in modern America, and since they readily adapt to humans and their noise and disruptions, killdeers are plentiful.

Spotted Sandpiper *(Actitis macularia).* This small sandpiper (7 to 8 inches in length) breeds from the southern edge of the Arctic south to Nevada, Missouri, and Virginia; it winters in the southern United States and in Central and South America. In its spring breeding plumage, the spotted sandpiper has a strikingly spotted breast. It is usually seen "teetering," bobbing its rear end up and down as it forages along a stream bank. When startled, it flies away low over the water, using rapid bursts of wing flapping interspersed with short, stiff-winged glides.

The spotted sandpiper feeds on a variety of animal matter and occupies many habitats near water, including shorelines of remote rivers and lakes, wetlands, and ponds and pools in cities and on farms. Its mating system, known as polyandry, is a mode of breeding unusual among birds. In polyandry, many of the sex roles are reversed. Females are more aggressive and active in courtship than males, and males take the primary parental role. A female may mate with up to four males and lay four sets of eggs; she will share incubation chores with only her final suitor, and in the other broods, the males take over incubating the eggs and feeding the young. In other cases, spotted sandpipers exhibit monogamous breeding, with females and males sharing in brood rearing.

Females lay three to five eggs (usually four) in a grass- or moss-lined depression in the ground. The buff-colored eggs are heavily blotched with dark brown. Incubation takes about three weeks. The young leave the nest soon after hatching and begin making their first flights after seventeen to twenty-one days. Spotted sandpipers breed statewide in Pennsylvania. Nowhere are they common, but the state's *Breeding Bird Atlas* survey, conducted by volunteers during the 1980s, found evidence of breeding in all sixty-seven counties.

Upland Sandpiper *(Bartramia longicauda)*. At 11 to 12 inches long, the upland sandpiper (formerly called the upland plover) is larger than the spotted sandpiper and the killdeer. It has a short bill, rounded head, and streaked buffy-brown plumage. The species breeds from Alaska south through the prairies of Canada and the United States and east through Pennsylvania and southern New England. It winters in southern South America—where, of course, it enjoys summer conditions in January and February.

Upland sandpipers eat a range of insects, including grasshoppers, crickets, beetles, and larvae. Individuals forage by walking jerkily through the grass, darting their bills out to pick up insects from the ground or vegetation. The species nests in late May, often away from the water, in what one biologist calls "the agricultural counterpart of prairie in Pennsylvania": pastures, golf courses, airports, fallow fields, and surface-mined areas that have been reclaimed and planted with grass. Upland sandpipers sometimes nest in loose colonies. Nests are on the ground, hidden in dense grass. The female lays four eggs, which she and her mate incubate. Upland sandpipers increased in Pennsylvania during the 1800s, when much land was cleared for farming. However, the population has fallen since the early twentieth century, as abandoned farms have grown up in brush or woods, pasture acreage has been reduced, and intensive agriculture has reduced the open, grassy habitats that upland sandpipers prefer.

Other Shorebirds. Other species that migrate through Pennsylvania include the black-bellied plover, American golden-plover, semipalmated

plover, American avocet, greater yellowlegs, lesser yellowlegs, solitary sandpiper, pectoral sandpiper, least sandpiper, dunlin, and semipalmated sandpiper. Presque Isle State Park on Lake Erie is a good place to see a variety
of shorebirds.

Gulls and Terns. Gulls and terns get blown inland from the Atlantic coast
by storms, drift into Pennsylvania from the Great Lakes, and shift into the
state during winter. More of these birds show up in Pennsylvania during
winter than at any other time of the year. Look for gulls along lakes and
rivers, in parking lots, in plowed fields, and at dumps. Both gulls and terns
are seen most often along Lake Erie and on the lower Delaware and
Susquehanna rivers.

The ring-billed gull *(Larus delawarensis)* is found in inland settings more
often than any other gull species. Adults have gray and white plumage,
greenish yellow legs, and a black ring around the bill near the tip. Ring-
billed gulls breed on lakes and rivers from central Canada south to Colorado, in the Great Lakes area, and east to Labrador. They often winter in
Pennsylvania. In early spring, large flocks may build up in southeastern and
northwestern Pennsylvania; along Lake Erie, thousands gather to feed on
winter-killed alewives and gizzard shad that have washed ashore. In eastern
Pennsylvania, ring-billed gulls range along the Susquehanna and Delaware
rivers and may gather on large reservoirs.

Herring gulls *(Larus argentatus)* breed in the Great Lakes region and
along the Atlantic south to South Carolina. They are year-round residents
in Pennsylvania, and a few pairs have nested in the state since 1996. Look
for herring gulls along the major rivers and on large reservoirs, where they
may feed in mixed flocks with ring-billed gulls. Much larger than ring-
billed gulls, herring gulls have bodies that are 2 feet long and wingspans of
55 inches. The plumage is quite variable; it takes young herring gulls two
years to achieve their adult coloration.

The largest gull usually seen in Pennsylvania is the great black-backed
gull *(Larus marinus),* whose dark wings span 65 inches. Great black-backed
gulls winter along the shores of Lake Erie, on the lower Delaware, and on
the Susquehanna as far north as Northumberland County. They often
perch on rocks and islands. They eat winter-killed fish and, if given the
chance, prey on waterfowl, grebes, and other gulls, particularly on sick or
weak individuals.

Bonaparte's gulls *(Larus philadelphia)* and laughing gulls *(Larus atricilla)*
can be abundant during spring and fall, when they migrate through the

state. Two fairly common winter visitors are the Iceland gull *(Larus glau-coides)* and the glaucous gull *(Larus hyperboreus)*, pale-colored gulls that typically join flocks of herring gulls.

In general, terns are smaller than gulls. The four tern species most likely to be seen migrating through Pennsylvania are the Caspian tern *(Sterna caspia)*, common tern *(Sterna hirundo)*, Forster's tern *(Sterna forsteri)*, and black tern *(Chlidonias niger)*. The black tern breeds on marshes in Crawford County and is considered to be an endangered species in Pennsylvania.

WOODCOCK AND SNIPE

The American woodcock and the common snipe are upland-dwelling shorebirds, members of family Scolopacidae, a group of sandpipers whose eighty-plus species are distributed over the world. Both woodcock and snipe breed in Pennsylvania. The woodcock is the more common of the two in the state and favors a drier habitat. Both species have long, thin bills with prehensile tips; they feed by probing with their bills in soft soil and seizing invertebrate prey.

American Woodcock *(Scolopax minor)*. The woodcock is known by a host of folk names: timberdoodle, night partridge, big-eye, bogsucker, mudbat. The plumage is an overall mottled russet and brown. The breast and sides are beige; black bars run from side to side across the forehead and crown. The short tail is brick red and black, tipped with gray. The feet and toes, which are small and weak, are colored a grayish pink.

A woodcock is 10 to 12 inches long (a little longer than a bobwhite quail) and has a standing height of 5 inches and a wingspread of up to 20 inches. The bird has a chunky body, short neck, and large head. The wings are short and rounded. Females average a bit heavier than males (7.6 versus 6.2 ounces). The female's bill is 2.75 inches or slightly longer, while the male's bill is usually shorter than 2.5 inches. Sensitive nerve endings in the lower third of the bill help a woodcock locate earthworms, its main prey. The bird is able to open the tip of its upper bill, or mandible, while it is underground. The long tongue and the underside of the mandible are both rough-surfaced for grasping prey. The large, dark eyes are set well back and high on the sides of the head: this positioning lets the woodcock watch for danger in all directions—behind, above, and to the sides, as well as straight ahead—while probing for food. When a woodcock flushes from the ground, air passing through the rapidly beating wing primaries produces a whistling. The bird flutters up from cover, dodging branches and twigs, then levels off and flies from 10 to several hundred yards before setting down again.

Earthworms, high in fat and protein, make up about 60 percent of the diet. An additional 30 percent is insects (ants, flies, beetles, crickets, caterpillars, grasshoppers, and various larvae), crustaceans, millipedes, centipedes, and spiders. About 10 percent is plant food, mostly seeds of grasses, sedges, and weeds. An adult woodcock may eat more than its own weight in earthworms daily.

In spring, migrating woodcock arrive in Pennsylvania as early as late February, but most don't show up until the last two weeks in March; migration is complete by mid-April. Males establish personal territories known as "singing grounds," woodland clearings spotted with low brush, or open fields adjacent to woods or brush. Singing grounds vary in size, but a quarter acre seems adequate. Woodcock do their courting in the dim light of evening and morning. The male sounds a nasal, buzzing *peent* while sitting on the ground. He takes off and flies to as high as 300 feet on twittering wings, then spirals or zigzags back to earth, singing a liquid, warbling *pee chuck tee chuck chip chip chip*. Females seek out males on the singing grounds, and one male may mate with several females. In Pennsylvania, most breeding takes place from early March to mid-May.

Hens usually nest within 150 yards of the singing ground where they mated. The male plays no role in nest selection, incubation, or rearing the young. Favored nesting habitats include brushy woods near water, hillsides above moist bottomlands, old fields with low ground cover, brier patches, edges of shrub thickets, and young conifer stands. There may be little overhead cover (as in the case of old fields) or up to 50 feet of vegetation (as in hardwood stands); the average cover height is 12 feet. The nest is a slight depression on the ground in dead leaves. The nesting hen is very difficult to spot, so completely does she blend in with the ground. Woodcock nest from March into June. Although they are solitary nesters, hens may share nearby feeding grounds.

The eggs, usually four, are pinkish buff to cinnamon, covered with light brown blotches overlaid with darker speckles. Incubation begins with the last egg laid and lasts nineteen to twenty-two days. If disturbed early in her incubation period, the hen may abandon the nest. The longer she sits on the eggs, however, the less likely she is to desert them; toward the end of her setting, she may sit tight even when touched by a human hand. Nest predators include domestic dogs and cats, skunks, opossums, raccoons, crows, and snakes. Hens that lose their first clutch may renest, laying only three eggs. Eggs hatch from early April until mid-June, with a peak during the last week of April.

The eggs split lengthwise, unique among birds. Chicks leave the nest within a few hours of hatching. They are covered with fine down colored pale brownish or buff, with brown spots and stripes in a camouflaging pattern. From the day of hatching, chicks freeze when threatened or in response to the hen's alarm call. The hen broods her young frequently, especially during rain and cold weather. At first, she finds worms for them, but after a few days, they can probe by themselves. After two weeks, the young can fly short distances, and at the end of four weeks, they are almost fully grown, fly strongly, and look like adults. The family breaks up when the juveniles are six to eight weeks old.

Woodcock migrate south in autumn. Birds from farther north may start passing through Pennsylvania in late September; the migration peaks in late October and early November, with stragglers until the end of November. Northwest winds and cold nights may launch large numbers of woodcock on their travels. They fly at low altitudes, around 50 feet, moving by night and resting and feeding during the day. They migrate alone or in loose flocks, generally covering 20 to 50 miles per night, although some individuals may fly as far as 300 miles in one night. Woodcock winter in coastal lowlands from the Carolinas south and west to eastern Texas, with concentrations in Louisiana and Mississippi. In spring, they return north, homing strongly to the areas where they were hatched.

Woodcock are hardy and seem able to recover from injuries that would kill most other birds. If a woodcock reaches adulthood, its life expectancy is about 1.8 years. Banded wild birds seven years old have been recovered. Woodcock perish from accidents, many occurring during night flight; from hunting (nearly two million are taken legally by hunters throughout the bird's migratory range); from predation, mainly by hawks and owls; and

The common snipe, top, and the American woodcock both have long, thin bills with prehensile tips, which they use to extract earthworms, grubs, and other prey from the ground.

from bad weather, when individuals returning north too early are caught by late-season blizzards and hard freezes that seal off their food supply.

Habitat requirements for woodcock change through the year. In spring, they need areas for courtship and nesting; in summer, for brood raising; during the fall and spring migrations, for feeding and resting; and they require wintering habitat in the southern states to which they migrate. Populations in the Northeast have been declining for more than twenty years, mainly because of habitat loss here and excessive hunting on the southern wintering range. Habitat is destroyed when people drain damp lowlands and put in highways, houses, and commercial developments, and when maturing forests close off singing grounds and nesting and feeding areas. The life of a woodcock cover that meets three-season habitat requirements is about twenty to twenty-five years in Pennsylvania. As the cover matures, different tree species take over, and the area grows less suitable for woodcock. Overmature aspen and alder tracts can be cut or burned, and the resulting shoot growth will restore good habitat.

Woodcock nest in every county in Pennsylvania. They are able to use islands of habitat in otherwise unsuitable locales, such as small, brushy areas surrounded by farmland or deep woods. Compared to many other birds, woodcock have a low potential productivity. A female raises only one brood per year, and each brood consists of only four (and sometimes only three) young. Fortunately the species has a high nesting success rate, 60 to 75 percent, and juvenile mortality is low.

Common Snipe *(Gallinago gallinago)*. Snipe look like slim woodcock: they have a long, slender bill (a bit longer than the woodcock's bill) and a cryptic brown and drab coloration. Snipe are 10 to 11 inches long, with a wingspread to 17 inches; adults weigh around 3.5 ounces, with females slightly heavier than males. Different subspecies, or races, breed in various parts of the world, including South America and northern Eurasia. I became familiar with snipe and their curious ways when our family spent a summer in an abandoned farmhouse on the coast of Iceland, surrounded by boggy moors where many snipe bred. In North America, snipe breed from Alaska to Labrador, south to Pennsylvania and the Upper Midwest. They favor habitats that are low and wet: bogs for breeding, marshes during migration and wintering. North American snipe winter as far south as Colombia and Venezuela.

Snipe feed by probing with their bills in damp soil, capturing fly and beetle larvae, earthworms, snails, and small crustaceans. They also consume a small amount of plant matter. When flushed from the ground, a snipe gives a rasping *scape scape* cry while fleeing in an erratic twisting and turning pattern; then it levels off in straightaway flight that can reach 60 miles per hour. In the spring, the male "winnows" to proclaim a breeding territory and to attract females. He climbs high into the sky and dives at a 45-degree angle with his tail spread and his wings beating. Air currents from the wings interact with the vibrating tail feathers, making a hollow, ascending *hu-hu-hu-hu-hu-hu*. (My wife said it sounded like elves laughing.) The North American subspecies makes a similar but slightly higher-pitched sound than the Eurasian race, because its wings and tail are of a different size and shape.

Female snipe nest from mid–April until early June, on the ground at the edge of a bog or a marsh. The nest is a small cup lined with fine grasses, sometimes sunk in sphagnum moss, occasionally overtopped with plants that form a protective canopy. The four eggs are light buff to dark brown, heavily marked with darker brown. The female incubates the clutch for eighteen to twenty days. Evidence exists that the male and female may split up the brood, with each adult rearing one or more chicks. After about three weeks, the young can fly for short distances, and their flight feathers are fully grown after thirty to thirty-five days.

Pennsylvania is on the southern fringe of the snipe's breeding range. A 1980s survey found snipe nesting in two northwestern counties, Erie and Crawford; in McKean, Potter, Tioga, and Bradford counties in the northern tier; and at scattered sites elsewhere in the state.

Sheldon, W. G. *The Book of the American Woodcock*. Amherst, MA: University of Massachusetts Press, 1967.

A Landowner's Guide to Woodcock Management in the Northeast (free booklet from the Moosehorn National Wildlife Refuge, RR1, Box 202, Baring, ME 04694).

MOURNING DOVE AND ROCK DOVE

The mourning dove and the rock dove belong to family Columbidae, a large group with over three hundred species worldwide. The mourning dove is native to North America. The rock dove, better known as the pigeon, is a native of Europe, Africa, and Asia that has been widely domesticated and carried by humans to many parts of the globe. Both of these birds are swift, strong fliers that feed almost exclusively on seeds and cereal grains. Prolific breeders, they nest repeatedly and raise several broods in a season.

Mourning Dove *(Zenaida macroura)*. An adult mourning dove weighs 3.5 to 5 ounces and measures 10 to 13 inches from bill to tail tip. It has a long, pointed tail and tapering wings that spread 17 to 19 inches. A dove is smaller and more streamlined than a pigeon; its neck is long, its head is small, and its beak is slender, short, and black. A dove's wings are gray, and its back, rump, and middle tail feathers are a grayish olive. The lateral tail feathers are bluish gray with black crossbars and white tips that flash when the bird is flying. Both sexes have a similar coloration, although the male's colors are brighter and more iridescent, with hints of pink and purplish bronze on the neck. In both sexes, the breast is a reddish fawn color.

Mourning doves can fly at speeds approaching 50 miles per hour, and they can swerve and dive suddenly to evade predators. The species' call earns the "mourning" half of the bird's name: a hollow, plaintive *ooah, cooo, coo, coo.* Depending on distance, only the last three notes may be discernible. The calling comes from males trying to attract females; after mating, calling helps to preserve the pair bond and to ward off other males. Females may coo in response, but their calls are weak and scarcely audible. Another distinctive sound made by the mourning dove is the whistling produced by the wings of the bird in flight.

Some 99 percent of the diet consists of weed seeds and waste grains. Doves also eat a few insects, snails, and slugs. They do not cling to stalks or

Mourning doves consume huge numbers of weed seeds,
which they pick up from the ground.

scratch for food, but feed by walking about in areas of low or scanty vegetation and picking up seeds from the ground. Favored weed seeds include
croton (also called doveweed), foxtail, smartweed, ragweed, lamb's-quarter,
pokeweed, and seeds of various grasses and sedges. Doves eat corn, wheat,
oats, barley, rye, grain sorghum, millet, and buckwheat left on the ground
by mechanical harvesters. Weed seeds can be very small; single doves have
been found with sixty-four hundred foxtail seeds in their crops. Doves pick
up grit, which is stored in the crop, a muscular chamber in the gullet; the
grit helps grind up seeds and prepare them for digestion. Doves seen along
roadsides are often picking up grit in the form of gravel, cinders, glass, or
other small, hard materials. In addition to food and grit, doves need water
each day. Ordinarily they fly to a stream, creek, or pond early in the morning and again in the evening.

Some doves live year-round in the Northeast; others are migratory.
Small bands of doves return to Pennsylvania in March and April; most of
these birds have wintered along the southeastern seaboard, mainly in North
and South Carolina. Doves breed in farmland broken up by fencerows, in
woodlots, in orchards, and in shrubby areas. They shun dense forest,
although they will breed in open woods and along forest edges. Mourning
doves also nest in towns and suburbs. They breed statewide in Pennsylvania.

In the spring, the male selects a territory and defends it by flying at
other males and pecking at them. He sits on a prominent perch—a snag,
treetop, fence, or utility wire—and coos to attract a female. He performs a
nuptial flight, gliding and vigorously clapping the wings together. The pair
select a nest site, and over the next four to six days, they build a nest. Some
doves use a vacant catbird, robin, or grackle nest as a platform. The male
collects small sticks; he stands on the female's back and offers the sticks to
her, and she weaves them into the nest. The eggs may be visible from the
ground through the loosely woven twigs, but the nests are surprisingly
strong despite their fragile appearance. They are built as high up as 50 feet,
although more often they are between 10 and 25 feet above the ground,
usually in the crotch of a branch in a conifer; sometimes they're constructed
in tangles of shrubs or vines or even on the ground.

The female lays two eggs, which are glossy white and unmarked. The
male incubates the eggs during the day, and the female incubates them at
night. After fourteen or fifteen days, the eggs hatch. The nestlings, called
squabs, are blind and covered with cream-colored down. For the first four
to six days, the adults feed the squabs a nutritious liquid called "pigeon's
milk" or "crop milk," secreted by the lining of the adults' crops. The milk,

a chalky mixture of cells and fluid, is given to the young by regurgitation. Gradually, seeds begin to compose the bulk of the squabs' diet. At fourteen days, the young are fully feathered and fledge from the nest, although they may remain nearby to be fed by their parents for another week or two.

The nesting cycle takes just over one month. Adults are monogamous during the breeding year and may make up to five nesting attempts over the course of the summer, finishing in August. About half of all nestings succeed, resulting in an average of four to six young produced annually by each adult pair. Spring and summer storms with high winds blow nests, eggs, and young out of trees; heavy rains and hail may kill adults as well as nestlings. Nest predators include blue jays, crows, squirrels, snakes, house cats, and others.

Juveniles complete their feather development two weeks after leaving the nest; they gather into small flocks to feed and roost. Migration of all age groups begins in August and continues into November. Flocks of a few to over twenty birds travel together, flying in the morning, resting and feeding around noon, flying again in the afternoon, feeding in the evening, and roosting at night; they average about 15 miles per day. Many doves do not migrate and spend the entire year in the north, especially if the winter stays mild. The winter range is generally south of Interstate 80 across North America. Wintering birds move between roosting sites, such as woodlots with dense trees to break the wind, and feeding areas, including picked grain fields.

Adults are preyed on by hawks (particularly the accipiters) and owls. The average life span is one year, and about half of all adults perish in any given year. The oldest wild dove on record lived to age nineteen. Because they feed on agricultural crops, there are far more doves in North America today than at the time of European settlement. The mourning dove is among the ten most abundant bird species in the United States, with a population estimated at 475 million.

Rock Dove *(Columba livia).* In the early seventeenth century, English settlers brought the first rock doves to North America as domesticated birds raised for the table. The species became feral and now breeds across the continent from southern Canada to Central America. It is found in every county in Pennsylvania, living near humans in farming country, towns, cities, and suburbs.

Rock doves are about 11 inches long and weigh 12 to 13 ounces. Their basic colors are bluish gray, bluish black, rusty red, pale gray, and white;

these colors intermix to produce individuals with various plumage patterns, highlighted with bright patches of iridescent feathering, particularly on the nape and throat. Rock doves eat grain, seeds, fruits, and nuts. Their flight is strong and direct. They are permanent residents in an area and do not migrate. Pairs are monogamous and are believed to mate for life.

When courting, the male spreads his tail, puffs up his chest, and struts about, bowing and cooing. He brings food to his mate. After breeding with her, he may make a display flight in which he claps his wings together in an exaggerated upstoke for three to five wingbeats and glides with his wings held in a deep V. In the wild, rock doves roost and nest in crevices and caves in rocky cliffs; in farming and urban areas, they use barn lofts, beams in old buildings, bridge supports, window ledges, and floors of abandoned houses. The male brings twigs, straw, and grass, which the female assembles into a shallow, flimsy nest. The two eggs are white and unmarked; both sexes incubate them for seventeen to nineteen days until they hatch. The squabs fledge after about six weeks, and the adults begin another breeding cycle, often using the same nest. A study in Kansas documented more than six broods per year.

Major predators of rock doves are humans, raccoons, opossums, hawks, great horned owls, and crows. The peregrine falcon is able to take the swift-flying rock dove on the wing, and pigeons are the main prey for peregrines breeding in cities in the Northeast.

Goodwin, D. *Pigeons and Doves of the World*. London: British Museum of Natural History, 1983.

CUCKOOS

Shy, sluggish birds of forested areas, cuckoos are heard more often than they're seen. When gypsy moth caterpillars denuded our woods in the early 1980s, I often noticed cuckoos feeding on the hairy pests or flashing across our clearing. An outbreak of gypsy moth or tent caterpillars may draw many cuckoos to a local area, with the birds eating huge numbers of the caterpillars. Both yellow-billed and black-billed cuckoos are slender, fairly large birds (11 to 12 inches) with rounded wings, long tails, and down-curved bills. Plumage is dark olive-brown above and whitish below. The yellow-billed species has a yellow lower mandible and prominent white spots on its undertail surfaces; the black-billed cuckoo has a dark bill and a narrow red eye ring. Within each species, the sexes look alike.

Most of the 140 species in the family Cuculidae live in the Old World. The European cuckoo and its call provide the model for the traditional cuckoo clock; the European cuckoo is a nest parasite, laying its eggs in the nests of other birds, who then rear the cuckoo's young. Our two eastern species occasionally lay their eggs in other birds' nests. Another family member is the roadrunner of the desert Southwest.

Black-Billed Cuckoo *(Coccyzus erythropthalmus)*. Black-billed cuckoos inhabit woods edges, groves, and moist thickets. They forage by climbing about on branches and in dense shrubs, taking caterpillars, beetles, cicadas, grasshoppers, and other insects. They eat fruits, including mulberries, elderberries, and wild grapes. The black-billed cuckoo is more apt to inhabit deep woods at higher elevations than the yellow-billed cuckoo, and the black-billed is the more common species in northern Pennsylvania. The call is a rapid, rhythmic triplet, *cucucu,* repeated thirty-five to fifty times per minute.

Black-billed cuckoos arrive in Pennsylvania in May. They are late breeders whose young do not hatch until ample insect food is available, usually in June. Males feed females during courtship. Black-billed cuckoos generally build their nests about 6 feet above the ground in the thick foliage of deciduous or evergreen trees. The nest is a loose assemblage of twigs lined with catkins, plant fibers, dry leaves, or pine needles. The female lays two or three (occasionally up to five) greenish blue eggs, which hatch after about two weeks. The young begin to fly when they are about three weeks old. Black-billed cuckoos have been observed laying eggs in nests of yellow-billed cuckoos, yellow warblers, chipping sparrows, cardinals, gray catbirds, wood thrushes, and other birds; ornithologists speculate that cuckoos become ready to lay before their nests have been finished and must quickly find places to deposit eggs. (The yellow-billed cuckoo uses at least eleven different species as hosts, most frequently the American robin, gray catbird, and wood thrush.)

Black-billed cuckoos leave Pennsylvania in August and September. Mostly they migrate at night. Their wintering range is in South America, centered on Colombia, Ecuador, and northern Peru.

Yellow-Billed Cuckoo *(Coccyzus americanus)*. This handsome bird usually inhabits open woods at lower elevations than the black-billed cuckoo. It is also found in overgrown pastures, reverting fields, roadside thickets, orchards, groves along streams, and suburban areas. The species has expanded its breeding range northward into Pennsylvania during the last fifty years, and today it

The yellow-billed cuckoo inhabits open woods and
feeds mainly on insects, especially caterpillars.

is more widespread in the state than the black-billed cuckoo. Yellow-billed cuckoos feed mainly on large insects: caterpillars (particularly gypsy moth larvae, tent caterpillars, and fall webworms), cicadas, beetles, katydids, grasshoppers, and crickets. They also eat berries and snap up an occasional treefrog. The song is less musical than that of the black-billed cuckoo, a hollow, wooden *ka ka ka kow kow kowp-kowp-kowp-kowp,* never given in distinct triplets. A folk name for the bird is rain crow, based on a belief that cuckoos start calling just before it rains.

The onset of breeding is linked to an adequate food supply. Yellow-billed cuckoos have an astonishingly fast breeding cycle, with only seventeen days passing between egg laying and the fledging of young. In Pennsylvania, most nesting takes place in mid-June. The yellow-billed cuckoo builds a slightly looser, messier nest than its black-billed relative, usually 4 to 10 feet above the ground. The female lays three to four pale greenish blue eggs. Incubation is by both sexes and takes nine to eleven days. A week after hatching, the young begin to leave the nest and perch on nearby branches. The male may feed the first fledglings, and the female may oversee later-departing young. Yellow-billed cuckoos quietly take their leave from the Northeast in late September and early October. They migrate through Central America and the West Indies to a wintering range that includes Venezuela, Brazil, Uruguay, and Argentina.

OWLS

Owls are birds of prey, occupying by night the hunting and feeding niches that hawks hold in the day. Owls are superbly adapted to find, catch, and kill prey under cover of darkness. They've been doing it for eons: owl fossils found in the midwestern United States date back sixty million years. Eight species breed in Pennsylvania or visit the state in winter. The barn owl, great horned owl, eastern screech-owl (see the following chapter), barred owl, long-eared owl, and northern saw-whet owl breed and are permanent residents; the short-eared owl is mainly a winter visitor, with a few individuals breeding in the state; and the snowy owl is sometimes seen in winter, especially in the northern counties. In general, owls are not strongly migratory, although some species shift southward in winter.

Taxonomists divide these raptors into two families: Tytonidae, the barn owls; and Strigidae, which includes all other owls. The barn owl ranges over most of the world, with related species in South America, Europe, Africa, Asia, and Australia. The Strigidae also have a near-worldwide distribution.

Dense, soft plumage makes owls look heavier than they really are. Their drab-colored feathers blend with the background of shaded daytime roosts. The feathering on an owl's legs provides insulation and protects against defensive bites of their prey. Both sexes are colored essentially alike; females are usually larger and heavier than males of the same species.

A suite of adaptations helps owls hunt. Their huge eyes have extremely large retinas that make their vision fifty to one hundred times more efficient than human sight at distinguishing small objects in dim light. Owls' retinas are packed with rods, which are light-gathering cells, as opposed to color-discerning cone cells. The eyes point forward to provide stereoscopic vision, promoting depth perception. (The eyes are fixed in the skull; to look to one side, an owl must move its head. Some species can twist their necks more than 270 degrees—almost all the way around.) An owl's head is large and broad to accommodate the two widely spaced, highly developed ears. In some species the ear openings are asymmetrical, an arrangement that helps the owl pinpoint sound sources. Rocking its head may let an owl locate a sound in the vertical as well as the horizontal plane. Even in total darkness, a barn owl can find and catch prey by using its hearing alone.

To dampen any sound made when the bird flies, the leading edges of the first primary feathers are soft and serrated, and the trailing edges are fluffy. Lightweight wings with a large surface area buoy an owl up and let it fly with a minimum of flapping. Because its flight is so quiet, an owl can pick up sounds made by prey while it glides in silently. Attacking, an owl spreads the four toes of each foot: the toes are arranged two to the front and two to the rear in an X shape, giving a wide coverage. Each toe ends in a sharp, strongly hooked claw, and the underside of the foot is ridged for extra gripping ability.

An owl swallows small prey whole. On larger creatures, the owl holds the kill down with its talons, tears the carcass apart with its hooked beak, and bolts the pieces. An owl's stomach absorbs nutritious matter and forms indigestible items—hair, feathers, bones, claws, insect chitin—into oblong pellets that are regurgitated about seven hours after a meal. The pellets, also called castings, can be found under daytime roosts and nighttime feeding stations. The larger the owl, the larger the pellets. The castings can be broken apart and the hard bony contents separated from fur and feathers; closely examining these objects will reveal the owl's diet.

Most owls call to proclaim individual territories and to attract members of the opposite sex. They also call softly for short-range communication between mates or between parents and offspring. When cornered or frightened, owls hiss or make clattering noises by snapping their mandibles together. Owls do not build nests; they take over abandoned crow or hawk nests, or they use holes in trees or earthen banks. Sometimes they line their nests with a few feathers or plumes of down. Owls tend to be early nesters. Some species lay eggs in late winter: when owl fledglings leave the nest, abundant immature offspring of other wildlife make easy prey for the inexperienced owlets. Plus, the young owls have a longer period to learn how to hunt before the next winter arrives.

Owl eggs are rounded, white, and unmarked, usually three to five per clutch. The female does most of the incubating, and the male brings food to her. After the eggs hatch, both parents feed the young. Nestlings are covered with pale down. Young found in the same nest are invariably of different sizes, since the female starts incubating as soon as she lays her first egg, which therefore hatches first. As many as two weeks may pass between the hatching of the first and last eggs. Because they cannot compete with their larger, older siblings, late-hatching owlets will die if their parents cannot find enough food around the nest. This natural check helps balance the predator population with the prey supply.

By day, most owls rest in tree cavities, in the dense foliage of trees (often conifers), or in barns. They hunt mainly at night and occasionally at dusk or on cloudy days. Most quarter the ground in silent flight or watch from a perch. Owls generally kill what is easiest to catch or to find. Beneficial birds, they prey on rodents, including many, such as rats and house mice, that are pests to humankind.

Barn Owl *(Tyto alba).* The barn owl is a long-legged, light-colored bird with a white, heart-shaped face. It is sometimes called the monkey-faced owl. A barn owl is 15 to 20 inches in length and has a 44-inch wingspan; females weigh about 24 ounces, males up to 20 ounces. Both sexes have whitish or pale cinnamon underparts and buffy or rusty upper plumage. A barn owl displays neither of the two characteristics often associated with owls: upright tufts of feathers on the head, and hooting-type calls. Barn owls make long, drawn-out whistles, loud hisses, and snoring sounds.

Barn owls nest on beams and ledges in barns, silos, and church belfries. As wooden barns fall down and are replaced by modern pole buildings, barn owl nesting habitat shrinks.

Barn owls nest in barns, church towers, hollow trees, old buildings, silos, and caves. They do not build nests, although their castings may form a base that the eggs lie upon. They usually nest in March, April, or May and lay from three to eleven eggs (generally five to seven) at two- to three-day intervals. Incubation takes about thirty-three days. After the eggs hatch, both parents feed the young. A nestling barn owl can eat its weight in food every night. The young fledge when they are nine to twelve weeks old.

Barn owls hunt in open fields, flying low over the ground. Ornithologists studied 200 disgorged pellets from a pair of barn owls that nested in a tower of the Smithsonian Institution building in Washington, D.C. The pellets contained 444 skulls, including those of 225 meadow mice, 179 house mice, 20 rats, and 20 shrews—all caught in the city. Other studies have confirmed mice, meadow voles, and shrews as important prey. Barn owls occasionally take small birds, flying squirrels, young rabbits, and insects.

Pennsylvania is near the northern limit of the species' range. Some individuals stay here through the winter, but others—particularly juveniles—move south in the fall. Barn owls favor low, open country and are not found in forested mountains. In Pennsylvania, most breeding has been documented in the southeast, particularly in Lancaster and Lebanon counties. A shift from pasture to row crops and a loss of nesting sites (old barns torn down and replaced by modern pole buildings) are major threats to the barn owl, which has been listed as a vulnerable species by the Pennsylvania Biological Survey. Barn owls readily use nest boxes, which can be mounted on trees, the sides of buildings, or beams inside barns and silos.

Great Horned Owl *(Bubo virginianus)*. This large owl is sometimes called the tiger of the air; it is our most powerful and aggressive owl. It weighs up to 3.5 pounds, is almost 2 feet long, and has a wingspan of 4 to 5 feet. A great horned owl has soft brown plumage above, mottled with grayish white; undersides of light gray barred with dark brown; a "collar" of white feathers on the upper breast; a rust-colored face; and prominent ear tufts, the so-called "horns," up to 2 inches in length. The species is known as the hoot owl for its call: three to eight (usually five) deep, booming, uninflected hoots, *hoo-hoohoo hoo hoo.*

Great horned owls hoot to stake out breeding territories in October and November. The hooting tapers off in January, and the owls hoot very little when they are incubating eggs and raising nestlings from February to April. Great horned owls are believed to mate for life. They nest in crow, hawk, and heron nests, in tree cavities, on rocky ledges, and in hollow stumps. A mated pair may clean debris from an appropriated nest, and the female may then add a few feathers. She lays a total of two to six eggs, usually in late February; often she becomes temporarily covered with snow while incubating her clutch. Horned owls defend their nests and young fiercely and may even attack humans who get too close. The eggs hatch after about a month; the owlets are covered with white down and at first are weak and blind. After five or six weeks, they may leave the nest and clamber about in nearby branches. The young can fly after nine to ten weeks, and their parents feed them for several weeks thereafter.

Great horned owls kill rabbits, rats (both native woodrats and introduced Norway rats), mice, voles, hares, domestic poultry, grouse, ducks, geese, smaller owls and hawks, squirrels, fox pups, opossums, skunks (the skunk's defensive spray apparently does not deter *Bubo virginianus*), porcupines (often with fatal results for both predator and prey), house cats, weasels, muskrats, snakes, frogs, insects . . . in short, most animals other than the large mammals.

Great horned owls live in forests, woodlots interspersed with farmland, remote wilderness, suburbs, and cities and towns. The species ranges throughout the Americas. In Pennsylvania, the great horned owl is a common resident in all seasons. It was heavily persecuted in the past, with bounties paid to people who destroyed owls, which were believed to kill large amounts of game. Today it is illegal to shoot this or any other owl.

Snowy Owl *(Nyctea scandiaca)*. Rare, irregular visitors to the Keystone State, snowy owls show up mainly from November to January. They breed on the arctic tundra. If food becomes scarce there, large numbers of owls may shift south. Population ebbs of lemmings and hares, and the accompanying owl migrations, seem to occur at four- or five-year intervals. Immature snowy owls, which are darker in color than adults, go farther south. About every ten years, an even greater number of snowy owls leave the north and migrate as far south as Georgia.

The white plumage is faintly barred with black, with the feet and legs heavily feathered. The full, soft feathering keeps the bird warm while it rests between hunting forays, and the white color blends it into a snow-covered setting. The snowy owl is slightly larger than the great horned owl, with a 24-inch body, a 60-inch wingspread, and a body weight of up to 5 pounds. In Pennsylvania, the snowy owl seeks out open fields that resemble its tundra home. It often perches on a fence post to look for mice and voles. In areas near water, it may take ducks and geese. The snowy owl is largely crepuscular (active in the twilight) but is forced to hunt by day during the arctic summer, when darkness is almost nonexistent; in Pennsylvania, the snowy owl continues these habits and often hunts during daylight. It does not call south of its breeding grounds, where it announces itself with deep hoots. Because snowy owls are not persecuted on their remote breeding range, they often let people approach them quite closely.

Barred Owl *(Strix varia)*. The barred owl is a large bird of the deep woods. It has a rounded head with no "horns" and brown eyes (it is the only brown-eyed Pennsylvania owl, other than the barn owl; the others all have yellow eyes). The barred owl ranges over the eastern United States, its distribution often coinciding with that of the red-shouldered hawk. A barred owl weighs up to 2 pounds, with a 44-inch wingspan and a body length of 20 inches. It has gray-brown plumage and pale undersides barred crosswise on the breast and streaked lengthwise on the belly.

The barred owl is the most vocal of our owls. Its hoots have a more varied inflection than those of the great horned owl, although they are not as deep or as booming. The barred owl's call is eight hoots in two groups of four: *hoohoo-hoohoo . . . hoo-hoo-hoohooaw* (usually described as "Who cooks for you, who cooks for you all?"). It calls early in the night, at dawn, and sometimes on cloudy days. Paired barred owls call back and forth to each other. Once on a winter camping expedition I was yanked out of a sound sleep when a barred owl opened up, at volume, from the hemlock under which I'd pitched my tent.

Barred owls almost always nest in hollow trees. In March, they lay two to four eggs that hatch after twenty-eight to thirty-three days. Pairs may show a strong attachment to the same nest area and return to it year after year. Barred owls favor dense woods, wooded ravines, bottomland forests, and the edges of swamps and marshes. They hunt mainly at dawn and dusk, taking mice, voles, squirrels (including flying squirrels), rabbits, opossums, shrews, birds, frogs, salamanders, snakes, insects, and crayfish.

Long-Eared Owl *(Asio otus)*. This slender, crow-size owl has long wings that make it look larger in flight than it really is. A long-eared owl has a 16-inch body and a 40-inch wingspan and weighs about 11 ounces. It gets its name from two prominent ear tufts. Although it looks like a smaller version of the great horned owl, the long-eared owl can be told from its larger relative by a streaked rather than a barred belly and by its closer-set horns. The call is a moaning, dovelike *hoo, hoo, hoo.*

Long-eared owls often nest in dense conifers, recycling an old crow, raven, or hawk nest. The four to six eggs are incubated by the female; the eggs hatch after twenty-six to twenty-eight days, and the oldest owlet may be eight to ten days old when the last egg hatches. *Asio otus* preys mainly on small mammals, including voles, mice, and shrews, and on small birds, bats, and snakes. An ideal habitat includes dense trees for nesting and roosting, with open county for hunting. Its secretive nature and strongly nocturnal habits have kept the long-eared owl one of the most poorly known nesting birds in the state.

Short-Eared Owl *(Asio flammeus).* Also called the marsh owl, the short-eared owl visits Pennsylvania mainly during the winter, although a few pairs have bred here. The short-eared owl is 13 to 17 inches long, has a 42-inch wingspan, and weighs about 1 pound. The upper plumage is buffy brown, and the pale breast is heavily streaked with brown. The short-eared owl's ear tufts are small and hard to discern, but its ear openings are large and its hearing is excellent.

The short-eared owl is mainly nocturnal but is also active during the day. It hunts over open country, including marshes and wet pastures, its irregular, flopping flight resembling that of a nighthawk or a huge moth. The short-eared owl is not particularly vocal. It sounds a barking, sneezy *kee-yow!* It preys mainly on voles and also takes lemmings, mice, shrews, rabbits, bats, and muskrats. When hunting, it flies low over the ground and often hovers before dropping to its prey. It can be thought of as the nocturnal counterpart of the harrier, or marsh hawk.

In recent years, short-eared owls have nested on reclaimed strip-mine lands in western Pennsylvania, particularly in Clarion County. They nest on the ground, in slight depressions sparsely lined with grasses, weeds, and feathers; bushes or sedge clumps often hide the nests. The female lays four to eight eggs and incubates them for twenty-four to thirty-seven days. The species ranges from the arctic tundra south to Ohio, Pennsylvania, and New Jersey in the East. *Asio flammeus* is on the endangered species list in Pennsylvania. Ornithologists hope to keep this owl breeding in the state by establishing and protecting areas of undisturbed grassland on revegetated strip mines.

Northern Saw-Whet Owl *(Aegolius acadicus).* With a body length of 8 inches and an 18-inch wingspan, the saw-whet is Pennsylvania's smallest owl. Its plumage is a dull chocolate brown above, spotted with white, and its undersides are white dappled with dark reddish brown. Juveniles are a deep brown over most of their bodies. The species does not have ear tufts. The call is a mellow whistled note repeated mechanically by the male between 100 and 130 times per minute: *too, too, too, too*. Settlers named the bird after its song, which reminded them of the rasping of a whetstone sharpening a saw.

Saw-whet owls are strongly nocturnal and seldom seen. They inhabit moist woodlands with dense undergrowth of conifers or brushy shrubs. They breed from March through July, nesting in deserted woodpecker holes (especially those of pileated woodpeckers and flickers), squirrel dens, hollow trees and stumps, and artificial nesting boxes. The female lays four to seven eggs. She incubates them almost continually until they hatch, after about four weeks; the male brings her food. The young leave the nest after four to five weeks. They stay in the nest's vicinity and are fed, mainly by the male, for another month or longer. The female may possibly seek out another male and raise a second brood.

Saw-whet owls are quite tame, and some pairs nest on the edges of yards and in towns. They sit perched on low branches and glide down to take mice, voles, shrews, young squirrels, small birds, and large insects; in turn, they are preyed on by barred and great horned owls. The species breeds, apparently in low densities, in wooded habitat throughout the state. As a cavity nester, it benefits when loggers leave dead snags in the woods and when people put up nest boxes.

Walker, L. W. *The Book of Owls.* New York: Alfred A. Knopf, 1974.

SCREECH-OWL

The eastern screech-owl, *Otus asio*, is the smallest of our ear-tufted owls. It is common and widespread across Pennsylvania and the Northeast, wherever there are trees with cavities for daytime seclusion and for nesting and rearing young. In late summer, I often waken in the night to the sound of screech-owls calling: a haunting, melodic tremolo (not a screech, although these diminutive predators sometimes make screeching sounds) given by parent birds to keep the newly fledged owlets with the family group.

Biology. The eastern screech-owl is 7 to 10 inches long (about the size of a robin) and has a 21-inch wingspread. Adults weigh 5 to 8 ounces, with females averaging 17 percent heavier than males. The body is stout, and the wings have rounded tips. The large eyes are yellow. The head has two prominent earlike tufts.

Screech-owls are clad in a mottled plumage that resembles tree bark. The species comes in two color morphs, or variations: brownish gray and reddish brown. The morphs are not related to age, sex, or season; birds of both colors readily interbreed, and both colors can show up in a single brood. More gray birds are found in the northern part of the species' range. Merrill Wood was for many years professor of ornithology at Penn State. (I took his class; a small, plump, highly tanned individual, Professor Wood reminded me of a brown creeper.) The professor noted in his book *Birds of Pennsylvania* that 75 percent of Pennsylvania screech-owls are gray and 25 percent are red.

Eastern screech-owls hunt by sitting on a perch and scanning the ground, then gliding down to attack. If nothing offers after a minute or two, they move to another perch. They hunt during the evening, at night, in early morning, and occasionally in broad daylight. A pair will hunt over 15 to 100 acres. During the breeding season, most food is taken within 100 yards of the nest. Screech-owls prey on cardinals, blue jays, grackles, doves, flickers, sparrows, and many other birds. They catch mice, voles, flying squirrels, and young rats and cottontails. Much of the diet may consist of insects: beetles, moths, crickets, grasshoppers, cicadas. Screech-owls eat earthworms, take fish and crayfish from shallow water, and capture frogs, toads, and small snakes. Screech-owls are themselves eaten by great horned owls, barred owls, and various land predators.

During the day, adults roost separately in tree cavities or in dense foliage, often in conifers. If approached by a potential predator, a screech-owl will close its eyes, erect its ear tufts, and stretch upward so that it looks like a bark-covered stub. Smaller birds that discover a resting screech-owl may "mob" it, scolding and flitting around it from branch to branch; blue jays often mob screech-owls and even the cavities where the owls have nested in the past.

Screech-owls are monogamous and are thought to mate for life. They do not migrate. In late winter, the male begins calling, using a high, single-pitch trill to advertise two or three nest cavities: natural hollows in trees, such as abandoned flicker or pileated woodpecker nests, generally 5 to 20 feet high; artificial nesting boxes; or crevices in buildings. Alternative nest sites are important, since pairs usually switch sites if a first nesting fails. After mating, the female lays a clutch of two to seven (usually four) eggs in one of the cavities. The eggs are elliptical to nearly round and are pure white in color.

The female incubates the eggs, which hatch after about thirty days. She starts incubating with the first egg laid, and the young hatch at intervals. The youngest and smallest owlets may starve or suffocate, and in about a quarter of the broods, particularly the largest ones, they are killed and eaten by their older siblings. The male roosts a few yards away from the nest. He supplies food for the female throughout her incubating and continues to bring food after the eggs hatch, with the female dismembering the prey and feeding it to the chicks. About ten days after the eggs hatch, the female joins her mate in hunting to feed the growing owlets. Eggs and young may be eaten by raccoons, opossums, black rat snakes, and squirrels. Replacement clutches are smaller than original ones—usually just three eggs.

After about four weeks, the young climb out of the cavity and into branches in the tree, or they jump out onto the ground and struggle into a nearby tree. They can fly weakly after two or three days; after another three to five days, they follow the adults in flight. Adults defend their young

Eastern screech-owls hunt by sitting on a perch,
such as a dead snag, and scanning the ground for prey.

aggressively, flying at intruders, including humans, and raking them with sharp claws. They may sound an alarm screech: a loud, piercing call. The parents keep the family together using the monotonic trill: the mellow quavering notes, given mostly in a single pitch, carry 75 to 150 yards.

For the first two weeks, owlets stay near each other while ranging farther from the nest. They begin catching insects on the ground. They stay in their parents' range for about ten weeks and continue to be fed by the adults. In early fall, the young disperse. A study in Texas found the average movement to be 1.7 miles, along woodland corridors, with males venturing farther than females.

Although able to live for at least fourteen years, the average female screech-owl survives for only one year, including a single breeding season. Because screech-owls habitually drop from perches and fly along low to the ground, many are killed by cars. Others die from collisions with limbs and house windows and from eating insects poisoned by biocides (another good reason not to have your lawn periodically doused with toxic chemicals).

Habitat. The eastern screech-owl lives in forests, woodlots, old orchards, city parks, cemeteries, towns, and shady suburbs where trees are large enough to have cavities (about 8 inches in diameter, at cavity level). Extensively forested areas are less attractive than woods broken by clearings and fields. The Texas study referenced above found that nests near suburban houses were more often successful than those in rural areas. The investigator surmised that the more open suburban setting had fewer shrubs where screech-owls' predators and competitors could hide.

Population. Because they are shy and nocturnal, screech-owls may live unnoticed in an area for years. In Pennsylvania, *Otus asio* is widely distributed in every county. Putting up nesting boxes can protect screech-owls in areas where the cutting of mature trees threatens to destroy their nesting sites.

Gehlbach, F. R. *The Eastern Screech Owl*. College Station, TX: Texas A&M University Press, 1994.

NIGHTHAWK AND WHIP-POOR-WILL

The common nighthawk and the whip-poor-will belong to a group of nocturnal and crepuscular birds known as the nightjars; eighty species are found around the world. Nightjars have large heads and eyes and exceedingly wide mouths, which they use as scoops for catching insects in midair. Their broad wings and large tails contribute to a buoyant, maneuverable flight. Their legs are short, and their feet are small and weak. Most spend the day resting on the ground or roosting in trees, perched lengthwise on limbs. *Nightjar* seemingly refers to the birds' nocturnal habits and the jarring, grating aspect of their calling. The nightjars are also known as "goatsuckers," from an erroneous belief that the birds use their expansive maws to steal milk from goats and other livestock.

Common Nighthawk *(Chordeiles minor)*. The name nighthawk is a misnomer, since this bird is not related to the hawks. A nighthawk is about 9 inches long, with a wingspread of almost 2 feet; individuals weigh from 2.5 to 3.5 ounces. The plumage is a mix of dark gray and brown. The long wings have a crook about halfway out and then taper to a point. The tail has a white band; white brightens the chin and throat; and a white "bandage" is clearly visible from below, marking the flight feathers on each wing.

Nighthawks fly mainly at dusk and at dawn. During warm summer evenings, flocks of nighthawks can be seen flying high above towns and farms. Unlike whip-poor-wills, which sit and wait and then sally forth to catch individual insects, nighthawks remain on the wing for extended periods, flapping, gliding, stalling, and swerving as they chase and arrest their prey. More than fifty insect species have been reported as prey, including flying ants, June bugs, mosquitoes, moths, mayflies, caddisflies, wasps, and grasshoppers. Nighthawks drink on the wing, skimming the surface of lakes and streams. They do not fly during heavy rain, strong winds, or cold weather.

The call is a loud, nasal *peent,* which, according to one source, resembles the word "beard" whispered loudly. As part of his breeding display, the male also makes a booming sound, which is produced by air rushing through the primary wing feathers after a sudden downward flexing of the wings during a dive. While camping in the Badlands National Monument in South Dakota, I once sat enraptured by nighthawks "booming" above the prairie through the extended twilight of a June evening and on into the moonlit night.

Chordeiles minor has a large breeding range: from the Yukon Territory to Labrador and south to Florida, Texas, and Central America. The birds nest in open fields, gravel beaches, rock ledges, burned-over woods, grasslands, and the flat, graveled roofs of buildings. The female nighthawk does not build a nest; she lays her two eggs directly on the ground. Egg laying peaks around the first of June. Nighthawk eggs are creamy or pale gray, dotted with brown and gray. The female does most of the incubating (the male may spell her at times), and the eggs hatch after about eighteen days. The hatchlings' eyes are open, and they are able to move feebly in the nest. Females may feign injury to draw predators away. Both parents feed the chicks by regurgitating insects to them. After around eighteen days, young nighthawks make their first flights. They can fly capably by thirty days, and by fifty days they are fully developed. Nighthawks raise only one brood per year. They migrate south through Pennsylvania from August to October.

The average life span is estimated at four to five years; banded birds as old as nine years have been recovered. Since the 1960s, the number of breeding and migrating nighthawks has fallen noticeably. This decline may stem from indiscriminate use of pesticides, increased predation, or changes in habitat, either in the northern breeding range or in the southern wintering area, which includes South America and about which little is known. Nighthawks seem to be abandoning traditional rural nesting sites; in Pennsylvania, most nesting takes place on building roofs in urban areas.

Whip-Poor-Will *(Caprimulgus vociferus).* The whip-poor-will lives in moist woods across the eastern and southern United States. It is about the size of a common nighthawk, but its wingspan is not as great and its wings are broader and more rounded. On each side of the bill, a vertical row of hair-like bristles flares toward the front; these bristles funnel insect prey into the mouth. The plumage is a mix of camouflaging browns. Both sexes have a white neck band, and the male has white outer tail feathers.

Whip-poor-wills perch on branches or sit on the ground or along roadsides, where the birds' eyes gleam red or bright orange in the glare of automobile headlights. (This "eyeshine" is caused by a reflective layer at the back of the retina called the tapetum, which amplifies small amounts of light by passing them back through the retina a second time.) Its soft feathering lets a whip-poor-will fly almost as quietly as an owl, helping the bird intercept moths, many of which can detect, through tympanic membranes, sounds made by potential predators. From their waiting places, whip-poor-wills fly up to take sphinx moths, noctuid moths, and the big silk moths—cecropia, luna, and polyphemus—as well as mosquitoes, gnats, June bugs, crane flies, and other insects. The whip-poor-will's sit-and-wait foraging

*The whip-poor-will sallies forth from the ground to catch
flying insects. The bird's broad bill is fringed with hairlike feathers
that help funnel prey into the bird's mouth.*

strategy is less energy-expensive than the common nighthawk's in-flight foraging and may be what allows *Caprimulgus vociferus* to arrive earlier on the
northern breeding grounds and to survive periods of cold weather and
accompanying prey shortages.

The whip-poor-will is named for the male's repetitive nocturnal calling. The *whip* is sharp, the *poor* falls away, and the *will*—the highest note in
the sequence—is a bullwhip snapping in the night. The call carries about
half a mile. Listeners close to the calling bird may hear a soft verbal *knock*
sound before each repetition. In Pennsylvania, whip-poor-wills start calling
in late April or early May, when males arrive from the south; the calling
continues through June and dwindles in July. Whip-poor-wills call mainly
at dawn and dusk. An elderly friend of mine who lived on our road, after
hearing a whip-poor-will start up outside his house, counted for four con-
secutive minutes, recording fifty-five, fifty-six, fifty-six, and fifty-seven repetitions. He noted the time, then sat reading. The bird kept singing,
without changing position or tempo, for ninety-one minutes. Figuring an
average of fifty-six calls per minute, my neighbor arrived at a total of more
than five thousand *whip-poor-will* calls in an hour and a half.

The calling attracts females. Whip-poor-will courtship involves head
bobbing, bowing, and sidling about on the ground. After mating, the
female lays two eggs on the ground in dry open woods, often near the edge

of a clearing. Most egg laying occurs between mid-May and mid-July. The eggs are off-white, speckled with tan, brown, or lilac; they blend in with the dead leaves, as does the adult who incubates them. Several times I have almost stepped on whip-poor-wills incubating eggs or brooding young. In one case, the incubating bird was a male. On another occasion, the adult, a female, flew directly at my face, then fell to the ground and tried to distract me by feigning an injury.

The eggs hatch after about three weeks of incubation. The reproduction of whip-poor-wills may correlate with the lunar cycle: males sing longer on moonlit nights, and hatching may occur when the moon is waxing (on its way to being full), so that the increasing light makes foraging easier for the adults, who must now feed nestlings as well as themselves. Parent birds feed their young by regurgitating insects to them. The fledglings can fly about twenty days after hatching. Whip-poor-wills begin leaving the Northeast in August and September, with stragglers into October. The species winters in the southeastern states, in areas where the related chuck-will's-widow *(Caprimulgus carolinensis)* breeds in summer. (The chuck-will's-widow withdraws to Central and South America in winter.) Some whip-poor-wills migrate to Central America and the West Indies.

Whip-poor-wills reach their greatest numbers in young brushy woods, abandoned farms (sometimes called "whip-poor-will farms"), and woodland edges, where rank plant growth supports large numbers of insects. The birds hunt in forest clearings and around water, orchards, and gardens. In Pennsylvania, the population remains strongest in the Ridge and Valley province in the south-central counties. The growth of suburbs and cities has eliminated this species from much of southeastern Pennsylvania. Whip-poor-wills cease breeding in areas where woods become too mature. The species has declined over much of the East during the last three decades.

RUBY-THROATED HUMMINGBIRD

The ruby-throated hummingbird, *Archilochus colubris,* is the only hummingbird that breeds in the Northeast. The hummingbirds live in the New World, with most of the three-hundred-plus species inhabiting the tropics. These birds hover at flowers and feed on nectar; many also consume insects. Although small and dainty-looking, hummingbirds defend their territories as aggressively as larger birds do, and some species—including the ruby-throated—undertake long, strenuous migrations.

Biology. The ruby-throated hummingbird breeds from southern Canada south to the Gulf coast and west to the Great Plains. Adults are about 3 inches long and weigh 0.1 ounce, which is less than a penny. Both sexes have glistening green-bronze backs and pale bellies. The male sports a bright metallic red gorget, or throat patch; on the female, this area is gray-ish white. The bill is long and thin. The feet and legs are small and weak and are used for perching rather than for walking or hopping.

Hummingbirds have the largest breast muscles, relative to body size, of all birds. They are unique in their ability to hover in place for extended periods and to suddenly fly backward, sideways, or up and down. In flight, a ruby-throated hummingbird beats its wings about fifty-three times per second, and as rapidly as eighty times per second when moving forward. Hummingbirds have flexible shoulder joints that let their wings move in a pattern like a figure eight laid on its side, with both forward and backward strokes generating lift. Minute changes in the angle of the wings let the bird control its speed and course. Scientists have calculated that hovering requires 204 calories per gram of body weight per hour, compared with 20.6 calories needed by the bird at rest. A hummingbird's heart beats more than ten times per second during activity. The bird must eat almost constantly to fuel its high-speed metabolism.

Hummingbirds insert their bills into flowers to feed on nectar; in the process, they act as pollinators for many plants. They are especially attracted to bright red blossoms; scientists believe that some plants, including a woodland vine known as the trumpet creeper, evolved red, tubular flowers especially to attract hummingbirds. Ruby-throated hummingbirds take nectar from more than thirty species of flowers, including wild bergamot, bee-balm, spotted jewelweed, honeysuckle, and cardinal flower. When the structure of a plant permits it, the bird may perch while feeding; otherwise it hovers. A hummingbird does not suck in nectar but rather laps it up using its long, grooved tongue. Hummingbirds also take sap from trees, vis-iting rows of "sap wells," small holes that yellow-bellied sapsuckers exca-vate in birches, maples, and other trees; the sap contains sucrose and amino acids. Up to 60 percent of an individual's diet may be insects, including mosquitoes, gnats, fruit flies, and small bees. Hummingbirds pluck spiders out of their webs and glean aphids, small caterpillars, and insect eggs from the leaves and bark of trees.

When sleeping, a hummingbird retracts its neck, points its bill slightly upward, and keeps its body feathers fluffed up to reduce heat loss. At times,

A ruby-throated hummingbird inserts its bill into a flower,
then laps up nectar using its long, grooved tongue.

it may enter a torpid state: its temperature drops and its metabolism slows, letting the bird get through the night, or through a cold snap, without starving.

Ruby-throated hummingbirds arrive in Pennsylvania in late April and early May, with males preceding females by a week or two. Males stake out individual territories of about 0.25 acre and defend them vigorously against other hummingbirds, both male and female. Males ruby-throats give a string of chipping calls from a perch in the center of their territory. If food sources are abundant, only 50 feet may separate neighboring males.

Hummingbirds are solitary, and males and females get together only for courtship and mating. When confronted by a female, the male does a series of U-shaped looping dives with an arc length of 3 feet or more; these maneuvers may actually be a part of the male's territorial defense. Once the female perches, the male changes his displaying to a series of side-to-side arcs, which show off his colorful throat patch. Most breeding occurs in June, and one male may mate with several females.

The female picks a nest site, usually in a deciduous tree in dense woodland, 5 to 30 feet above ground. Near the tip of a downward-sloping branch, she constructs a platform of thistle and dandelion down, attached to the branch with spider silk. She uses plant down for the nest's side walls, binding the material with spider webbing or pine resin and cementing bud scales and lichens to the outside. The finished nest is a soft, flexible, well-camouflaged cup about 2 inches wide and 1.5 inches high. Some females refurbish old nests.

The female lays two oval eggs (occasionally one and rarely three), each about 0.5 inch long by 0.33 inch wide and weighing 0.02 ounce. The eggs are white. Egg laying in Pennsylvania runs from May to July or August. The female incubates her clutch for fourteen to sixteen days; the young are naked and dark gray in color. Their mother feeds them nectar and insects, inserting her bill into that of a young bird and pumping the food down its gullet. The female broods the nestlings almost constantly, except when foraging. The young hummingbirds' eyes open after nine days, and the female ceases brooding and starts bringing them whole insects clasped in her bill. The young fledge after eighteen to twenty days; when they leave the nest, they weigh more than their mother. The female may continue to feed them for four to seven days as they learn to forage. She may mate again and raise a second brood.

After breeding, hummingbirds start building up body fat for migration. An individual's weight can double in as few as seven to ten days. Males begin leaving the breeding range in early September; females and juveniles (whose plumage matches the females') may stay until early October. Ruby-throated hummingbirds join many other birds in migrating along ridge tops. They can be seen in good numbers when a strong cold front ushers in a north wind. The hummingbirds winter in southern Florida, Louisiana, Texas, and Central America south to Panama. Although migratory routes are poorly documented, it's known that some individuals follow the coast and others fly nonstop across the Gulf of Mexico.

Hummingbirds are preyed on by house cats, American kestrels, merlins, sharp-shinned hawks, loggerhead shrikes, great crested flycatchers, and even frogs and bass; blue jays have been seen killing and eating nestlings. But probably more hummingbirds succumb to accidents: crashing into windows and cars and, at night during migration, colliding with telecommunication towers. Females have been documented to live for nine years and males for five years.

Habitat. Ruby-throated hummingbirds inhabit open woods, woods edges, gardens, and orchards. They are found in cities, unbroken forest, and extensive farming areas, but they are less common in these places than they are in mixed wooded and open land. Floodplain forests and areas along streams offer nesting sites and abundant flowers. In their wintering range, hummingbirds feed on nectar and insects.

Population. In Pennsylvania, *Archilochus colubris* breeds almost statewide. The population here and in the rest of the species' range seems to be stable.

Stokes, D. W., and L. Q. Stokes. *The Hummingbird Book: The Easy Guide to Attracting, Identifying, and Enjoying Hummingbirds.* Boston: Little, Brown, 1989.

BELTED KINGFISHER

Paddling down a stream, I have often startled—and been startled by—a kingfisher. The bird takes off from its perch, sounding an alarm call that rattles down the wooded corridor. It flashes downstream, making two or three strokes of its blue-gray wings, followed by a short glide, then more wing pumping, sometimes skimming so low that its wingtips seem to brush the water. When the bird reaches the end of its territory, it quietly loops around behind me. Frequently I'm scolded by another kingfisher at the next bend in the stream.

The belted kingfisher, *Ceryle alcyon,* belongs to the family Alcedinidae. Six species of kingfishers live in North and South America, and eighty species inhabit other parts of the world: Australia's laughing kookaburra is a well-known family member. In North America, the belted kingfisher breeds from Alaska to Labrador and south to Florida, Texas, and California. Most king-fishers winter in the lower forty-eight states where open water remains available, and some individuals go as far south as northern South America.

Biology. A kingfisher has a stocky body and a large head with a ragged-looking double-pointed crest. The beak is sturdy and sharply pointed, the tail is short, and the feet—especially when considered along with the out-size head—appear to be oddly small. Adults are 11 to 14 inches in length and weigh 5 to 6 ounces. The white neck ring and breast stand out against the blue-gray body plumage. The female has an area of rusty feathers adorning her sides and breast, which the male lacks.

Kingfishers live along the banks of streams, rivers, and lakes, where they catch fish near the surface or in shallow water. They mainly take fish that are 4 or 5 inches long or shorter. Kingfishers hunt from perches—branches, utility wires, pilings, and bridge supports—or hover above the water while scanning it for prey. A kingfisher dives into the water with its eyes closed and uses its bill to grab its prey. After catching a fish, the bird flies back to its perch. It stuns the fish by whacking it against the perch and swallows it headfirst. Kingfishers take whatever types of fish inhabit a given waterway, from bullheads to sticklebacks to trout. When heavy rains make stream waters cloudy, kingfishers may turn to crayfish. They also eat mollusks, insects, reptiles, amphibians, and the occasional small bird or mammal. After feeding, a kingfisher coughs up a small pellet of indigestible matter, such as bones and fish scales.

People often hear kingfishers before seeing them. The rattle call is given freely, both as an alarm signal and during territorial disputes. Mated pairs use a softer version of the same call to communicate with each other. Kingfishers become active just before sunrise, when they forage and patrol their territories; they do most of their feeding between seven and ten in the morning and are less active at midday. At night, they roost in trees. Kingfishers are solitary except when breeding. Both males and females defend individual territories, calling stridently and flying at and attacking intruding kingfishers. A territory may include 1,000 yards of stream or lake bank.

In spring, migrating kingfishers return to Pennsylvania in March and April (others may have stayed through the winter, if streams did not freeze over). The male establishes and defends a breeding territory; once a female is attracted and the two pair up, she joins in defending the territory. During courtship, the male feeds the female. After mating, the male, followed by the female, may soar and then dip close to the surface of the water. Breeding reaches a peak in early May.

Kingfishers nest in burrows that they dig into steep banks above streams, in road cuts, and in sand and gravel pits. Often the burrows are a few feet below the top of the bank, where topsoil gives way to sandier subsoil. Burrows are usually near or along the water, but sometimes they're a mile or farther away. Both birds excavate the burrow, a task that may take three days to two weeks. The tunnel is 3 to 4 inches in diameter, slopes upward, extends a yard or two into the bank, and ends in an unlined chamber 8 to 12 inches across and 6 or 7 inches high. Before entering, an adult will land on a convenient perch, give the rattle call, and fly straight into the burrow opening. To tell whether a burrow is in use, look for twin grooves on the outer lip made by the kingfishers' feet. During breeding, kingfishers are sensitive to disturbance by humans and may desert an area if bothered too frequently.

On the dirt floor of the nest chamber, the female lays five to eight white eggs. Both sexes incubate the clutch, with the female setting at night. The eggs hatch after about twenty-four days. The altricial young have pink flesh, and their eyes are shut. The female broods them continuously for

Kingfishers' territories often center on
stream riffles where fish are abundant.

three to four days after hatching. The adults regurgitate fish to the young and, as the hatchlings grow and strengthen, begin bringing them whole fish as frequently as once every twenty minutes. After defecating, the young use their bills to peck or scratch at the nest chamber's walls, so that dirt covers up their waste. When two weeks old, the young may crawl from the nest chamber into the burrow. They leave the nest four weeks after hatching; the parents hold fish in their bills, sit on a nearby perch, and coax the young into flying from the entry. The adults feed the fledglings for about three weeks, as the young learn how to fish. Parents may teach their offspring to dive by dropping insects into the water beneath the youngsters' perch.

Skunks, minks, raccoons, and black rat snakes kill some young in the nest; after they fledge, juveniles are vulnerable to hawks. Kingfishers escape from predators by diving into the water. Individuals breed during their first year after hatching. In the northern parts of its range, *Ceryle alcyon* raises one brood per year. After the mating season, pairs break up and individuals settle on and defend smaller territories. Kingfishers migrate south from September into December. Most kingfishers in the Northeast are partial migrants, able to stay on and survive the winter if the streams stay unfrozen and the birds can find fish. When migrating, kingfishers follow rivers, lake shores, and coastlines.

Habitat. Kingfishers inhabit streams, rivers, ponds, lakes, and estuaries. Individual territories often center on stream riffles, which are good fishing spots. Kingfishers prefer open running water that is not turbid. On lakes, they use sheltered coves and shallow bays. For nesting, they need earthen banks in which to excavate their burrows. In winter, they resort to rocky coastlines, swamps, brackish lagoons, oxbows, bayous, and shores of rivers and reservoirs.

Population. Pennsylvania is veined with streams, and kingfishers are widely distributed across the state. The birds are absent from places such as southern Clearfield County, where acid mine drainage has polluted long sections of waterways. Stream channelization destroys the vertical banks needed for nesting. Biologists believe that breeding densities reflect the number of suitable aquatic foraging sites, especially stream riffles. A study in Ohio found five pairs of kingfishers nesting along 6 miles of river shoreline; another survey in New Brunswick documented ten pairs in 1 mile.

WOODPECKERS

A drum roll at dawn, a bird in undulating flight through the forest, wood chips littering the ground at the base of a tree—all signal the presence of a woodpecker. Woodpeckers belong to the family Picidae, with over 20 species in the United States and 215 species worldwide. Seven species breed in Pennsylvania and the Northeast.

Woodpeckers are well adapted for climbing and hammering on trees. They drill holes into trees to uncover insect food and create nesting and roosting shelters; they drum to communicate with each other. A woodpecker has a sharp, stout bill with a chisel-like tip for chipping and digging into trunks and branches. In pecking out wood, the bird aims its blows from alternating directions, much like a human woodchopper does. Bones between the beak and the unusually thick skull are not as rigidly joined as they are in other birds, and spongy, shock-absorbing tissues connect these flexible joints. Strong neck muscles provide the force for hammering, and bristly feathers shield the nostrils from dust and flying chips.

The tongue of most woodpeckers is rounded, horny, rich in tactile cells, and bathed in sticky saliva. The tongue's tip is pointed and barbed. After chopping exposes the cavity of a wood-boring insect, the long, flexible tongue feels out the prey, impales it, and draws it out. The tongue is nearly twice as long as the woodpecker's head and winds around inside the back of the skull when retracted. To get purchase on trees, a woodpecker has short, muscular legs and sharp claws on its feet. In most species, two toes point forward and two backward; this opposed, "yoke-toed" arrangement lets the bird climb vertical surfaces with ease. Stiff, pointed tail feathers catch on rough tree bark to brace the body during hammering and to provide an extra prop while the bird hitches its way upward. When a woodpecker molts and replaces its feathers, the two middle tail feathers (the stiffest and strongest ones) do not fall out until the other ten tail feathers have grown back in to support the bird's weight.

When a woodpecker launches itself from a tree trunk, it pumps its wings for four or five strokes, then folds them against its body; during this brief pause, the bird loses a few feet of altitude. Then comes another flurry of wingbeats, another pause, and so on, resulting in a distinctive undulant flight.

Woodpeckers feed mainly on wood-boring grubs, insects (often with an emphasis on ants), insect eggs, and pupae. The various species also consume sap, nuts, and the fruits of trees and shrubs. Woodpeckers search for food using their eyesight, probe with their bills into crevices, fleck off bits of bark, and excavate into dead wood. Hollow sounds—echoes from the woodpecker's exploratory tapping—may reveal the location of a wood borer's channel; the bird then delivers up to a hundred strokes per minute to uncover the morsel. Even in winter, woodpeckers have no trouble locating insects, which eliminates the need to migrate.

Most woodpeckers drum on limbs, hollow trunks, drainpipes, garbage can lids, tin roofs, and anything else that yields a resonant sound. Drumming designates an individual's territory and can attract a mate. Soft tapping may be a means of bonding or communicating between mates or between parents and offspring. Courtship and nesting habits are similar in most species. Much of the rivalry between males is confined to noisy, chattering pursuit. After pairs form, the male and the female together excavate a nest cavity in a branch or tree trunk. The female lays her clutch directly on wood chips left in the bottom of the cavity. The eggs are white. Both sexes incubate them, with the more aggressive male often staying on the eggs to protect them overnight. Young are altricial; for two to three weeks, they remain in the nest and are fed whole or predigested food by their parents. In the southeastern states, woodpeckers may raise two broods, but in Pennsylvania, most bring off one brood per year.

Woodpeckers excavate holes in living trees but rarely in healthy ones. By stripping the bark from dead and dying trees and cleaning up resident wood borers and carpenter ants, they prevent those pests from spreading to nearby healthy trees. Their hole-making activities also provide nesting and resting habitats for owls, bluebirds, tree swallows, nuthatches, chickadees, and squirrels. Cooper's hawks, sharp-shinned hawks, goshawks, red-tailed hawks, and the larger owls are the major predators on woodpeckers. Raccoons, opossums, weasels, squirrels, and black rat snakes take eggs and young.

Red-Headed Woodpecker *(Melanerpes erythrocephalus)*. Length is 8 to 9 inches; wingspread, 18 inches. The adult's head is a brilliant, eye-catching scarlet; that of the juvenile is brown. Body plumage is black and white, with a prominent white wing area that flashes when the bird flies. The call is a raucous *kwrrk*. The red-headed woodpecker inhabits open forests, farm woodlots, groves, orchards, and shade trees in towns and parks. Highly omnivorous, it feeds on beetles, ants, grasshoppers, caterpillars, other insects, spiders, acorns, wild fruits, apples, and corn. It gathers nuts in the fall and caches them in holes for use as winter food. Like the flicker, it often forages on the ground. Red-headed woodpeckers catch insects in flight, and sometimes they eat the eggs and nestlings of other birds.

Red-headed woodpeckers usually site their nests 8 to 80 feet above the ground in bare dead trees, and occasionally in fence posts. The female lays four to seven eggs (usually five), with a twelve- to thirteen-day incubation period. The young fledge after about four weeks. In Pennsylvania, the red-headed woodpecker inhabits wide valleys and agricultural areas; it is most common in the northwest and in the valleys of the south-central region, and it is rare in the deep woods of the northern tier and the Poconos. The species nests in much of the United States east of the Rockies. Starlings competing for nesting sites may be causing a general decline that has been documented in the red-headed woodpecker population.

Red-Bellied Woodpecker *(Melanerpes carolinus)*. Adults are 8 to 9 inches long and have a 17-inch wingspread. This shy bird shows a "ladder-back" plumage pattern, with black and white banding like a ladder's rungs on its back. The crown and the nape of the neck are red in the male; in the female, only the nape is red; and the immature's entire head is brown. The belly is tinged a very light red. The red-bellied woodpecker feeds on acorns, beechnuts, hickory nuts, wild grapes, mulberries, and fruits of poison ivy and dogwood; animal foods include beetles, wood-boring larvae, ants, treefrogs, and the eggs of smaller birds. Red-bellied woodpeckers inhabit coniferous and deciduous woods, groves, woodlots, orchards, and yards. Listen for the species' contact call, one to four *churr* notes, or a repeated *wick-uh* given during territorial disputes.

The nest is a cavity excavated in dead wood, usually less than 40 feet above the ground. The female lays three to eight eggs (generally four or five). Both parents incubate the clutch for twelve to fourteen days. The young leave the nest around twenty-four days after hatching; the parents may continue to feed their offspring for another six weeks or longer. During a recent scrap outside my office window, a pair of flickers outcompeted two red-bellied woodpeckers for a cavity in which the red-bellied pair, or another pair of the same species, had nested the previous year; later I saw the male red-bellied woodpecker quietly leaving the cavity with a flicker's egg in his beak. Subsequently, the flickers abandoned the nest.

Except during the April-to-August breeding season, red-bellied woodpeckers move about within a local area gleaning food. The species ranges from Pennsylvania south to Florida and west to Minnesota, Kansas, and Texas. *Melanerpes carolinus* breeds almost statewide in Pennsylvania, although its distribution is scattered in the mountainous north-central region and in the northeastern counties. Red-bellied woodpeckers do not often live above an elevation of 2,000 feet; perhaps harsh winter weather excludes the birds from the uplands.

Yellow-Bellied Sapsucker *(Sphyrapicus varius)*. Length is 7 to 8 inches; wingspread, 14 inches. Plumage is variable within the species, but the narrow longitudinal wing stripes (visible when the bird is perched) and the finely mottled back are good field marks; the back coloration blends well with tree bark. The belly is tinged yellow, and the head is red, black, and white. The call is a cat- or jaylike mewing note. Individuals also tap or drum in a distinctive staccato rhythm.

Sapsuckers inhabit moist forests, woodlots, orchards, and clear-cuts where some trees remain. They drill sap wells, parallel rows of tiny holes in live trees (up to thirty holes per day) and return later to drink sap and to catch small insects attracted by the sweet liquid. The sapsucker has a brush-tipped tongue for soaking up sap. Sapsuckers also eat beetles, ants, caterpillars, insect eggs, and spiders; the cambium, or the layer beneath the bark, of maple, aspen, birch, fir, hickory, beech, pine, oak, and other trees; and fruits and seeds. Warblers, ruby-throated hummingbirds, other woodpeckers, chipmunks, and squirrels also drink from sapsuckers' sap wells.

Pennsylvania is on the southern fringe of the species' breeding range; the yellow-bellied sapsucker nests mainly across the northern tier of counties in hardwood forests at elevations above 1,500 feet. The nest is a gourd-shaped cavity excavated 8 to 40 feet up in a tree; aspens and other trees afflicted with fungal disease are often chosen as nest sites, because the fungus creates a soft center that is easily dug out. The female lays four to seven eggs (usually five or six), with a twelve- to thirteen-day incubation period. The parents feed insects, sap, and fruit to the young.

Sphyrapicus varius is the most migratory of our woodpeckers, wintering in the southeastern states, the West Indies, and Central America. In the winter, they inhabit mixed woodlands and prefer areas with conifers.

Downy Woodpecker *(Picoides pubescens)*. With a length of 5 to 6 inches and a wingspread of 11.5 inches, the downy is the smallest woodpecker in eastern North America. It is also the most common. A downy woodpecker resembles a small hairy woodpecker and has a similar white back stripe and white breast. The male has a red patch on the back of the head. Bill length of the downy is less than the width of its head, while that of the hairy is equal to or greater than the width of its head. The downy woodpecker's outer tail feathers are barred with black (in the hairy, these are solid white). The downy feeds on wood-boring larvae, moths, beetles, ants, aphids, gall wasps, spiders, poison ivy and dogwood fruits, berries, corn, apples, and acorns. Calls include a soft *pik* and a rattling sound.

The downy woodpecker lives in open forests of mainly deciduous growth, woodlots, orchards, and parks, from wilderness areas to suburbs. Courtship begins in April. Both sexes share in excavating a nest in rotting wood 12 to 30 feet above the ground, often in a dead stub on the underside of a limb. The three to six eggs (usually four or five) are incubated by both parents for twelve days. The downy woodpecker is a common resident in all seasons. In fall and early winter, males and females have separate feeding areas. In winter, downy woodpeckers forage for weed seeds in fields, search and excavate in tree bark for dormant insects, and visit backyard suet feeders. They often join into mixed flocks with chickadees, nuthatches, and titmice.

Downy woodpecker, above, and northern flicker.

Hairy Woodpecker *(Picoides villosus).* Length is 8 to 9 inches; wingspread, 15 inches. This woodpecker has a vertical white stripe down the center of its back, black wings stippled with white, white feathers forming the outer edge of the tail, and a white breast. The sexes are similar, but the female lacks the male's small red patch on the back of the head. Hairy woodpeckers live from Alaska to Panama and from Newfoundland to Florida. Primary habitat is forested land and wooded swamps with large trees. Hairy woodpeckers feed on beetle larvae, ants, caterpillars, spiders, seeds, nuts, and berries. Males tend to forage higher in trees than females.

The hairy woodpecker is more furtive than its close relative, the downy woodpecker. Males and females may live on separate territories in early winter; pairs form in midwinter, with the female's home territory becoming the hub of the breeding area. A nest cavity is excavated 5 to 60 feet up in a tree. The female lays three to six eggs (commonly four), and both she and her mate incubate the eggs for around fourteen days. After about a month, the young leave the nest, and the parents continue to feed them for several weeks. Hairy woodpeckers require mature trees for breeding and benefit when loggers leave large dead snags in the woods.

Northern Flicker *(Colaptes auratus)*. Length is 8 to 10 inches; wingspread, up to 20 inches (about the size of a blue jay). Also known as the yellow-shafted flicker or yellowhammer, this species has a brown back, no white on the wings, a prominent black band high on the breast, bright red at the nape of the neck, and yellow underwings. The male has a black "mustache" marking extending back from the bill. In flight, the white rump shows prominently. The species ranges from Alaska to Nicaragua. Flickers breed across Pennsylvania, and some individuals winter in the southern counties.

Flickers favor partially open country—woodlots, woods edges, orchards, and yards—rather than deep woods. They often land on the ground or on sidewalks to eat ants, which may constitute up to three-quarters of the diet. They also hop about eating beetles, grasshoppers, crickets, termites, and other insects. In fall and winter, flickers feed on poison ivy fruits, berries, nuts, sumac seeds, and corn. Males defend their territories aggressively, calling *flick* or *flicker* and displaying their bright underwings and tail feathers. Both sexes work to clean out a cavity in a dead tree, a broken-off stub, a fence post, or a utility pole, and occasionally in the side of a building; the excavation may take up to two weeks. Flickers also use artificial nesting boxes. Starlings are a particular scourge to flickers, often driving them out of newly excavated nests. The female lays three to ten eggs (usually six to eight), which hatch following eleven to sixteen days of incubation. In contrast to many other woodpeckers, which bring whole food to their young, flickers feed their offspring by regurgitating insects. The young fledge after about a month.

Pileated Woodpecker *(Dryocopus pileatus)*. Length is 12 to 17 inches; wingspread, to 27 inches: crow-size, but with a long, slender neck. Also called the Indian hen and log cock, this is the largest eastern woodpecker except for the closely related and almost certainly extinct ivorybill. The pileated has a solid, dull black back and tail and a conspicuous red crest for which it is named *(pileus* is Latin for "cap"). The female is similar to the male but lacks red cheek patches and has less red in the crest. Pileated woodpeckers fly strongly, using periods of flapping broken by moments of soaring, creating an undulating flight. The wings' undersurfaces flash white during flight. Pileated woodpeckers drum loudly and rapidly, starting off fast, then trailing off at the end. The call, *wick-uh wick-uh wick-uh,* carries a great distance through the woods.

Digging deep into rotted wood in search of carpenter ants (up to 60 percent of the diet), a pileated woodpecker may hammer out fist-size chunks of wood; the bird twists its head and beak as it strikes to gain leverage, leaving a large rectangular hole. Pileated woodpeckers also feed on beetles, wasps, wood-boring larvae, nuts, and wild fruits, including those of dogwood, wild grape, greenbrier, and black gum: I've watched these big birds dangling from the springy branches of the gum tree next to my office, snapping up the deep blue fruits about 5 feet away from the window.

Pileated woodpeckers inhabit mature coniferous and deciduous forests, remote mountainous territory, valley woodlots, and city parks. The nest is a new hole excavated each year, in a tree from 15 to 70 feet up. The cavity is 10 to 24 inches deep. The female lays one to six eggs (usually four), incubated by both sexes (the male has night duty) for eighteen days. Both parents feed the young, by regurgitation. Fledglings leave the nest after about four weeks and may stay with their parents for up to three months. When the original eastern forests were cleared in the 1700s and 1800s, the pileated woodpecker population dropped sharply. Since 1900, the species has slowly come back, and today this magnificent woodpecker is again common. It breeds from Alaska to Florida in suitable wooded habitat and is found in every county in Pennsylvania.

Bent, A. C. *Life Histories of North American Woodpeckers.* Bloomington, IN: Indiana University Press, 1992.

Winkler, H., D. Christie, and D. Nurney. *Woodpeckers: A Guide to Woodpeckers of the World.* Boston: Houghton Mifflin, 1995.

FLYCATCHERS

The tyrant flycatchers—family Tyrannidae—are found only in the New World. The family name stems from the aggressive, almost tyrannical, behavior of some of the birds in this large group of over four hundred species, most of which live in the tropics. Pennsylvania has ten species. Flycatchers are often hard to identify, even for veteran birders; since the birds are drab and tend to stay among thick foliage, they end up being distinguishable only by their songs. The sexes are colored alike. Flycatchers are perching birds, members of order Passeriformes, whose feet have three toes pointing forward and one pointing backward, letting them perch easily on branches. (All of the species described from here to the end of the "Birds" section, except for the chimney swift, are perching birds.)

Flycatchers catch and eat flies and many other insects, particularly flying ants, bees, and wasps. In forested areas, large flycatchers may specialize in larger insects, medium-size flycatchers may take slightly smaller prey, and small flycatchers may zero in on the smallest insects. Such feeding stratification reduces competition and lets several species use the same area. Also, different species prefer subtly different habitats, with varying amounts and densities of undergrowth and degrees of canopy shading.

When foraging, a flycatcher sits upright on a perch, scanning its surroundings while waiting for an insect to approach. The bird darts out in swift, maneuverable flight, snatches its prey out of the air with its beak, and swallows it on the spot or returns to the perch to eat its meal. Several adaptations help a flycatcher take insects. Its drab plumage makes the waiting bird hard to see, not just by its prey, but also by hawks that hunt for flycatchers and other small birds. The bill is flat and wide, suggesting somewhat the bills of swallows and nightjars, although not nearly so compressed or gaping. Bristles at the corner of the mouth may function as "feelers," letting a flycatcher make last-second adjustments before snapping its bill shut on prey. Keen eyesight lets a flycatcher spot insects and judge distances accurately. In addition to catching insects on the wing, flycatchers sometimes hover near foliage and pick off insects and spiders clinging to the vegetation. Some species land and catch prey on the ground. Most of our flycatchers occasionally eat berries and seeds.

Of our ten breeding species, most build open cups anchored to small branches of trees and shrubs. One, the yellow-bellied flycatcher, builds an enclosed nest on the ground. The familiar eastern phoebe plasters its nest against a rock wall or on a building rafter. And the great crested flycatcher uses a tree cavity. In most cases, the female does most or all of the incubating, while the male defends the nesting territory and helps to feed the young.

Flycatchers advertise their home territories using their voices; some employ a special "dawn song" given just before sunrise and rarely sung later in the day. Because many flycatchers are so similar in appearance, individuals probably recognize their own species mainly by sound. Biologists believe that in at least several types, the distinctive song is innate, not learned, as is the case with most other birds, which learn to sing by listening to adults of their kind.

Since they depend on eating insects, flycatchers must vacate northern areas in winter. They migrate at night. The various species winter in open and forested habitats along the Gulf coast, on the Caribbean islands, and in Central and South America. In South America, an estimated 10 percent of all birds belong to the tyrant flycatcher family. In much of their wintering range—which is probably their original or ancestral home, from where populations expanded their breeding ranges eons in the past—flycatchers are vulnerable to habitat loss as large forested tracts are logged or converted to agriculture.

Olive-Sided Flycatcher *(Contopus cooperi)*. Although once fairly common in Pennsylvania, this species may or may not breed in the state today. Its white throat and breast contrast with its dark olive sides. A fairly large (7 to 8 inches long), big-headed bird, the olive-sided flycatcher inhabits cool coniferous forests, generally near water. The male sounds a repetitive *pip pip pip,* plus a song that has been rendered as *hic-three-beers.* Individuals sit high in dead snags or on branches, sally forth to catch prey—mainly wasps, winged ants, and bees—and return to the perch to eat. Olive-sided flycatchers place their cup-shaped nests in trees 5 to 70 feet above the ground, among dense twigs or needles; three young are usual. The main breeding range is in Canada. The species migrates north through Pennsylvania in late May and early June, leaves again in August and September, and spends the winter months in the rain forests of South America. This long-range migration has earned it the nickname "peregrine of flycatchers."

Eastern Wood-Pewee *(Contopus virens)*. The eastern wood-pewee breeds throughout eastern North America from southern Canada to the Gulf of Mexico. It is found in all counties in Pennsylvania. To locate this drab, olive-gray, sparrow-size bird, listen for the male's oft-repeated namesake call—*pee-o-wee*—which is given throughout the day and particularly at dawn and dusk. Pewees use almost every woodland habitat, including woodlots, woods edges, mature forests (both deciduous and mixed), parks, and urban areas with shade trees. They perch in one place for an extended period, flying out to intercept passing insects; one study found an average perching height of 35 feet above the ground. Pewees eat flies, beetles, small wasps, and moths. They also consume elderberries, blackberries, and fruits of dogwood and pokeweed.

Males defend breeding territories of 2 to 6 acres. Pairs begin nesting in late May. The nest is a compact cup woven of plant matter, hairs, and spider silk, its outer surfaces studded with lichens; it looks like a larger version of the ruby-throated hummingbird's nest. The three eggs are incubated by the female and hatch after twelve or thirteen days. Both parents feed the young, which make their first flights at fourteen to eighteen days. Blue jays are major predators, taking both eggs and young. Most perching birds stop singing regularly in late summer, but male wood-pewees keep up their chanting until the autumn migration. The species departs from Penn's Woods in August and September, with a few individuals hanging on until October. Wood-pewees winter in the tropics from Panama to Bolivia, in shrubby woods and along forest edges.

Yellow-Bellied Flycatcher, Acadian Flycatcher, Alder Flycatcher, Willow Flycatcher, and **Least Flycatcher** *(Empidonax* species). These small flycatchers, around 5 inches in length, have olive-colored backs and heads, pale breasts, and pale eye rings and wingbars. They spend much of the day hunting from a perch. When perched, they occasionally flip their tails up and down. Extremely difficult to identify in the field, they are usually distinguished by voice and habitat.

The yellow-bellied flycatcher *(Empidonax flaviventris)* lives in the deep shade of coniferous woods and cold bogs. A shy bird, it inhabits remote uplands in a scattering of our northern counties. The call is a quiet, ascending *chu-wee.* The cup-shaped nest is built of rootlets and mosses and is hidden on or near the ground, in a cavity among the roots of a fallen tree, in a hummock of sphagnum moss, or at the base of a conifer. The species nests mainly in Canada, as far west as the Yukon Territory, with all individuals apparently migrating through the East. The yellow-bellied flycatcher is considered a threatened species in Pennsylvania.

The Acadian flycatcher *(Empidonax virescens)* nests mainly in the Southeast, and Pennsylvania is near the northern limit of its range. The type, or first, example of the species was discovered near Philadelphia in 1807 by the Scottish-born ornithologist Alexander Wilson. The species is misnamed, since it does not inhabit Acadia, the former French colony centered on Nova Scotia. The Acadian flycatcher lives in moist woods near streams and requires large areas of contiguous forest. The male sounds a low, sharp *spit-chee!* The Acadian often chooses a beech tree in which to build its hammocklike nest; stems and grasses dangle from the nest, giving it an unkempt appearance. Acadian flycatchers winter mainly in the rain forests of Colombia and Ecuador, where they sometimes follow mass movements of army ants and prey on insects set to flight by the creeping columns.

The alder flycatcher *(Empidonax alnorum)* and the willow flycatcher *(Empidonax traillii)* were, until the 1970s, considered to be one species— Traill's flycatcher, named by John James Audubon for Dr. Thomas Traill, one of his supporters. However, the two flycatchers have different voices, use slightly different habitats, build different kinds of nests, and are reproductively isolated. The alder sings *fee-bee-o* and the willow *fitz-bew;* the alder builds a loose cup for a nest, usually within a few feet of the ground, while the willow flycatcher's nest is compact and felted, and often sited higher above the ground. Both alder and willow flycatchers nest in thickets of willows, alders, and other shrubs, but the willow flycatcher tends to use drier, more open sites than its lookalike relative. In Pennsylvania, alder flycatchers nest mainly in the north; willow flycatchers nest statewide, with the fewest records coming from the north-central region.

The least flycatcher, *Empidonax minimus,* is the smallest and probably the most common of the eastern *Empidonax* flycatchers. It lives along woodland edges and often perches in the open. The male calls out an emphatic *chebeck!,* accented on the second syllable. The least flycatcher eats small wasps, winged ants, midges, flies, beetles, caterpillars, grasshoppers, spiders, and berries. Pairs sometimes nest in loose colonies. The nest, a neat cup, is usually placed in a vertical fork of a branch in a small tree or sapling. The three to five eggs are incubated for thirteen to fifteen days. The least flycatcher's breeding range stretches from western Canada to Nova Scotia and south in the Appalachians to Tennessee and North Carolina. *Empidonax minimus* is fairly common across much of Pennsylvania, except for the southeast and the lower Susquehanna drainage, where it is absent. In autumn, adults migrate ahead of juveniles to wintering grounds in Mexico and Central America.

Eastern Phoebe *(Sayornis phoebe).* Anyone who has spent time at a woodland cabin probably has come to know this jaunty, medium-size (6.5 to 7 inches) flycatcher. Phoebes breed statewide in Pennsylvania, except in heavily urbanized areas. They eat a variety of insects, including small wasps, bees, beetles, flies, and moths. They often take prey from vegetation and from the ground, and they eat seeds and berries. The female builds a nest out of mud, moss, leaves, grass, and hair, tucking the cup-shaped structure into a sheltered spot beneath a rock ledge or house eave, against a stone wall, on a bridge beam or barn or porch support. A pair may use the same nest several years in a row.

The female lays four or five eggs and incubates them for around sixteen days. Both parents feed the nestlings, which fledge about sixteen days after hatching. Eastern phoebes typically rear two breeds per summer. One of the harbingers of spring at our home is the arrival of the first male phoebes, around the middle of March; they announce themselves with repeated *fee-bee* calls and the species' characteristic up-and-down tail flicking. I often wonder how (or if) they survive when a late snowfall sends insects back into dormancy. Under our garage, phoebes nest within 20 feet of American robins, and each species tolerates the other's presence; perhaps there is little overlap in the foods they eat. In the Northeast, populations have risen since European settlement, with phoebes taking advantage of nest sites created by human construction. The species winters in the Gulf states and in Mexico.

Great Crested Flycatcher *(Myiarchus crinitus).* At 8 to 9 inches, this is our largest flycatcher. The great crested flycatcher sports a yellow belly, a gray breast, and rusty-red tail and wing feathers. When agitated, it erects a prominent head crest. The species breeds in mature woods throughout Pennsylvania and eastern North America and can also be found in wooded suburbs, farm woodlots, and orchards. Great crested flycatchers feed among the treetops, hopping from limb to limb taking caterpillars, katydids, crickets, beetles, and spiders, and flapping out into openings and clearings to seize moths, butterflies, beetles, bees, and wasps. Great crested flycatchers eat wild fruits in late summer and fall.

The call is a loud, insistent *wheep!* Great crested flycatchers defend their territories against intrusions by squirrels and other birds. They nest in tree cavities, including old woodpecker holes, as well as hollow fence posts and artificial nesting boxes. (One nest was even found in the barrel of a cannon in Gettysburg National Military Park.) Both male and female bring in grass, weeds, bark strips, rootlets, and feathers, often building up this cushion as high as the entry hole. They have the curious habit of placing a shed snakeskin or a scrap of cellophane among the nest material; some ornithologists speculate that the crinkly foreign matter may deter nest predators. Great crested flycatchers depart from Pennsylvania in September en route to wintering grounds in southern Florida and from Mexico to Colombia.

Eastern Kingbird *(Tyrannus tyrannus).* This bold, aggressive flycatcher breeds in open country across North America. Look for kingbirds in scattered trees along roads and streams, orchards, fencerows, and forest clearings. The bird gets its name because it dominates other birds, including many larger than itself, driving them away from its territory. Of all the flycatchers, kingbirds are among the easiest to locate and observe. They are

Flycatchers, like this eastern kingbird, forage by sitting on perches, waiting for insects to fly past, and darting out to seize their prey.

about 8 inches long and are dark gray and white, with a white-tipped tail and a small red streak on the head. Roger Tory Peterson describes the species' call as "a rapid sputter of nervous bickering notes." Kingbirds feed on beetles, wasps, bees, winged ants, grasshoppers, honeybees, and many other insects.

Kingbirds often attack crows, hawks, and owls, flying high in the air, getting above the larger birds, and diving at them repeatedly. After driving off an adversary, a kingbird may perform a display known as "tumble flight," in which it glides back to the earth in stages, sometimes tumbling in midair. After mating, the female does not let the male help her build the nest and may actually drive him away until after the eggs hatch. The nest is a bulky cup 7 to 30 feet up in a shrub, tree, or snag. The two to five eggs hatch after sixteen days. Both parents feed the nestlings, which can fly after around seventeen days; they may be fed by their parents for a month after fledging, with family members sounding rapid *kitterkitter* calls back and forth. Kingbirds have a very different lifestyle on their wintering range in South America, where they coexist in flocks and switch to a diet of berries.

SHRIKES

Shrikes are called "butcher birds" for their habit of hanging their prey on thorns or barbed-wire fencing as a means of storing it. Shrikes are classified as songbirds, and they belong to the family Laniidae, mainly an Old World group with only two species in the Americas. These birds have hooked bills, which they use to kill small birds, small rodents, and large insects. The two North American species, the northern shrike and the loggerhead shrike, have similar appearances, with black masks, gray-and-white bodies, and white-edged tails. They perch on bushes, fences, and utility lines watching for their prey, then fly in swiftly to attack it.

Northern Shrike *(Lanius excubitor)*. These robin-size predators breed in the far north from Labrador to Alaska. They inhabit sparse stands of spruce and willow and alder thickets, preying on voles, other rodents, and insects. In winter, northern shrikes migrate south, with the vanguard often reaching Ohio, Pennsylvania, and Virginia in the East; these winter visitors stay in open and brushy areas. In some years, a fair number of northern shrikes winter in Pennsylvania, while in other years few come this far south.

Loggerhead Shrike *(Lanius ludovicianus)*. Slightly smaller than the northern shrike, the loggerhead shrike breeds across the South and Midwest. It favors pasture or cropland with scattered trees or shrubs on which to perch while scanning for prey. Grasshoppers, crickets, beetles, wasps, and other insects are eaten, as well as mice and other rodents (especially in winter), small birds, lizards, frogs, snakes, spiders, and snails.

In the Northeast, the population of loggerhead shrikes has declined in recent years, and scientists are not sure why. It's believed that shrikes spread into the region in the early 1800s after forests were cut down, and it may be that reforestation has made the area less suitable; pesticide contamination may also be playing a role. Loggerhead shrikes apparently are gone from New England. Virginia and Maryland have shrikes. During the *Breeding Bird Atlas* survey conducted in Pennsylvania during the 1980s, volunteers spotted shrikes in several south-central counties, but they could not confirm that the birds had nested. More recently, several pairs have nested in Adams and Franklin counties. The loggerhead shrike is listed as an endangered species in Pennsylvania.

VIREOS

The more than fifty species of vireos live only in the New World. About 5 to 6 inches long and olive or gray in color, they keep mainly to the treetops and are heard—thanks to the males' incessant singing—much more frequently than they're seen. Five species breed in Pennsylvania, including the red-eyed vireo, perhaps the most abundant bird of mature hardwood forests in the Northeast. Another species, the Philadelphia vireo, migrates through our state. The word *vireo* comes from Latin and means "green bird."

Vireos feed mainly in the midlevel and upper canopies of trees and in understory shrubs. They glean insect prey while walking along or hopping among branches, hovering near leaf surfaces, making short flights, and inspecting bark furrows. Vireos also eat berries, especially in fall and winter. The best time to look for them is in early spring, after migratory birds have arrived and before the leaves come out fully. During breeding season, males sing throughout the day, even during hot noon hours, and they keep on singing into late summer after most other birds have quieted.

Vireos breed in May and June. Males perform stylized posturings in front of prospective mates. They may spread their tails and fluff up their feathers while weaving their bodies from side to side or up and down; both males and females may flutter their wings. Pairs are thought to be monogamous. They nest among the foliage of trees and shrubs. The typical nest is a cup made out of plant matter held together by spider or caterpillar silk, hanging hammocklike in the fork between two twigs. In most species, both males and females work at building the nest. Vireos' eggs are white, marked with brown or black spots. The usual clutch is three to five, with an average of four eggs. Both parents share in incubating the eggs and bringing food to the nestlings. In Pennsylvania, most vireos rear one brood per summer. Brown-headed cowbirds lay eggs in, or parasitize, many vireos' nests. After the eggs hatch, the vireo adults feed the cowbird young, and the baby cowbirds grow faster than the baby vireos, who generally do not survive. Fragmented forests in the Northeast may give both cowbirds and nest predators increased access to vireos' nests. Several vireo species also face problems on their wintering ranges in Central and South America, where thousands of acres of tropical forest have been logged into oblivion. Vireos are preyed on mainly by accipiter hawks.

White-Eyed Vireo *(Vireo griseus).* This smallish vireo sings *chick-oh-per-weeoh-chick:* the sharp notes at the couplet's beginning and end are distinctive among vireos. The bird has yellowish eye rings and white eyes. The species inhabits woods edges, overgrown pastures, brushy swamps, swales, glades, and alder tangles. White-eyed vireos feed actively in the low branches and foliage of dense cover, taking moths, butterflies, caterpillars, beetles, wasps, ants, bees, flies, and many other insects, as well as spiders and snails. As summer wanes and during migration, they eat fruits and berries.

White-eyed vireos nest throughout the East. Pennsylvania is near the northern limit of the breeding range, with most nesting reported from the state's southwestern and southeastern corners, including suburbs around Philadelphia and Pittsburgh. The nest, usually 2 to 6 feet above the ground, is slightly cone-shaped, distinguishing it from the more rounded, cuplike nest of the red-eyed vireo. The nest consists of small pieces of soft wood and bark held together with cobwebs, with an inner lining of dry grass and fine stems. The eggs are incubated for about two weeks, and the young fledge nine to eleven days after hatching. In some areas, brown-headed cowbirds parasitize nearly half of all nests. White-eyed vireos winter in the southern United States, Mexico, and Central America.

Yellow-Throated Vireo *(Vireo flavifrons)*. This is the most colorful of our vireos, with a bright yellow throat and breast. The male's song is a string of short, buzzy, robinlike phrases given twenty to thirty-five times per minute. Yellow-throated vireos live along the edges of bottomland and mature upland forests, and in open wooded places such as orchards, parks, and shady areas in towns. The species avoids coniferous woods and the unbroken forest interior. The bill, typical for vireos, is sturdy, slightly curved, and has a hook at the tip, useful for nabbing and tearing apart caterpillars, the single food item most prevalent in this species' diet. Yellow-throated vireos also feed on many other insects and eat the fruits of multiflora rose, sassafras, wild grape, pokeberry, and other plants. The nest, a thick-walled cup made of strips of inner bark and grasses, is generally 20 to 40 feet up in a tree, and within 20 inches of the trunk. *Vireo flavifrons* breeds throughout the eastern United States and in southern Canada, and winters in Mexico, Central America, and South America.

Blue-Headed Vireo *(Vireo solitarius)*. Also known as the solitary vireo, this bird has a blue-gray head and white eye rings. It is our earliest spring vireo, arriving in April and May. The song, a series of short, whistled phrases, is among the most mellifluous of all vireos' calls. The species thrives in a wide range of forested settings, particularly in open woods where pines or hemlocks predominate. In Pennsylvania, most blue-headed vireos nest at elevations above 1,000 feet. The species breeds widely in northern Pennsylvania and is absent during the summer from the state's southeastern and southwestern corners. Blue-headed vireos feed almost exclusively on insects and spiders, foraging among the leaves, branches, and twigs in the upper zones of trees. The nest is an open cup made of grasses, inner bark, and other plant materials, its outside adorned with lichens or papery scraps from old hornets' nests. It is usually placed less than 10 feet up in a tree.

Warbling Vireo *(Vireo gilvus)*. This drab, grayish olive bird has a whitish breast and a faint pale stripe above the eye. Roger Tory Peterson terms the song "a single languid warble unlike the broken phraseology of the other vireos." Males sing from late April until mid-September. Warbling vireos eat many caterpillars, plus insects ranging in size from aphids to dragonflies. In late summer, they turn to fruits of dogwood, pokeberry, sumac, elderberry, and other plants. Warbling vireos breed across much of North America in open, mixed, or deciduous woods, fencerows, roadside trees, shade trees in open country, woodlot edges, and trees along streams and rivers. Nests are built in trees, higher above the ground than those of most other vireos: 20 to 90 feet up. Both sexes build the neat cup out of bark strips, leaves, grasses, feathers, and plant down. Male warbling vireos are such persistent singers that they even give voice while helping to incubate eggs. In Pennsylvania, the population of warbling vireos seems to be concentrated in the state's four corners and along the major river systems.

Philadelphia Vireo *(Vireo philadelphicus)*. Its name notwithstanding, the Philadelphia vireo does not nest in the vicinity of Philadelphia or anywhere else in Pennsylvania. (The type, or first, specimen of this bird was collected near Philadelphia.) The Philadelphia vireo lacks wingbars and has a yellow-tinged breast. Experienced birders may spot this uncommon migrant in May, and again in September and October, as it passes through the Keystone State, shifting between its northern breeding grounds—primarily New England and southern Canada—and its wintering range in southern Central America. Philadelphia vireos often join flocks of migrating warblers.

Red-Eyed Vireo *(Vireo olivaceus)*. Although this bird is common and abundant throughout the forests of the Keystone State, its greenish leaf-matching coloration, countershading (the pale belly, when seen from below, blends with sun-dappled foliage and the sky), and treetop habits combine to make

*The red-eyed vireo inhabits the forest canopy. It is one of
the most abundant woodland birds in the Northeast.*

it an unfamiliar bird to most Pennsylvanians. The song is a series of robinlike phrases (ornithologists have noted around forty of these locutions), often repeated for an hour or longer without cease. Males continue to sing into the summer, even during hot afternoons. Red-eyed vireos breed in every county in Pennsylvania. They use a variety of woods settings, including second-growth forest, woodlots, mature deciduous or mixed woodlands, and shade trees in cities and towns. An ideal habitat would be an extensive stand of mature moist forest with an understory of shrubs and smaller trees.

Male red-eyed vireos feed in the high canopy, while females forage lower down. Red-eyed vireos hover while picking insects from leaves and flowers. They feed on caterpillars (gypsy moth, tent caterpillars, fall webworms, and many others), beetles, bugs, flies, walkingsticks, cicadas, and treehoppers. They also eat the fruits of Virginia creeper, dogwood, sumac, and other plants. In spring, males establish 1- to 2-acre territories. Unlike our other vireos, the male red-eyed vireo does not help the female build the nest, which is a deep cup 2 to 60 feet (on average, 5 to 10 feet) above the forest floor. Chipmunks and red squirrels may eat eggs and nestlings. Up to half of all red-eyed vireo nests may be parasitized by brown-headed cowbirds; Hal Harrison, author of *A Field Guide to Birds' Nests,* found one red-eyed vireo female incubating four cowbird and no vireo eggs in its nest, the female cowbird apparently having punctured or pitched out the vireo's eggs.

In unparasitized nests, three to five young hatch after eleven to fourteen days' incubation. Both parents feed the brood. The juveniles leave the nest when ten to twelve days old. Their parents feed them for another several weeks; the adults quit defending a home territory and lead the young about, searching for food. Red-eyed vireos usually raise one brood per year. The species winters in northern South America, including the Amazon River basin, where the birds feed mainly on fruit. Despite the wholesale cutting of tropical forests and the fragmenting of woods in the Northeast, the red-eyed vireo seems to have a stable population, perhaps because it can adapt to different wooded habitats.

CROWS AND RAVEN

Two species of crows and the common raven breed in Pennsylvania. All belong to family Corvidae, in the large order of the Passeriformes, or perching birds, the dominant avian group on earth today. Crows and ravens are intelligent, alert, and adaptable. Their bodies, feet, and bills remain unspecialized, letting them exploit many different sorts of food in a range of habitats. Adults of all three species are black from their beaks to the tips of their tails; juveniles show varying amounts of dark brown in their plumage.

American Crow *(Corvus brachyrhynchos).* Crows breed across most of North America, except for hot deserts. Adults are 17 to 19 inches long, have a 3-foot wingspan, and weigh 16 ounces. In level flight, crows can reach 25 to 30 miles per hour. The species name *brachyrhynchos* means "short beak"; actually, the crow's sturdy beak is fairly large, up to 2.5 inches long, but short compared with that of the raven. Crows have keen senses of sight and hearing. Wary birds, they usually post sentries while they feed; the sentries watch for danger and warn the feeding birds with sharp alarm caws.

Crows can be seen year-round throughout Pennsylvania. That doesn't mean that the same individual birds are present at all times: many crows that breed here migrate south in late September or early October and are replaced by birds from farther north. Northern migrants remain here over winter, while crows from our state may fly as far as the Gulf coast. Crows are gregarious. During most of the year, they flock in groups ranging from family units to several hundred birds. In winter, crows may gather by the tens of thousands in farming areas, roosting at night in groves of trees; flocks range widely to feed, up to 30 miles a day. Throughout the year, crows eat grasshoppers, caterpillars, grubs, other insects, worms, grain, berries, fruits, the eggs and young of other birds, carrion, garbage—just about anything edible that they can find or overpower. They feed mainly on the ground and sometimes in trees. Indigestible matter that is accidentally swallowed, such as hair, feathers, and bones, gets coughed up later in pellet form.

The large winter flocks break up in early spring, with the birds regrouping in two basic units. One unit is a small wandering flock of non-breeding birds. The second social unit is a family flock organized around a previously mated pair and sometimes including one or more offspring from past years. Young crows do not mature sexually until they are two years old or older and may remain with their parents for up to four years.

During courtship, the male faces the female, fluffs up his body plumage, spreads his wings and tail, and bows while sounding a brief, rattling song. Pairs touch bills and preen each other's feathers, a practice known as allo-preening. Some authorities believe crows mate for life. Pairs are quiet and secretive near their nests. Crows begin nesting as early as late March in Pennsylvania. They place their nests in the crotches of trees—oaks and conifers are preferred—10 to 70 feet above the ground. A crow's nest is about 2 feet across, made of twigs, sticks, bark, and vines, with a central cup lined with moss, shredded bark, grass, fur, or feathers. The female lays three to eight eggs (usually four to six); the eggs are bluish green, blotched and spotted with brown and gray. The female does most or perhaps all of the incubating and is fed on the nest by her mate. The young hatch after about eighteen days. Both parents, sometimes aided by nest helpers (young from previous years), feed the nestlings, which fledge four to five weeks after hatching.

Crows are both predators and prey. They rob the nests of songbirds and waterfowl, stealing or breaking eggs and eating their contents, and killing and eating young. Crow eggs and nestlings are eaten by raccoons, opossums, and tree-climbing snakes. Hawks and owls kill fledgling and adult crows. Crows are especially vulnerable to night attacks by great horned owls. Should crows spot an owl during the day, they will mob it: swoop in, call excitedly to attract other crows, and if the opportunity presents itself, use their bills to strike the larger, less maneuverable raptor. I have been close to a band of crows mobbing a great horned owl, and the unbridled hatred they displayed nearly made my hair stand on end. Crows also mob hawks and ravens. In turn, crows are mobbed by smaller birds, especially kingbirds and red-winged blackbirds.

Before Europeans settled North America and farming and logging opened up the wooded wilderness, crows were rarer than they are today.

Even today, crows breed more readily in semiopen habitats in lowlands and valleys than they do in heavily wooded mountainous terrain. Crows are most numerous in agricultural districts offering a variety and a reliable supply of food. One habitat necessity is an adequate number of trees for roosting and nesting. Crow populations have grown since the 1960s, perhaps because crows are not hunted as intensely as they once were; because they have begun to breed more readily in towns and cities; and because large numbers of road-killed animals provide ample winter food.

The large, heavy bill of the common raven distinguishes it from the American crow.

Fish Crow *(Corvus ossifragus).* The fish crow is slimmer and has a narrower beak and slightly shorter legs than the American crow. In the field, the two look almost identical. The fish crow, however, is typically found near water: along the Atlantic coast and, in Pennsylvania, in the drainage basins of rivers flowing into the Atlantic. One way to identify a fish crow is by hearing it call: a short, nasal *car* or *cuh-cuh,* as opposed to the distinct, ringing *caw* of the American crow. The fish crow, it is sometimes said, sounds like "a crow with a cold."

As its name implies, this bird feeds on fish. Along beaches and estuaries, it captures small crabs, steals food from smaller gulls and terns, and scavenges for carrion. Inland, it eats insects, crayfish, berries, seeds, nuts, birds' and turtles' eggs, and garbage. Fish crows usually forage in flocks, walking along shores or in shallow water. If herons or ducks are frightened off their nests, fish crows may eat the eggs.

The nest and eggs of the fish crow are similar to those of the American crow. In Pennsylvania, fish crows breed along the Delaware, Schuylkill, Lehigh, and Susquehanna rivers. Recently, the population has been expanding up these river systems and spreading from bottomland forests into nearby farms and woodlots.

Common Raven *(Corvus corax)*. Ravens live in northern regions of both the Old and New World. They range from Alaska to Labrador and south to Nicaragua in the West and Georgia in the mountains of the East. In Pennsylvania, the common raven breeds mainly in remote parts of the north-central counties on the Allegheny High Plateau. Ravens are also fairly abundant in the wooded mountains of the Ridge and Valley region of south-central Pennsylvania. Before settlement, ravens probably were more common in the state than they are today, but by the late nineteenth century, they were so rare that ornithologists believed the species to be nearing extinction. Today ravens are showing a greater tolerance of human-caused change and seem to be increasing in numbers.

Ravens are 20 to 25 inches in length, with a wingspread of 4 feet—about the same size as a large hawk. The average weight is 32 ounces, twice that of a crow. The plumage is entirely black, with green and purple iridescence, and the feathers of the throat are shaggy. Ravens' beaks are noticeably larger and heavier than those of crows. Ravens are generally quite wary. Their numerous calls include guttural croaks, gurgling notes, and a sharp, metallic *tock*. Outside my house, I'll sometimes hear a raven calling; when I look up, the bird will be no more than a tiny black dot soaring against the firmament.

Ravens are skillful fliers, and their courtship display flight is spectacular: the male soars, swoops, flies upside down, and tumbles in midair. After mating, a pair will seek out an isolated nesting spot, usually at least a mile away from other ravens. Ravens nest from mid-February through May, on rock outcrops and in the crowns of large trees. Of seventeen raven nests found in a Pennsylvania study, thirteen were on cliffs, three were in hemlocks 45 to 80 feet up, and one was 85 feet up in a white pine. Ravens often build a new nest on top of their nest of the previous year. The nests are loosely constructed out of large sticks, twigs, and grapevines. The outside diameter is 2 to 4 feet; the inside diameter is 1 foot; and the depth of the central hollow is 6 inches. The central hollow is lined with deer hair, moss, shredded bark, and grass. The nests are often messy and stained with excrement. The female lays three to seven eggs (usually four to six), which are greenish and covered with brown or olive markings; the eggshells are rough and dull looking. Incubation, mainly by the female, lasts three weeks. The young leave the nest five to six weeks after hatching.

Ravens eat rodents, insects, frogs, lizards, the eggs and young of other birds, grain, fruit, and garbage. They consume much carrion, especially in winter. In northern Pennsylvania, ravens are often seen along roadways, where they feed on highway-killed animals. That ravens are adjusting to humans and their developments is shown by the family group I observed recently, two parents and four slightly smaller young, all foraging for insects on a grassy lawn near a shopping center in State College. A contact suggested that perhaps these were the ravens that had nested in the steelwork of the Penn State University football stadium.

Angell, T. *Ravens, Crows, Magpies, and Jays.* Seattle: University of Washington Press, 1978.

BLUE JAY

The blue jay, *Cyanocitta cristata,* belongs to the family Corvidae, the Corvids, whose other members include crows, ravens, and magpies in North America. Corvids have the largest cerebrums, relative to body size, of all birds, and scientists believe them to be the smartest. Corvids are social birds, and many species live in flocks when not nesting. The bold, colorful blue jay breeds from southern Canada south to Florida and west to the Rocky Mountains.

Biology. The blue jay is 11 to 12 inches in length (larger than a robin) and has a blue back marked with black and white; its underparts are off-white, and it bears a prominent blue crest on its head. The sturdy beak is straight and sharp, suited for a variety of tasks, including hammering, probing, seizing, and carrying.

Blue jays live in wooded and partly wooded areas. About three-quarters of their diet is vegetable matter: acorns, beechnuts, seeds (including sunflower seeds from feeding stations), corn, grain, fruits, and berries. The remaining 25 percent includes insects—ants, caterpillars, beetles, grasshoppers, and others—along with spiders, snails, frogs, small rodents, carrion, and the eggs and nestlings of other birds. In the spring, blue jays eat caterpillars of the gypsy moth and tent moth, major forest pests. In autumn, jays cache many acorns by storing them under the leaf duff in forest clearings and meadows. They retrieve some of the nuts in winter; others they forget about, and the resulting sprouts help forests to regenerate, particularly on cut-over and burned lands. Confronted with abundant nuts and seeds, a jay may fill its expandable throat; later, it will disgorge the food and cache or eat it. To open an acorn, the bird grips the nut in one foot and hammers the shell apart with its bill.

Blue jay.

Blue jays are quite vocal. They sound a raucous *jaay* to attract other jays and as an alarm call. A bell-like *toolool* is given during courtship, as is a *wheedelee* call, sometimes referred to as the "squeaky hinge" call. Blue jays often mimic the *kee-yer* calls of hawks.

Blue jays have an interesting social courtship. In early spring, from three to ten males (thought to be yearling birds) shadow one female, bobbing their bodies up and down and sounding *toolool* calls. Aggressive displaying apparently scares off the competitors one by one until a single male is left as the female's mate. Ornithologists believe that older jays, ones that have bred in the past, pair up earlier and do not participate in courtship flocks. Once paired, birds move about quietly, with the female giving *kueu kueu* calls to the male when he brings her food. The female may make several preliminary or "dummy" nests, using twigs brought by the male; this behavior may help the pair bond. Later, the female, with the help of the male, assembles the breeding nest, often in a dense conifer or shrub, 5 to 50 feet above the ground. The nest is 7 to 8 inches across, built of twigs, bark, mosses, and leaves, with a 4-inch central cup lined with rootlets. In a survey of blue jay nests in Pennsylvania, twenty were in white pine, eighteen in hemlock, two in red spruce, two in fir, twelve in white oak, and five in alder, with others in black gum, viburnum, pitch pine, and dogwood.

In May or June, the female lays three to six eggs, pale olive or buff spotted with brown or gray. Both sexes incubate. Blue jays are silent and furtive around the nest; one year a pair nested in a white oak next to our house, and I hardly knew they were there once egg laying and incubating commenced. If necessary, blue jays strongly defend their nest, calling loudly and diving at hawks, owls, crows, squirrels, and ground predators. Yet they will allow other jays to land quite near the nest. The eggs hatch after seventeen to eighteen days. Both parents feed the young, bringing them insects, other invertebrates, and carrion. Adult blue jays often raid the nests of smaller birds, including vireos, warblers, and sparrows, eating eggs and nestlings. Biologists believe that forest fragmentation gives jays greater access to the nests of woodland birds.

The young leave the nest after seventeen to twenty-one days. The family stays together for another month or two, with the fledglings clamoring for food and their parents obliging them, even when the juveniles are almost adult-size. In the North, blue jays raise one brood per summer; jays in the South may rear two broods. When the adults molt in July and August, their new plumage comes in a deep, beautiful blue. In fact, the blue of birds' plumage is not caused by pigmentation but by structure: small voids in the feathering do not absorb the blue part of the light spectrum and instead cause it to scatter, giving an appearance of blue. Grind up a jay's feather, and you'll be left with a blackish powder.

In late summer and early fall, family groups merge into larger foraging flocks. As the weather grows colder, these groups fragment again into smaller bands. Birds from Canada shift southward in September and October, and juveniles from the northern United States also drift to the south. In some years—perhaps when nuts are scarce—blue jays move in large numbers; accipiters, particularly sharp-shinned and Cooper's hawks, accompany the flocks, picking off unwary members. The longevity record for the species is sixteen years. Among adults, the annual survival rate is estimated at 55 percent.

Habitat. Blue jays inhabit extensive woods, farm woodlots, and developed areas. They avoid strictly coniferous forests. They thrive where nut-bearing oak and beech trees are plentiful. Although primarily forest birds, blue jays have adapted to breed in cities, where they nest in parks and along tree-lined streets.

Population. The population in Pennsylvania and the Northeast is healthy. On a continental scale, the species is expanding to the northwest in Canada. Biologists estimate two or three breeding pairs of blue jays per 100 acres of suitable habitat.

HORNED LARK

Named for twin feather tufts adorning its head, the horned lark *(Eremophila alpestris)* is the only lark native to North America. The species is Holarctic, present in both North America and Eurasia. Horned larks breed across Canada into Alaska and in much of the United States, especially the prairie states; they winter south to North Carolina. The species is present in Pennsylvania year-round.

Biology. Horned larks have a streaked brown and tan plumage and a yellow face marked with black; the black "horns" can be hard to distinguish in the field. Ground-dwelling birds that favor open or barren terrain, horned larks walk about (they do not hop) searching for grass and weed seeds, waste grain left by harvesters or scattered when manure is spread, and insects, including grasshoppers, beetles, and caterpillars. In autumn, seeds make up 60 to 70 percent of the diet; in winter, 80 to 100 percent.

Larks are early breeders: the season runs from February to June, with a peak in March. The male proclaims his territory and tries to attract females by flying high into the air, hovering and circling while sounding a sweet tinkling song, and finally diving suddenly to the ground. The female selects the actual nest site, often near or partly beneath a grass tussock. Over two to four days, she builds a shallow cup of weeds and stems lined with fine grasses. She lays two to five (usually three or four) eggs; they are pale gray, peppered with brown. The female does all of the incubating, leaving the nest for short periods to feed. The eggs hatch after ten to twelve days. The young are altricial; both parents feed insects to them. Falcons, hawks, owls, weasels, skunks, raccoons, and house cats prey on adults and young. Meadow voles, deer mice, and shrews are nest predators.

Young horned larks leave the nest eight to ten days after hatching; their parents continue to feed them. As with many other ground-nesting birds, the young of *Eremophila alpestris* walk long before they're able to fly. After about four weeks, fledglings can fly, and soon they become independent. Flocks of horned larks increase in size during summer and fall. In most places, horned larks raise two clutches, sometimes three. Individuals are permanent residents throughout most of the species' breeding range, although flocks shift about in search of food.

Winter is the best time of the year to see horned larks. Look for them in windswept fields, where they may gather in mixed flocks with Lapland longspurs and snow buntings, which breed farther to the north and winter in the Northeast. When snow lies deep, the birds forage along the edges of roads.

Habitat. Horned larks can be found in stubble fields, mowed areas around airstrips, golf courses, heavily grazed pastures, lake flats, and reclaimed strip-mined lands regrown with sparse vegetation. They avoid areas with trees and bushes. The species is losing habitat as former farmlands revert to brush and woods.

Population. Horned larks increased dramatically in the eastern United States during the late 1800s and early 1900s, as humans cut down forests to create fields and pastures. The species spread into Pennsylvania from the west and colonized the entire state between 1888 and 1927. Since the mid-1960s, the population has been declining by about 9 percent each year, as suggested by Breeding Bird Survey data compiled by the U.S. Fish and Wildlife Service. Horned larks are most common in western Pennsylvania, particularly on reclaimed strip-mined lands north of Pittsburgh.

CHIMNEY SWIFT, PURPLE MARTIN, AND SWALLOWS

Swifts, martins, and swallows are built for life in the air. They have long, tapering wings and lightweight bodies. Their short, wide bills open to expose gaping mouths for scooping up insect prey. The chimney swift has tiny, almost vestigial feet with four clawed toes facing forward, letting it cling to upright surfaces; the feet of the purple martin and the swallows have three toes forward and one to the rear, for perching on branches and wires.

Many of these birds are social and breed in colonies. Purple martins usually nest in artificial boxes with multiple chambers, put up by people wanting to attract these insect eaters; the other swallows build or occupy different sorts of nests, depending on their species. Most swallows do not defend territories. The males sing mainly to attract mates and to communicate with them. Both parents usually share in incubating the eggs and feeding the young. Swifts, martins, and swallows often forage in groups, soaring high above forests, farms, and urban areas. During wet weather, they hunt at lower altitudes, where insects fly under damp conditions. These birds undertake long migrations. The seven species that breed in Pennsylvania winter in the Gulf states, Central America, and South America.

Chimney Swift *(Chaetura pelagica)*. The common name combines the bird's favorite nesting habitat and the speed of its flight. The chimney swift is sooty gray, about 5 inches long, and has a 1-foot wingspan; the body looks stubby between the long, narrow wings. The swift spends most of the daytime hours in the air; its flight is batlike, with shallow wingbeats and erratic stalls and turns as the bird singles out insects or sweeps through clouds of prey. Chimney swifts eat flies, leafhoppers, flying ants, mayflies, stoneflies, beetles, leaf bugs, and other insects. They take spiders, mainly small ones floating on strands of silk borne aloft by air currents. Chimney swifts utter a loud clicking call in flight. They drink on the wing, skimming low over ponds, and even gather materials for their nests while in flight, using their feet to break tips off dead branches and carry them back to the nest site.

Chimney swifts are thought to be monogamous and to mate for life. Pairs sometimes glide in tandem with their wings raised in a V. In the past, chimney swifts nested in hollow trees and caverns. Today they use man-made structures almost exclusively: factory and house chimneys, silos, air shafts, and old wells, where they are protected from storms and predators. The nest is shaped like a half saucer and cemented to a vertical surface, the twigs held together by the adults' glutinous saliva, which solidifies as it dries. Females lay three to six eggs (four or five are usual), which are white and unmarked. Both sexes incubate for eighteen to twenty-one days. The newly hatched young are altricial and are fed regurgitated insects. Sometimes a third "parent," probably a yearling offspring of the adults, helps feed and brood nestlings. The young fledge a month after hatching and join feeding flocks. In late summer, swifts gather in the evening before flying into large factory and school chimneys, where they roost by the thousands.

Although originally forest-dwelling birds, chimney swifts are not common in the densely wooded parts of Pennsylvania, where trees may not be mature enough to offer cavities for nesting and roosting. Swifts arrive in the Northeast in May, raise a single brood in June and July, and head south in August and September. They winter mainly in the Amazon Basin. The average life span is four years.

Purple Martin *(Progne subis)*. At 8 inches in length, the martin is the largest North American swallow. Adult males are a glistening blue-black; females and yearlings are grayish with pale bellies. Both sexes have a notched tail. Martins, less maneuverable than the other swallows, glide in circles punctuated with short periods of flapping flight. Before Europeans came to the New World, Native Americans were hanging gourds around their villages to attract purple martins, which also nested in caves and hollow trees. Today the vast majority of martins nest colonially in compartmented boxes that people put up for them.

Martins inhabit meadows, farmland, and open areas near water. They feed on winged ants, wasps, bees, flies, dragonflies, beetles, moths, and butterflies. Males arrive first in the spring, followed by females. The call is a throaty, gurgling *tchew-wew.* One male may mate with more than one female. The four or five eggs are white and unmarked, laid on a nest of grass, twigs, and leaves inside the nest chamber. The female incubates them for fifteen to eighteen days. In late summer, large numbers of purple martins congregate on Presque Isle in Lake Erie. The species winters in the Amazon Basin.

Tree Swallow *(Tachycineta bicolor).* Tree swallows nest across Canada and most of the northern United States. They are 5 to 6 inches long, an iridescent green-black or blue-black above and white below. They nest in tree cavities, woodpecker holes, and bluebird houses put up by humans. The earliest of our swallows to return north, they arrive in late March and April; unlike the other species, tree swallows switch to eating berries and seeds to survive cold periods when insects become torpid. They often breed near lakes, ponds, and marshes, competing for nesting cavities with bluebirds, starlings, house sparrows, and house wrens. Ornithologists believe that individuals choose new mates each year. Tree swallows are more aggressive than other swallow species and defend an area within a radius of about 15 yards from the nest. The female lines the nest cavity with grass, weeds, rootlets, and pine needles; after the four to seven pinkish white eggs are laid, she often adds feathers (usually white ones) from other birds. Incubation takes fourteen to fifteen days. The young fledge three weeks after hatching. Tree swallows migrate in flocks to wintering grounds in the Gulf states and Central America.

Northern Rough-Winged Swallow *(Stelgidopteryx serripennis).* This small (body length about 5 inches), nondescript brown and white swallow is named for small serrations in its outermost wing feathers. The species breeds across the United States and in Central America. Rough-winged swallows often forage in flight above moving water. The call is a short, harsh *trit trit*. The birds nest in cavities in rock faces, quarries, and stream banks, frequently in abandoned kingfisher burrows, occasionally in drainpipes and culverts. At the end of a 1- to 6-foot tunnel, the birds heap up twigs, bark, roots, and weeds, and line a central cup with fine grasses. The four to eight white eggs hatch after about sixteen days of incubation. Rough-winged swallows nest throughout Pennsylvania. They winter along the Gulf coast and in Central America.

Bank Swallow *(Riparia riparia)*. About 5.5 inches long, this small, brown-backed swallow has a dark band across its pale breast. Although they have small feet and tiny bills, bank swallows usually dig their own burrows, up to 5 feet deep in dirt banks, piles of gravel or sand, and road cuts. Nest entries of neighboring pairs may be only a foot apart. Colonies—which often have hundreds of pairs—arise and die out as banks of suitable burrowing materials become available and then lose the qualities that bank swallows require, such as steepness and height. Bank swallows forage over fields and wetlands and along rivers and ponds, taking flies, beetles, wasps, winged ants, dragonflies, stoneflies, moths, and other insects. They nest from May until July. The typical clutch has four or five eggs. American kestrels prey on fledglings. In late summer, bank swallows may gather in large flocks before departing for wintering grounds in South America. The species also breeds in Europe and Asia, where it is known as the sand martin.

Cliff Swallow *(Petrochelidon pyrrhonota)*. Body length is 5 to 6 inches; a pale rusty or buff-colored rump distinguishes this species. From below, the tail looks squared-off. Cliff swallows eat flying beetles, flies, winged ants, bees, wasps, mayflies, lacewings, and many other insects. They build gourd-shaped nests out of pellets of mud, attached to cliffs, bridge supports, dams, and the walls of unpainted barns and derelict buildings, beneath eaves that shield against rain. A typical nest takes one to two weeks to build and requires more than a thousand mud pellets. Colonies can be dense: in one instance, eight hundred nests were clustered on the side of a barn. The adults line the inside of the nest with grass, hair, and feathers. The three to six eggs are white spotted with brown. Both sexes incubate for about fifteen days. A female cliff swallow will sometimes lay an egg in another swallow's nest or carry an egg in her bill to a neighboring nest. Cliff swallows winter in southern South America. The population is thought to be increasing in North America.

Barn Swallow *(Hirundo rustica)*. Barn swallows have long, deeply notched tails. The flight of these sleek, blue and buff swallows can look like an aerial ballet, with the birds sideslipping, stalling, twisting and turning low over water or fields in pursuit of their prey: houseflies, horseflies, beetles, wasps, bees, winged ants, and other insects. In bad weather, barn swallows may land and eat spiders, snails, berries, or seeds.

Pairs nest on their own or near a few other pairs. Barn swallows are abundant breeding birds in Pennsylvania and the Northeast. They build bowl-shaped nests out of mud and straw, fixing them to walls, beams, and eaves of barns and other outbuildings; in culverts and under bridges; and, rarely, on the cliff faces and caves that were the species' original habitat before Europeans began settling North America. Barn swallows often line their nests with poultry feathers. The adults scold human intruders and dive at them, zipping past their heads. Each female lays four or five eggs, which are white spotted with brown. During the day, both male and female take turns incubating, switching about every fifteen minutes. Young leave the nest three weeks after hatching. Some pairs raise a second brood. Barn swallows from eastern North America winter in Panama, Puerto Rico, and throughout South America. *Hirundo rustica* is the most widespread swallow species in the world, breeding in North America, Europe, and Asia.

Turner, A., and C. Rose. *Swallows & Martins, an Identification Guide and Handbook.* Boston: Houghton Mifflin, 1989.

CHICKADEES, TUFTED TITMOUSE, NUTHATCHES, AND BROWN CREEPER

These woodland birds are mainly year-round residents in their breeding areas. They become most apparent in fall and winter, when all four types may be seen together along with downy woodpeckers and kinglets in mixed-species foraging flocks. In these groupings, the greater number of eyes improve foraging efficiency and better detect potential predators. Our two chickadees and the tufted titmouse belong to the family Paridae. These omnivorous feeders cache excess seeds in holes or bark crevices, remember the locations, and return later to eat the food. The two nuthatches are in the family Sittidae. They glean insects from the trunks of trees and also eat nuts; their common name derives from the way they "hack" nuts apart using their stout pointed bills. Taxonomists place the brown creeper in the family Certhiidae, a group that includes only six species, the other five of which live in Europe and Asia.

Black-Capped Chickadee *(Poecile atricapillus)*. A black cap and bib, buffy flanks, and a white belly mark this lively 5-inch-long bird. Chickadees have short, sharp bills and strong legs that let them hop about in trees and cling to branches upside down while feeding. They fly in an undulating manner, with rapid wingbeats, rarely going farther than 50 feet at a time. The species ranges across northern North America, living in deciduous and mixed forests, forest edges, thickets, swamps, and wooded areas in cities and suburbs. Black-capped chickadees are common throughout Pennsylvania except for the state's southwestern and southeastern corners, where they are replaced by the similar Carolina chickadee.

About two-thirds of a chickadee's diet consists of animal protein: moth and butterfly caterpillars (including early growth stages of gypsy moths and tent moths), other insects and their eggs and pupae, spiders, snails, and other invertebrates. In late summer and fall, chickadees eat wild berries and the seeds of ragweed, goldenrod, and staghorn sumac. In autumn, chickadees begin storing food in bark crevices, curled leaves, clusters of pine needles, and knotholes. Sometimes I'll split a billet for the woodstove, and from beneath the loose bark will spill a half dozen sunflower seeds gotten from the bird feeder and cached away by chickadees—or perhaps by titmice or nuthatches, which also cache seeds. The birds rely on these hoards when other foods become scarce. Chickadees also eat suet from feeding stations and fat from dead animals.

In winter, chickadees live in flocks of six to ten birds with one dominant pair. Listen for the *chick-a-dee-dee-dee* calls that flock members use to keep in contact while foraging around a territory of 20 or more acres. A flock will defend its territory against other chickadee flocks. At night, chickadees roost individually in tree cavities or among dense boughs of conifers. A roosting bird tucks its head under a wing to conserve body heat. On cold nights, a chickadee's temperature drops from a normal 108 degrees Fahrenheit to about 50 degrees, causing the bird to enter a state of regulated hypothermia, which saves significant amounts of energy. Chickadees lose weight each night as their bodies slowly burn fat to stay alive; they must replace those fat stores by feeding during the next day.

Chickadees mate for life. In spring, the winter flocks break up as pairs claim nesting territories ranging in size from 3 to 10 acres. Chickadees nest

Chickadees often take seeds from a bird feeder and cache them to be eaten later.

in May and June. The usual site is a hole in a tree, dug out by both sexes. Birch is a favorite, since this tree's outer bark stays intact after the inner wood rots and becomes soft enough for easy excavating. Chickadees also clear out cavities in aspen, alder, willow, and cherry trees, and use abandoned woodpecker holes; along our road one year, a pair nested in a hollow steel gate post. The cavity is usually 4 to 10 feet above the ground. The female assembles the nest by laying down a base of moss, then adding soft material such as animal fur or plant fiber. House wrens compete for nest cavities and may destroy chickadee eggs and broods; raccoons, opossums, and squirrels raid nests. Chickadees will renest if a first attempt fails. Only one brood is raised per year.

The five to nine eggs are white with reddish brown dots. The female incubates them, and the male brings her food. The eggs hatch after twelve days. Both adults feed the nestlings, which fledge about sixteen days after hatching. Three to four weeks after fledging, the young disperse, moving off in random directions. As winter approaches, they join feeding flocks. Some become "floaters," moving between three or more flocks, ready to pair with an opposite-sex bird should its mate die.

Chickadees are taken by many predators, including sharp-shinned hawks, American kestrels, eastern screech-owls, saw-whet owls, and domestic and feral cats. Sometimes chickadees mob these enemies while sounding *zee-zee-zee* alarm calls. The average life span for a chickadee is two and a half years, and the current longevity record is twelve years, nine months. Every few years, long-distance movements take place within the population, "irruptions" that may be launched by the failure of seed crops or by high reproductive success.

Carolina Chickadee *(Poecile carolinensis)*. Similar to the black-capped chickadee in appearance and life history, this species lives in milder climates across the southeastern United States. The Carolina chickadee is smaller than the black-capped chickadee, and its call is pitched higher and sung more rapidly. Where their ranges meet, the two species hybridize. The Carolina chickadee breeds in southeastern and southwestern Pennsylvania.

Tufted Titmouse *(Baeolophus bicolor)*. This trim bird has gray and white plumage, a prominent head crest, and black shoe-button eyes. The species ranges through eastern North America into southern New York and New England. It has extended its range northward over the last half century, perhaps because of climatic warming and an increase in bird feeding by humans. In the early 1900s, the tufted titmouse was absent from northern Pennsylvania; today it breeds statewide.

Titmice eat insects (caterpillars, wasps, bees, sawfly larvae, beetles, and many others, as well as eggs and pupae), spiders, snails, seeds, nuts, and

berries. Like the chickadee, the titmouse forages by hopping about in tree branches and often hangs upside down while inspecting the underside of a limb. To open a nut or seed, the bird holds the object with its feet and pounds with its bill. Titmice cache many seeds; with sunflower seeds, the birds usually remove the shell and hide the kernel within 120 feet of the feeding station, in cracks or furrows in tree bark, on the ground, or wedged into the end of a broken branch or twig.

Winter flocks are often made up of parents and their young of the previous year. Titmice are early breeders: males start giving their *peter peter* ter-

To open a nut or seed, a tufted titmouse holds the morsel with its feet and pounds it with its bill.

ritorial song in February. In Pennsylvania, pairs begin building nests in
late March and early April. Titmice
are believed not to excavate their
own nest cavities; instead, they use
natural cavities or abandoned woodpecker holes. Breeding territories
average 10 acres. The female lays five
or six eggs, which are white with
dark speckles, and incubates them for
two weeks. The young fledge about
eighteen days after hatching. Sometimes yearling birds stay on in their
natal territory and help their parents
raise the next year's brood.

Brown creeper, left,
and white-breasted nuthatch.

White-Breasted Nuthatch *(Sitta
carolinensis)*. The white-breasted nuthatch has a slate gray back, a white
breast and face, and a cap that is
black in the male and ashy gray in
the female. Nuthatches live in deciduous forests throughout Pennsylvania
and the East. They climb around in trees, walking in a herky-jerky manner
up and down and around the trunks, along branches and the undersides of
limbs. Both sexes sound a nasal *ank ank* call. Pairs live in home territories of
20 to 35 acres.

White-breasted nuthatches feed on insects and spiders in summer and
on nuts and seeds in winter. They relish suet at feeding stations and carry
away sunflower seeds for caching. Sometimes they forage on the ground.
Nuthatches wedge acorns and hickory nuts into tree bark and then hammer
the shells apart with blows from their awl-like beaks.

During courtship, the male bows to the female, spreading his tail and
drooping his wings while swaying back and forth; he also feeds her morsels.
Before building the nest, the birds rub or sweep crushed insects back and
forth over the inside and outside of the nest cavity. Ornithologists speculate
that this sweeping behavior leaves behind chemicals that may repel predators or nest competitors. The female builds a nest inside the cavity (commonly a rotted-out branch stub or an abandoned squirrel or woodpecker

hole) using twigs, bark fibers, grasses, and hair. She lays five to nine white, brown-spotted eggs and incubates them for twelve to fourteen days while her mate brings her food. Both parents feed insects and spiders to the young, who fledge after two or three weeks, usually in June.

Red-Breasted Nuthatch *(Sitta canadensis).* In Pennsylvania, this species is found mainly in the northern part of the state; it ranges through New England and across Canada. Slightly smaller than the white-breasted nuthatch, the red-breasted has a rusty tinge to its breast and a prominent black eye stripe. It lives in areas where evergreens are plentiful and often nests in pine plantations.

Red-breasted nuthatches feed on insects and on seeds, particularly those of conifers. They nest in tree cavities 5 to 40 feet above the ground. Five or six young are produced in a single annual brood. In some autumns, large numbers of red-breasted nuthatches show up south of their normal range; biologists believe that these movements are driven by poor cone production in northern forests.

Brown Creeper *(Certhia americana).* Brown creepers are inconspicuous birds whose intricately patterned backs help blend them in with the tree bark that is their near-perpetual home. Brown creepers breed across a huge range extending from Alaska and Newfoundland south to Nicaragua. They favor mature forests with many large trees. The species is found in much of Pennsylvania, although numbers are lower in the state's southeastern and southwestern corners. Braced by their long, stiff tails, brown creepers climb slowly up tree trunks following a spiral course. They inspect bark furrows and use their decurved bills to tease out insects, pupae, and eggs. They also eat spiders and seeds.

The call is a long, thin *seeee;* the male also voices a subtle breeding song. The species nests under peeling bark, often in a shagbark hickory or a dead or dying tree, less frequently in a cavity. A hammocklike twig nest is built to fit the available space. The female lays four to eight eggs, which are whitish and dotted with reddish brown. Incubation takes fourteen to seventeen days, and young leave the nest two weeks after hatching. Brown creepers from the Northeast may migrate to Florida and the Gulf coast. In winter, brown creepers sometimes mix in with foraging flocks of chickadees; perhaps these are resident creepers or northern birds that have shifted southward.

Smith, S. M. *The Black-Capped Chickadee, Behavioral Ecology and Natural History.* Ithaca, NY: Cornell University Press, 1991.
Harrop, S. *Chickadees, Tits, Nuthatches & Treecreepers.* Princeton, NJ: Princeton University Press, 1995.

WRENS

Wrens are small, active birds, basically brown in color, that often perch with their tails held straight up in the air. They forage on or just above the ground in thick brush, forest undergrowth, or marsh vegetation. Wrens belong to the family Troglodytidae, with about seventy species in the New World, most of them in the tropics. Only one species lives in the Old World: the winter wren, which likely spread from Alaska to Siberia and extended its range westward until, eons in the past, it reached Britain and Iceland.

Some wrens nest in cavities; others build roofed structures out of plant matter. The males of several species build preliminary or "dummy" nests in tree cavities, woodpecker holes, and nest boxes, less frequently in odd enclosed spaces like tin cans, pockets of clothing hung outdoors, hats, boots, flower pots, and drainpipes. Later a female will choose one of the preliminary nests, finish its construction, and lay eggs in it. Wrens often pester other birds and evict them from nest cavities, puncturing their eggs or pecking their young to death. They destroy nests in cavities and in the open; they also wreck other wrens' nests. Why such belligerence? Does an abundance of empty nests discourage predators from looking further and finding an active wren's nest? Or does killing its rivals' offspring reduce pressures on prey populations, making it easier for a wren to feed its own young?

Wrens eat insects and spiders. A few species also feed on berries and seeds. Owls, small hawks, and house cats take adult wrens; raccoons, opossums, minks, weasels, mice, squirrels, woodpeckers, and snakes raid wrens' nests. Some wrens migrate southward in winter, while other species remain as permanent residents on their breeding range. Five species are found in Pennsylvania.

Carolina Wren *(Thryothorus ludovicianus)*. The Carolina wren inhabits the eastern United States and Central America. It is a permanent resident wherever it breeds. At 5.5 to 6 inches long, and weighing 0.7 ounce, it is the largest of our wrens. Carolina wrens are colorful birds, with rust-brown upper plumage, a buffy or cinnamon breast, and a white stripe above each eye. They prefer moist or bottomland woods with moderate to dense shrubby cover; they also inhabit gardens and yards. Carolina wrens forage mainly on the ground, often near downed trees or brush piles, using their curved bills to lift up leaf litter and expose prey. They also climb up tree trunks like creepers or nuthatches. Carolina wrens catch caterpillars, chinch bugs, beetles, leafhoppers, grasshoppers, crickets, katydids, and many other insects. They eat seeds of poison ivy, sumac, smartweed, and other herbaceous plants, plus fruits and acorn mast.

Unlike the house wren, the Carolina wren is monogamous and mates for life. Pairs often forage together and defend a territory year-round. The species has a clear, ringing song, *tea-kettle, tea-kettle, tea-kettle.* Carolina wrens nest in tree cavities, birdhouses, crevices in stone walls, among exposed roots, and in cracks or crannies in buildings. Using leaves, twigs, and other plant materials, both sexes build a dome-shaped nest with a side entrance. The normal clutch is five or six eggs. Incubation is by the female and takes two weeks; the male feeds her on the nest. The young leave the nest about two weeks after hatching. Pairs usually raise two broods per year.

In the last century, the Carolina wren has been expanding northward. Pennsylvania is on the northern edge of the species' breeding range, which extends north after mild winters and ebbs south following harsh winters. Extended periods of ice and snow can devastate local populations. Bewick's wren *(Thryomanes bewickii)* is a similar-appearing species that bred in southern Pennsylvania until around 1976; since then, it has disappeared. Bewick's wren is listed as an extirpated species in Pennsylvania.

House Wren *(Troglodytes aedon).* The most common wren in Pennsylvania, this bird was named either for its penchant for nesting in birdhouses or because it often lives around humans' dwellings. A house wren is 5 inches long and weighs 0.33 ounce. Its overall color is gray-brown. House wrens live in open shrubby woodlands, small woodlots, woods edges, towns, suburban backyards, and city parks. They feed on insects, spiders, millipedes, and snails. The species breeds across southern Canada and the United States, south to the Carolinas in the East. Individuals from the East winter mainly in Georgia and Florida.

Males arrive on the breeding grounds in late April or early May. They establish territories of a half acre or larger and advertise for females with a rich, liquid song. Males build dummy nests out of twigs in tree cavities, nest boxes, or hollow fence posts; one male may construct up to seven such nests, defending them and the space around them. When building dummy nests, house wrens may destroy the nests and young of tree swallows, chickadees, bluebirds, and prothonotary warblers. Females either arrive later than the males or stay hidden in brush until they begin inspecting the males' territories. If a female finds a territory to her liking, she will finish one of the male's dummy nests by adding a lining of grass, plant fibers, rootlets, feathers, and animal hair.

Male wrens of several species (the house wren is shown here) often build preliminary nests in tree cavities, woodpecker holes, and nest boxes.

In May, the female lays five to eight eggs, which are white and speckled with reddish brown. She incubates them for twelve to fifteen days. After the eggs hatch, the male helps with feeding the young, bringing grasshoppers, crickets, caterpillars, and spiders to the growing nestlings. About two weeks after hatching, the young leave the nest. Females typically produce two broods per summer, rarely three. A female may abandon her first brood soon after the young have fledged, leaving the male to rear them; she may then move to another male's territory, mate again, and lay a second clutch. A male house wren may mate with two or more females in his territory, although he will usually help only the primary female raise her young. An unmated male may enter an established territory and try to drive away the resident male or mate with the female. If he succeeds in taking over a territory, he may destroy the female's eggs or young. At that point, she will usually renest.

Most house wrens leave the breeding range in September and early October. They migrate at night; some are killed when they collide with communication towers. On their southern wintering grounds, they forage in thick brush. The oldest house wren on record lived for seven years, but most individuals probably survive for only a year or two. The house wren benefits from forest fragmentation and does well in towns and residential areas. Ornithologists believe the species has been expanding southward since the period of European settlement began.

Winter Wren *(Troglodytes troglodytes)*. At just over 4 inches in length, this is our smallest wren. Its plumage is dark brown, and its tail is stubby. Look for the secretive winter wren in deep woods, particularly hemlocks, where it forages in brush piles and ravines—behaving "more like a mouse than a bird," notes the ornithologist Kenn Kaufman. The male's song is a series of warbles and trills. Foods include insects, spiders, small fish taken from stream shallows, and berries. In the East, winter wrens breed from Newfoundland south to Georgia in the Appalachians. In Pennsylvania, they breed in heavily forested mountainous areas. They nest in cavities, and a brood of five to six is the norm. Males may mate with more than one female. Winter wrens head south in early fall, although some remain in the north and winter along streams and in swamps.

Sedge Wren *(Cistothorus platensis)*. This small (4.5 inches), shy wren inhabits moist upland sedge meadows with little or no standing water. It was formerly known as the short-billed marsh wren. Sedge wrens often breed in small colonies. They may occupy a suitable habitat for several years, then disappear. Males sing a dry, rattling song; Hal Harrison once counted thirty-five to forty singing males on a 10-acre site. The nest is a ball of dried or green sedges woven into growing vegetation 2 to 3 feet above the ground. The usual clutch is seven eggs. Most females produce two broods per year, and males may mate with more than one female. The destruction of wetlands has harmed this species. Rare and declining in the Northeast, the sedge wren is listed as "threatened" in Pennsylvania.

Marsh Wren *(Cistothorus palustris)*. This is the typical wren of the cattail marsh. It is 4.5 to 5.5 inches long, its brown plumage marked with black and white stripes on the back and a white eye stripe. Marsh wrens arrive in breeding areas in late April or early May. The male's reedy, gurgling song lasts one to two seconds and is given up to twenty times per minute, by day and sometimes at night; not particularly musical, it reminded one naturalist of "air-bubbles forcing their way through mud or boggy ground when trod upon."

The marsh wren forages on the marsh floor, flitting up and clinging to stalks and leaves of cattails, bulrushes, and other plants while searching for prey. It takes aquatic insects and their larvae, terrestrial insects, spiders, and snails. Both males and females peck and destroy the eggs of other birds in their territory; red-winged blackbirds often attack marsh wrens on sight. Males typically build dummy nests—around six for each breeding nest used by a female. The female weaves an oblong nest out of cattails, reeds, and grasses, secured to standing vegetation. A short tunnel leads to a central cavity in which three to six eggs are laid. The female incubates the clutch for about two weeks. Fed by both parents, the young fledge after twelve to sixteen days. The adults care for them for another two weeks. Two broods are produced each year. Up to half of all breeding males may mate with two or more females. Marsh wrens winter along the Atlantic and Gulf coasts.

KINGLETS AND BLUE-GRAY GNATCATCHER

These tiny woodland birds flit about in trees and shrubs while searching for insects. Golden-crowned kinglets breed and winter in Pennsylvania. Ruby-crowned kinglets breed farther to the north, but individuals migrate through and winter in the Keystone State. Kinglets are about 3.5 inches long and weigh less than 0.33 ounce. They are dull olive-gray in color, with white wingbars; a good identifier is their incessant, nervous wing flicking. The blue-gray gnatcatcher breeds throughout the Northeast and is usually seen fluttering in the crowns of trees, darting out now and again to catch flying insects.

Golden-Crowned Kinglet (*Regulus satrapa*). The species nests from Labrador to Alaska and south in the Appalachian and Rocky mountains. Golden-crowned kinglets have brilliant markings on the tops of their heads: golden yellow in females, orangish in males. I've watched them in winter (the best time to observe kinglets, since they forage actively in a variety of habitats), when the only brightness in the somber woods seemed to emanate from their dazzling crowns. Golden-crowned kinglets favor dense stands of mature conifers, including native red and black spruce and eastern hemlock; they also inhabit planted stands of Norway spruce and red pine. During the March–April migration, and in winter, they may forage among deciduous trees. Golden-crowned kinglets eat small insects such as beetles, wasps, gnats, caterpillars, aphids, and scale insects. Often they feed in the crowns of conifers, hanging upside down on the tips of twigs. They eat spiders and drink sap from wells excavated by sapsuckers. In winter, they pick up tiny springtails and bark-hibernating insects, plus insect eggs. They join mixed flocks and feed alongside chickadees, titmice, nuthatches, and downy woodpeckers.

The male defends a breeding territory of about 4 acres by singing and erecting his colorful crown. The female builds the nest, typically in a spruce, 6 to 60 feet up (usually around 50 feet), a pendant mass of mosses and lichens attached to a horizontal limb near the trunk; sometimes ruffed grouse feathers are woven into the walls so that the plumes arch out over the cup. The female lays seven to ten eggs, white spotted with brown and gray; often they're arranged in two layers in the 2-inch-diameter cup. Incubation takes about two weeks. Both parents feed the nestlings, which leave the nest fourteen to nineteen days after hatching.

Golden-crowned kinglets are increasing in Pennsylvania as breeding habitats open up for them. Most of our state's mature conifers were logged off during the late 1800s; today spruce and pines planted on cutover land are maturing. During the state *Breeding Bird Atlas* survey conducted in the 1980s, observers spotted golden-crowned kinglets most frequently in plantations of 35- to 70-foot-tall Norway spruce and red pine at higher elevations on the Appalachian Plateau and in the Ridge and Valley region.

Ruby-Crowned Kinglet *(Regulus calendula)*. Males have bright scarlet crowns; females lack the brilliant head markings. Feeding habits are similar to those of the golden-crowned, except that the ruby-crowned kinglet hovers more frequently to take insects from foliage and flits out more readily to catch prey. The ruby-crowned kinglet breeds farther north and winters farther south than the golden-crowned species. In spring, ruby-crowned kinglets migrate through Pennsylvania in April and May, often in the company of warblers. In autumn, both the golden- and ruby-crowned species pass through the state in September and October. Some ruby-crowned kinglets winter in Pennsylvania.

Blue-Gray Gnatcatcher *(Polioptila caerulea)*. This slender 4.5-inch-long bird can be recognized by its long tail, held cocked up like a wren's tail and twitched from side to side. Its plumage is blue-gray above and whitish below—"like a miniature mockingbird," wrote one naturalist. The species ranges across the United States, breeding as far north as New England in the East. It is statewide in Pennsylvania in floodplain forests, open woods, and mature black willows bordering wetlands. It is common in Appalachian oak and mixed hardwoods, scarce and local in northern hardwoods, and favors edges over the interior of woods. The call is high-pitched and whining. I have spotted blue-gray gnatcatchers chasing after insects in the treetops, and I've also seen them plucking prey from shin-high huckleberry bushes. They eat leafhoppers, treehoppers, beetles, flies, small wasps, and spiders.

In Pennsylvania, most nests are on horizontal limbs of oaks, 20 to 50 feet up: these beautiful, precise cups are about 2.5 inches wide, and their outside walls are camouflaged with lichens. The female lays four to five eggs; the eggs are bluish white and freckled with reddish brown. Both adults share in incubating the clutch. The young hatch after about thirteen days and fledge from the nest two weeks after hatching. Blue-gray gnatcatchers leave the Northeast in September. They winter in the southern states, the Caribbean islands, and Central America.

Galati, R. *Golden-Crowned Kinglets.* Ames, IA: Iowa State University Press, 1991.

BLUEBIRD, ROBIN, AND THRUSHES

North America has fourteen thrush species, six of which breed in the Northeast, including the American robin and eastern bluebird; a seventh species, the gray-cheeked thrush, passes through our region during migration. Thrushes have strong legs and thin bills. They often forage on the ground, searching in leaf litter and on lawns for insects, spiders, earthworms, and snails; they eat berries in late summer, fall, and winter. Juveniles' spotted breasts help camouflage them. Hawks, falcons, owls, foxes, minks, and house cats prey on thrushes. Blue jays, grackles, crows, raccoons, weasels, squirrels, chipmunks, and snakes eat eggs and nestlings.

Many thrushes sing complex, mellifluous songs that delight human listeners. Most build open, cup-shaped nests secured to branches of low trees and shrubs. Robins may nest on building ledges and other flat surfaces; bluebirds choose tree cavities or artificial nesting boxes; hermit thrushes and veeries often nest on the ground. The females do most of the actual nest construction. The typical clutch is four or five eggs; all of the species breeding in the Northeast lay pale blue or blue-green eggs. Females do most or all of the incubating, and both parents feed the young.

Eastern Bluebird *(Sialia sialis).* This familiar species nests across much of the East and winters south to Nicaragua. A bluebird is 6 inches long and weighs about an ounce; it has a vivid blue back and wings and a ruddy breast. When not nesting, bluebirds wander in small feeding flocks. They favor semiopen habitats: orchards, pastures, hayfields, fence lines, cutover or burned areas, forest clearings, open woodlots, and suburban gardens and parks. The song consists of three or more soft, melodious notes *("tury, cher-wee, cheye-ley,"* as one observer has rendered it). Bluebirds eat crickets, grasshoppers, beetles, caterpillars, and many other insects, and they take spiders, centipedes, earthworms, and snails. Often they sit on a low perch, then flutter down to catch prey. In fall and winter, they turn to fruits, including those of sumac, dogwood, Virginia creeper, poison ivy, pokeweed, elderberry, wild cherry, bittersweet, honeysuckle, and wild grape.

The courting male sings to the female and flutters close to her with his wings and tail spread; he may pass food to her. Mated pairs preen each other's feathers. A study in New York found that bluebird territories used

Eastern bluebird.

for mating, nesting, and feeding averaged just over 5 acres. Bluebirds nest in abandoned woodpecker holes, tree cavities, hollow fence posts, and artificial boxes; often they face stiff competition for those sites from starlings, house sparrows, tree swallows, and house wrens, all of which have been known to kill adult bluebirds. Inside the cavity, the female builds a loose nest out of grasses and weed stalks, sometimes lining the central cup with feathers or animal hair. Early nesters, bluebirds lay their first clutches by late March or early April and their second clutches by early June. A typical clutch has four or five eggs. The female incubates them for about two weeks. Both parents feed the nestlings; after about eighteen days, the young fledge from the nest. A second clutch will usually have one fewer egg than a first clutch produced by the same pair.

Bluebirds are permanent residents in the southern parts of their range; in winter, individuals from northern areas may shift southward. In mild winters, I see many bluebirds in the agricultural valleys of central Pennsylvania, and on one frigid day in February, a flock of about a dozen descended on our house clearing and quickly cleaned up all the shriveled fruits still clinging to the black gum trees. Bluebirds arrive back on their breeding grounds in March and April, welcomed as harbingers of spring by winter-weary rural folk. Bluebirds nest statewide in Pennsylvania, avoiding deep woods and wooded ridges. The population of *Sialia sialis* probably peaked around 1900, when farmland covered two-thirds of the state; the number of bluebirds waned for many years thereafter, as unprofitable acres were abandoned and grew back up in forest. Bluebird numbers have risen over the last several decades, thanks to thousands of bluebird boxes put up by humans and perhaps to a decline in the house sparrow population.

Veery *(Catharus fuscescens).* Named for its call, this forest species has a reddish brown head, back, and tail and a faintly spotted breast. It breeds in southern Canada and in the northern United States, south in the Appalachians to Georgia; in Pennsylvania, where it arrives in May, it is most common in the northern half of the state, especially on the Pocono Plateau and from Potter County west to Crawford County. The veery favors damp deciduous forest with dense shrubs and ferns. Where its range overlaps that of the wood thrush and hermit thrush, the veery will be found in wetter, younger woods. Its song is a delicate, flutelike *da-vee-ur, vee-ur, veer, veer.* Mainly a ground forager, the veery feeds on insects (60 percent of its diet) and fruit (40 percent). In an Ontario study, individual territories averaged slightly more than half an acre.

The female builds a nest in a dense shrub near ground level or on the ground itself, often hiding it in vegetation at the base of a bush or small tree or in a brush pile. She lays three to five eggs and incubates them for ten to fourteen days. Brown-headed cowbirds lay eggs in the nests of veeries, which make no attempt to remove the eggs and end up raising the cowbirds along with their own young. Chipmunks sometimes prey on eggs and nestlings. The male helps to rear the brood, and the young leave the nest ten to twelve days after hatching. Veeries migrate at night. They winter in South America east of the Andes, mainly in Bolivia and Brazil.

Gray-Cheeked Thrush *(Catharus minimus)*. This elusive bird breeds in spruce forests and in alder and willow thickets in northern Canada and Alaska. Gray-cheeked thrushes pass through Pennsylvania in the latter part of May and again in late September and October. They forage on the ground in dense woods, and a birder must be stealthy and patient to glimpse one. The species winters in South America.

Swainson's Thrush *(Catharus ustulatus)*. A common migrant seen in woodlots and parks during spring and fall, this shy thrush nests but rarely in Pennsylvania, in a scattering of northern counties, with a concentration in 4,000 acres of old-growth forest in McKean and Warrens counties, including Heart's Content and Tionesta scenic areas in the Allegheny National Forest. The Swainson's thrush (also called the olive-backed thrush) breeds in New England, across Canada and Alaska, and in the U.S. Northwest. It can be identified by the bold buffy rings that surround its dark eyes. It has a melodious call with flutelike phrases. Swainson's thrush inhabits coniferous woods, generally spruce but also hemlock, where it nests in shrubby trees 2 to 10 feet above the ground. The species winters in tropical forests. Its name memorializes an English ornithologist.

Hermit Thrush *(Catharus guttatus)*. Many observers credit the hermit thrush with the loveliest of all bird songs, sometimes described as *Oh, holy holy-ah, purity purity, -eeh, sweetly sweetly.* The hermit thrush has a rufous tail and an olive head (in contrast with the wood thrush, which has an olive tail and a rufous head) and a spotted breast. When startled, a hermit thrush will usually fly to a perch and stare at an intruder while flicking its wings and slowly raising and lowering its tail.

The species' breeding range extends from Canada south into mountainous northern and central Pennsylvania. Hermit thrushes inhabit cool, damp mixed deciduous and coniferous woods. As quiet and unobtrusive as their name implies, they spend much time in the lower branches of shrubs and on the forest floor, where they forage for insects by hopping, then stopping and snatching with the bill. Animal matter makes up 90 percent of the diet in spring, 40 percent in winter. Hermit thrushes eat the fruits of elderberry, pokeberry, dogwood, greenbrier, Juneberry, sumac, poison ivy, and other plants.

Males arrive on the breeding range in April, in advance of females. Late snowstorms that cover up food sources may kill many early birds. Females usually build their nests on the ground (but also sometimes in trees 2 to 8 feet up), hiding them beneath boughs, weaving together twigs, bark fibers, ferns, mosses, and grasses, and adding a soft lining of conifer needles, plant fibers, and rootlets. The female incubates the three to four eggs for about twelve days, and the young are able to fly twelve days after hatching. Some pairs raise two broods over the summer.

Individuals have been known to survive more than eight years in the wild, but most do not live that long. The hermit thrush winters over much of the southern United States and south through Mexico to Guatemala. It is the only species in genus *Catharus* to winter in North America. For this reason, the rampant cutting of tropical forests has not hurt the hermit thrush population as much as it has harmed some other thrush species.

Wood Thrush *(Hylocichla mustelina)*. According to my journal, these are the dates in recent years when the calling of the first wood thrushes percolated through the woods around our Centre County home: April 15, May 3, April 8 ("but then they went away for 2 weeks"). The lilting, flutelike song is usually rendered as *ee-o-lay*, and it goes on increasingly through May, especially at dawn and dusk. Wood thrushes have reddish heads, olive backs and tails, and prominently spotted breasts; they are not as shy as other forest thrushes nor as bold as robins. Wood thrushes feed on beetles, caterpillars, crickets, ants, moths, and sowbugs, plus spiders, earthworms, and snails. They also eat many fruits and berries. Wood thrushes nest throughout eastern North America. They are statewide in Pennsylvania in moist lowland woods, dry upland forest, wooded ravines, orchards, city parks, and wooded suburbs. Territories range in size from 0.25 to 2 acres.

The female builds her nest on a branch or in a fork of a tree 6 to 50 feet above the ground (on average, 10 feet up), using grasses, moss, bark, and leaves cemented together with mud. An inner cup is lined with rootlets. The nest looks like a robin's nest but is smaller (a maximum of 5.5

inches in diameter, compared to the robin's 6.5 inches). Three to four eggs are usual for a first clutch; any later ones will have two to three eggs. The young hatch after two weeks and leave the nest around twelve days later. Brown-headed cowbirds may parasitize wood thrush nests, although in some cases, the foreign young do not affect the growth or success of the host's young. House cats, black rat snakes, flying squirrels, grackles, blue jays, weasels, and white-footed mice take eggs, nestlings, and young. In Delaware, a study of 378 wood thrush nests that did not fledge young found that 71 percent were lost to predation.

Wood thrushes quit singing in late summer but continue to sound *bwubububub* contact notes and *bweebeebeebee* alarm calls. They head south in August and September to forests from southeastern Mexico to Panama. The wood thrush population has declined since the 1980s, perhaps because fragmented forests in the Northeast are making thrush nests more accessible to parasites and predators. A study of 171 nests in Berks County revealed that predators wrecked 56 percent of the nests that were in small forest fragments, 22 percent in large forest fragments, and only 10 percent in unbroken woods. Wood thrushes have also lost crucial wintering habitats to widespread logging in the tropics.

American Robin *(Turdus migratorius)*. This adaptable songbird lives in towns, cities, farmland, cutover areas, woods edges, and deep woods. Early settlers named it after the European robin. The American species is about 10 inches long and has dark upperparts and a brick red breast, both colors more intense in males than in females, plus a white eye ring. Juveniles have paler colors and spotted breasts. Only the males sing, a hearty *cheeriup, cheerily, cheeriup* given repeatedly. Robins feed on beetles and other insects, earthworms, and fruits, both wild and cultivated, with fruit making up 60 percent of the annual diet. Robins often hunt for prey on lawns; they take earthworms that surface after the soil has been soaked by rain. Robins locate their prey by sight rather than by sound.

Robins arrive on their breeding territories in late March and early April; individuals may have wintered far to the south or close by in wooded or brushy swamps. Males home strongly to areas where they were born. They begin to establish territories that, as the breeding season progresses, resolve themselves into about a third of an acre. The territories of several males may overlap along their edges. Males may roost communally at night, then resume defending their territories during the day. Ornithologists have not discerned any specific courtship behavior; pairs seem to simply get

American robins locate their prey by sight.
They often hunt for insects and earthworms on lawns.

together. The male brings nest material to the female, and she weaves together grasses, weed stalks, and string, plastering them with mud and repeatedly forming a central cup with her own body. (Females often have a muddy band on the breast during nest building.) The cup is lined with fine grasses. Robins nest in trees (conifers for early broods, before deciduous trees have leafed out) and sturdy shrubs, and on porch supports, windowsills, and bridge and barn beams; sometimes robins repair and reuse their last year's nest.

The female lays three to seven eggs (usually four), colored the distinctive "robin's-egg" blue. Unlike many other thrushes, robins usually discern and eject the eggs of the brown-headed cowbird, a nest parasite. The female does all of the incubating and leaves the nest for about ten minutes per hour to feed herself. Male robins sing vociferously just before broods hatch, some twelve to fourteen days after the eggs are laid. Both parents feed the young, mainly on insects and earthworms, and the fledglings leave the nest after about fourteen days. The male may take over feeding the young while his mate begins a second nesting.

Pairs start to break up and communal flocks begin forming in July and August. The flocks move around to find trees and shrubs with good crops of berries, and in October, most of the flocks fly south. Although some robins winter in the North, most migrate to the southern states; a few go as far south as Guatemala. In farming areas, robins may share large winter roosts with European starlings, common grackles, and brown-headed cowbirds. The American robin is probably the most abundant bird species in Pennsylvania.

Eiserer, L. A. *The American Robin: A Backyard Institution*. Chicago: Nelson-Hall, 1976.

CATBIRD, MOCKINGBIRD, AND BROWN THRASHER

The gray catbird, northern mockingbird, and brown thrasher are among the most vocal of our birds. All belong to the family Mimidae, the "mimic thrushes" or "mimids," and they often imitate the calls of other species, stringing these remembered vocalizations into long, variable songs. The family Mimidae has over thirty species, which are found only in the New World, with most inhabiting the tropics. The mimids have long tails and short, rounded wings. Our three species in the Northeast are solitary (living singly, in pairs, and in family groups rather than in flocks), feed mainly on the ground and in shrubs, and generally eat insects in summer and fruits in winter. The sexes look alike. Adults are preyed on by owls, hawks, foxes, and house cats, and their nests may be raided by snakes, blue jays, crows, grackles, opossums, squirrels, and raccoons.

Gray Catbird *(Dumetella carolinensis)*. The gray catbird is 8 to 9 inches long, smaller and slenderer than a robin, and overall dark gray with a black cap. Individuals often jerk their tails—up, down, and in circles. The species is named for its mewling call, although catbirds also make other sounds. They migrate between breeding grounds in the eastern two-thirds of North America and wintering areas in the coastal Southeast and in Central America. Gray catbirds are abundant statewide in Pennsylvania, inhabiting hedgerows, woods undergrowth, regenerating cutover land, shrubby areas near water, woods edges, and suburban plantings. They avoid dense forests.

In summer, their diet is about 60 percent fruit; in spring, 20 percent fruit. Catbirds eat beetles, ants, caterpillars, grasshoppers, crickets, and other insects. They often forage on the ground, using their bills to flick aside leaves and twigs while searching for prey.

Gray catbirds feed and nest in hedgerows, woodland edges, and shrubby areas.

Although not as talkative as the northern mockingbird, the catbird is still a versatile vocalizer. Its ability comes in part from the structure of its syrinx, or voice box, whose two sides operate independently, letting the bird sing with two voices at the same time. A catbird calls out a rapid string of syllables—more than one hundred types in some individuals—including squeaks, chitters, whistles, whines, and songs swiped from other birds. The babble, which lasts up to ten minutes, is frequently punctuated by the familiar catlike mewl.

Catbirds are monogamous. They nest from May into July and usually raise two broods per year. The nest, substantial and deeply cupped, is placed in a dense thicket, brier patch, vine tangle, or shrubby tree, 3 to 9 feet above the ground. The female lays three to five eggs, which are a dark greenish blue and unmarked. Brown-headed cowbirds may lay their eggs in catbird nests, but catbirds almost always recognize the parasites' contrasting eggs (pale and dotted with brown) and eject them. Catbirds destroy the eggs and nestlings of some other birds, including wood-pewees, robins, and sparrows; biologists don't know whether this behavior represents an attack on competitors or a feeding strategy. Parents feed their young mainly insects and spiders. Incubation takes two weeks, and the young leave the nest ten or eleven days after hatching.

Northern Mockingbird *(Mimus polyglottos)*. The slender, robin-size mockingbird has a gray back, pale breast, and conspicuous white patches on the tail and wings. When foraging, a mockingbird will often stop and flick its wings, opening them to expose the white patches. The species lives year-round on its range, which overlays most of the lower forty-eight states and includes southern Canada, the Caribbean islands, and Mexico. Mockingbirds live in towns and cities, where they often forage on lawns, and in thickets, road margins, woods edges, cutover lands, and farms. They like a mix of low shrubs and open terrain. In Pennsylvania, mockingbirds are most common in the southeast, the south-central (although not in the mountains), and the southwestern regions.

About half of the diet consists of insects and other invertebrates, and the other half is native and cultivated fruits. Mockingbirds eat beetles, ants, bees, wasps, grasshoppers, and other insects; they also prey on spiders, earthworms, snails, and sowbugs. In fall and winter, mockingbirds eat elderberries, grapes, apples, barberries, hawthorn, and (a particular favorite) multiflora rose hips. Mockingbirds attack and sometimes even kill cedar waxwings, with whom they compete for fruit. Mockingbirds may visit feeding stations for seeds and suet, pugnaciously chasing other birds away.

Both male and female mockingbirds sing, but the males are the true virtuosos. They mimic snatches of other birds' songs, calls of crickets and frogs, dogs barking, and mechanical noises like squeaky hinges and squealing tires. A male's repertoire increases as the bird ages and may ultimately include more than 150 distinct song types. Usually an individual repeats one sound or song three to six or more times, then switches to another song, and so on, singing for minutes on end. (Brown thrashers usually repeat each song once, and catbirds do not repeat.) In the spring, male mockingbirds sing to establish territories and to attract mates, starting about an hour before sunrise. They sing in flight, on the ground, from perches, when building nest foundations, during and after copulation, while foraging—even with food clutched in their bills. Unmated males may sing during the night,

The northern mockingbird has large white patches on its wings and tail.

usually from a hidden perch. Mockingbirds sing from March to August (during the breeding season) and from late September into November (while establishing fall and winter feeding territories).

Mockingbirds are mainly monogamous. Courting males and females chase each other in flight. The nest is a bulky cup built in a dense shrub or a tree, usually 3 to 10 feet above the ground. The female lays three or four greenish to bluish gray eggs, blotched with brown. She incubates them for twelve to thirteen days. Both sexes feed the young, which fledge after twelve days, although they do not become strong fliers for another week. At fledging, the male may take over feeding the young while the female lays and begins incubating the next clutch. This division of labor lets mockingbirds produce two and sometimes three broods (up to four in the South) during each breeding season. Mockingbirds aggressively defend their nests, driving away predators and humans who come too close.

Some mockingbirds spend the year as a pair on a single territory, while others, particularly in the northern part of the range, use different breeding and wintering territories. In northern populations, some individuals may migrate south in winter. Young disperse up to 200 miles from where they hatched. The spread of multiflora rose (an invasive species once planted widely for wildlife habitat) and the planting of ornamental shrubs (especially *Pyracantha,* or firethorn) have provided mockingbirds with winter food and shelter, aiding the species in a northward expansion that has gone on for nearly a century.

Brown Thrasher *(Toxostoma rufum).* The largest of our three mimids, the brown thrasher has an 11- to 12-inch length, half of which is tail. The plumage is a rich reddish brown above, heavily streaked below. The name "thrasher" may come from the bird's habit of thrashing the ground litter, using its long, curved bill to sling aside leaves and dirt while foraging. Brown thrashers breed across the eastern two-thirds of North America. The species nests statewide in Pennsylvania, more commonly in the southern than in the northern counties. Brown thrashers prefer brushy, thorny places, including hedgerows, thickets, forest margins and clearings, and old fields overgrown with shrubs. Generally shyer than catbirds and mockingbirds, thrashers are less likely to live around people, and some will flee into escape cover at the sight of a human.

Brown thrashers feed on insects (more than half of the annual diet), berries, small fruits, seeds, and nuts, including many acorns. Occasionally they take crayfish, lizards, and small frogs. The best time to observe brown

The brown thrasher often "thrashes" the ground litter,
using its bill to sling aside leaves while searching for food.

thrashers is in April, before nest building has commenced, when males sing from high, exposed perches. The song is full of improvisation and the mimicry of other species, including flickers, titmice, cardinals, and thrushes; observers have reported over three thousand song types, the largest repertoire of any North American bird. The alarm call is a crackling note that sounds like a loud, smacking kiss. After mating, males continue to sing but in a quieter tone. Territories are 2 to 10 acres.

The nest, hidden in tangled cover, is built of sticks and twigs and lined with rootlets. Thrashers place their nests from 1.5 to 20 feet above the ground and occasionally on the ground itself. The female lays four eggs, which are pale blue and freckled with reddish brown. Both parents incubate the clutch. The eggs hatch after eleven to fourteen days, and the young leave the nest nine to thirteen days after hatching. They stay in the vicinity, and their parents bring them food. Two broods are usual; some thrashers switch mates between same-season broods. Nesting runs from early May to the end of July in Pennsylvania.

Brown thrashers in southern areas are permanent residents, but most individuals breeding in the Northeast leave the region in September and October and take up residence in thickets in the Gulf states. In Pennsylvania, the brown thrasher population seems to have decreased by about 4 percent a year since the mid-1960s, perhaps because of cowbird parasitism, nest predation, and the loss of brushy habitats.

Nickell, W. P. *Habitats, Territory and Nesting of the Catbird.* South Bend, IN: University of Notre Dame Press, 1965.
Doughty, R. W. *The Mockingbird.* Austin, TX: University of Texas Press, 1988.

CEDAR WAXWING

The cedar waxwing, *Bombycilla cedrorum,* is named for its penchant for eating the berries of cedar trees, and for the red tips on its secondary flight feathers (the part of the wings nearest the body), which resemble the sealing wax formerly used for securing envelopes. Cedar waxwings breed across North America and winter on the southeastern coastal plain and in Central America.

Biology. Cedar waxwings are sleek, olive-buff in color, with a pale yellow breast and a prominent head crest. The bill is small and black, and the tail is tipped with bright yellow. Unlike most other perching birds, cedar waxwings coexist in flocks and exhibit little territoriality, even during breeding season. They do not sing, but sound a thin, high *zeee.*

Waxwings are frugivorous birds, or fruit eaters: in the northeastern United States, over 80 percent of the diet consists of sugary fruit. In addition to the berries of eastern red cedars, waxwings eat serviceberries, mulberries, cherries, raspberries, strawberries, blueberries, pokeberries, and fruits of mountain ash, honeysuckle, crab apple, hawthorn, and firethorn. In summer, adults sit on perches and then fly out and intercept insects, including emerging mayflies and dragonflies, flying ants, beetles, and scale insects. Cedar waxwings generally feed in small flocks of less than thirty birds in summer and thirty to a hundred birds in fall and winter. Occasionally, flocks contain up to several thousand members.

Cedar waxwings arrive on the northern breeding grounds in late May. They are late nesters, and the hatching of their young coincides with the ripening of fruits in midsummer. Waxwings nest in loose colonies of ten or more pairs; breeding birds flock at nearby fruit sources, foraging side by side without showing aggression. Around the nest, individuals may guard their mates or defend nest materials. During courtship, paired birds often pass fruit to each other. They build their nest in a tree (often a cedar, white pine, apple, pear, hawthorn, or burr oak) up to 50 feet above the ground. It is loosely woven out of grasses, twigs, and weed stems and lined with fine plant matter. Females lay eggs in June and July; most clutches contain three to five eggs, which are pale gray dotted with brown. The altricial young hatch after twelve to thirteen days. Their parents feed them insects and, later, fruit. The young leave the nest about sixteen days after hatching. The adults, monogamous during the breeding season, usually rear a second brood.

Cedar waxwings move about during winter in search of fruit. Birds breeding in the same area may winter far apart: six birds banded in Michigan were recovered at places from Georgia to Mexico. Merlins, sharp-shinned hawks, and Cooper's hawks prey on adults, and blue jays eat nestlings; mockingbirds defending winter fruit sources may kill waxwings by knocking them to the ground and striking them with their bills. Waxwings also die from pesticide poisoning and from eating fermented berries. Biologists estimate the maximum life span of an individual to be seven years. The annual mortality rate is about 55 percent.

Habitat. Cedar waxwings nest in open woods, forest edges, old fields overgrown with shrubs and small trees, shady groves, and orchards. They avoid deep woods. They often forage above streams, marshes, and beaver ponds, perching in old snags and flying out to catch emerging aquatic insects. Their droppings disperse seeds and help fruiting plants to regenerate themselves.

Population. Over the last thirty years, cedar waxwing numbers have increased in much of North America, and the species seems to be expanding into new areas. In Pennsylvania, cedar waxwings are found statewide; they are least common in cities and in intensively farmed areas of the southeast. Some ornithologists believe the number of breeding waxwings tripled between the mid-1960s and the 1990s, an increase fostered by the widespread planting of fruiting trees and ornamental shrubs.

WOOD-WARBLERS

Like jewels strewn through the woods, our native wood-warblers appear in early spring, the males arrayed in gleaming colors. Thirty-one warbler species breed in Pennsylvania, and eight migrate through Penn's Woods headed for breeding grounds farther north. In central Pennsylvania, the first species begin arriving in late March. The great mass of warblers passes through between May 10 and 15, and then the migration trickles off until it ends in late May—by which time the trees have leafed out, making it tough to spot the canopy-dwelling species. In southern Pennsylvania, look for the migration to begin and end a few days to a week earlier; in northern Pennsylvania, it's somewhat later. In August, warblers start moving south again, with migration peaking in late September and ending in October, although stragglers may still come through in November. By now, most species have molted into cryptic shades of olive and brown: the "confusing fall warblers" of field guides.

Wood-warblers are small, lively birds that use a range of habitats. All of the North American species are migratory; almost certainly, most of them arose in the tropics and extended their ranges northward to exploit new breeding areas. The name warbler is a misnomer, because few species possess warbling voices, and many have thin, scratchy, unmusical songs. Males use two calls: a song to advertise territory and a shorter call to attract a mate and communicate with her.

Wood-warblers breed in May and June, in woods and brushland, in areas that may be dry, moist, or wet. The various species forage from ground level to the treetops and eat mainly small insects plus a few fruits; some warblers take flower nectar. Notes Scott Weidensaul in *Mountains of the Heart,* "Wood-warblers are among the most habitat-specific of North American birds." When several species inhabit the same area, their feeding strategies usually differ enough that the birds don't compete directly. Nesting habits vary widely. The prothonotary warbler *(Protonotaria citrea),* a rare breeder in wetlands and bottomland forests in Pennsylvania, builds its nest in a tree cavity, often using an old downy woodpecker hole. The Nashville warbler *(Vermivora ruficapilla)* is one of several species that nest on the ground. Some warblers nest exclusively in conifers; others use hardwoods. The northern parula *(Parula americana)* weaves its nest into hanging clumps of lichens, twigs, or pine needles. Most species are monogamous. Generally the female builds the nest. The eggs, usually two to five per clutch, are whitish with dark spots. The female does most or all of the incubating, and both parents feed the young.

We know less about warblers' habitat requirements and feeding activities on their winter range. Most species winter in Mexico, Central America, and South America, where they forage in mixed flocks that include several to many different species. Wood-warblers tend to shun lowland rain forests, instead preferring foothill and mountain woods. A few hardy species (the yellow-rumped warbler, *Dendroica coronata,* is one) stay in North America all winter. Warblers are small birds with limited fat reserves, and many perish from the rigors of migrating. A route followed by many species in spring requires a nonstop flight from the Yucatan Peninsula across the Gulf of Mexico to Louisiana, Mississippi, Alabama, and Florida; if migrating birds encounter headwinds, they may exhaust their strength, fall into the ocean, and drown. Tremendous numbers of warblers and other night-migrating birds die when they fly into communication towers and tall buildings, particularly on cloudy nights. Many individuals are killed by the smaller hawks and owls. Warblers have been documented to live for over ten years in the wild, but most die before reaching that age.

Some wood-warbler populations are holding their own. Those of others, such as the cerulean warbler *(Dendroica cerulea),* which breeds in mature forests, have declined in recent years. When woodlands are broken up into smaller patches by logging or home development, warblers lose habitat. In fragmented woods, birds and mammals—including blue jays, raccoons, foxes, squirrels, and house cats—can prey more easily on warblers and their nests. Brown-headed cowbirds, which live in open areas, may find greater access to warblers' nests: the female cowbirds surreptitiously lay eggs in the nests, and when the young cowbirds hatch, they are raised by the host adults, whose own smaller, slower-to-develop young may not survive.

The following is a closer look at eight common wood-warblers of Pennsylvania.

Yellow Warbler *(Dendroica petechia).* This showy, all yellow bird has a rufous-streaked breast. The male's song is a lively *weet weet weet weet tsee tsee.* Yellow warblers breed statewide in Pennsylvania. Look for them in low brush or shrubs, woods edges, orchards, parks, and gardens; they're often found along streams and near swamps. Caterpillars may make up two-thirds of the diet. Yellow warblers snatch up mayflies, moths, mosquitoes, beetles, damselflies, treehoppers, and other insects. They pluck their prey from twigs and leaves, hover to glean it from the undersides of foliage, and make short flights to intercept it. The nest is a neat, open cup built of plant materials and lined with plant down or fur. *Dendroica petechia* is often parasitized by cowbirds, whose foreign eggs cause some yellow warblers to desert their nests or build a new nest on top of the cowbird eggs. Yellow warblers arrive in Pennsylvania in April and May and head south again as early as July and August. They winter in Mexico, Central America, and northern South America.

The yellow warbler often breeds along streams and in wetlands.

Chestnut-Sided Warbler *(Dendroica pensylvanica).* In spring, both sexes sport a yellow crown, black face markings, and chestnut streaks on their sides. Chestnut-sided warblers are common migrants, and the species breeds across the northern half of the state, in the Allegheny Mountains, and elsewhere in scattered locations. This now-common species increased its numbers after Pennsylvania's virgin forests were logged. Chestnut-sided warblers inhabit brush and briers, cutover woods, and reverting fields. They forage for insects by hopping from branch to branch, darting out now and then to catch their prey in midair. The song is similar to the yellow warbler's and has been rendered as *please please please ta meetcha.* The nest is built in dense shrubs or blackberry tangles and is woven out of strips of cedar or grapevine bark, weeds, grasses, and roots, with a soft lining. Adults in autumn have a dull greenish plumage and look not at all like their bright spring selves. The winter range extends from Mexico through Panama.

Black-Throated Blue Warbler *(Dendroica caerulescens).* One of the handsomest birds in the forest, the black-throated blue warbler is aptly described by its name. (The black and slatey blue colors are set off by a white breast.) The species typically nests in dense shrubs in the deep woods, often in bottomland cove forests well-stocked with hemlocks, with a bubbling stream nearby and plenty of gnats, moths, crane flies, caterpillars, and other insects to prey on. Males forage higher in the understory than do females; black-throated blue warblers sometimes steal insects from spiderwebs. Males sing a buzzy, drawn-out *zur, zur, zree* (some people hear it as *I-am-so-la-zeee*). The nest is a bulky cup hidden in a rhododendron, laurel bush, or shrubby conifer. In Pennsylvania, the black-throated blue warbler nests in the mountainous northern tier and in the northern Ridge and Valley region. The species breeds across southern Canada and in the Appalachians south to Georgia; it winters in the Bahamas and the Greater Antilles.

Black-and-White Warbler *(Mniotilta varia).* This widespread, abundant bird acts more like a nuthatch or a creeper than a warbler, foraging methodically on tree bark, circling the trunks and limbs of trees while looking for insects and their eggs. Both males and females have zebra stripes on the back and crown. Next to the Louisiana waterthrush, the black-and-white warbler is the earliest spring migrant, arriving well before the leaves push out. Individuals often feed low in trees and usually nest in deciduous woods. The male sings a thin *weesee, weesee, weesee,* repeating the phrase at least seven times. The female builds a nest out of dry dead leaves and lines a central cup with grasses, strips of grapevine bark, rootlets, and weed fibers. The nest is sited at the base of a tree or tucked partway under a log, stump, or rock. Black-and-white warblers breed across most of Pennsylvania and in the eastern United States and much of Canada. They winter in Florida, the Gulf coast states, the West Indies, and from Mexico south into South America.

American redstarts feed in the treetops, catching insect prey.

American Redstart *(Setaphaga ruticilla)*. In Pennsylvania, the American redstart is rare in the highly agricultural southeast but common in the forested northern tier and in other wooded parts of the state. The male is an eye-catching mix of black, orange, and white; orange patches show on the wings and tail, which the bird often flashes open and shut. American redstarts inhabit sapling woods, river groves, forest edges, and tree-lined creek banks. A Wisconsin study found the species to be three times as common in woods of greater than 80 acres than in woodlots of less than 14 acres. Redstarts flutter about in treetops, hovering among the leaves, leaping up or darting out like a flycatcher to grab a passing insect. Bristles framing its mouth help a redstart catch flying prey. Redstarts eat insects, spiders, seeds, and berries. The song is a series of high-pitched, indistinct *tsee* notes. The female builds a cup-shaped nest in a tree fork 4 to 70 feet in the air. Some males breed with more than one female in their territories. Redstarts head south in August and September; they winter along the Gulf coast from Mexico south to South America.

Ovenbird *(Seiurus aurocapillus)*. This bird gets its name from the covered, dome-shaped nest that it builds on the ground, which reminded early observers of a Dutch oven. An ovenbird looks like a miniature thrush, olive-brown above and with a dark-streaked (rather than a spotted) breast and an orange, black-rimmed stripe atop the head. Ovenbirds prefer extensive tracts of dry mature deciduous woods, but they also inhabit other forest types, including swamplands. They feed on the ground, taking beetles, ants, caterpillars, bugs, worms, spiders, and snails. The song is an emphatic *Teacher! Teacher! Teacher!*, repeated about ten times at an increasing volume, with three to four sessions per minute. The species breeds across Canada and the Northeast; in Pennsylvania, it nests statewide, although it is absent from heavily farmed and urbanized districts. Ovenbirds arrive here in April and May and depart in September and October. They winter in Florida, Mexico, Central America, and the West Indies.

Louisiana Waterthrush *(Seiurus motacilla)*. In April, trout fishermen see this shy warbler walking on stones along the edges of streams, turning over wet leaves with its bill and flitting out over the water to catch insect prey. A Louisiana waterthrush looks like a thrush and acts like a sandpiper, teetering and dipping, elevated above slick rocks on its long legs and stabilized by its large, long-toed feet. Waterthrushes eat bugs, beetles, adult and larval mayflies, dragonflies, crane fly larvae, ants, caterpillars, and other insects, along with centipedes, small crustaceans, and snails. They breed from April to June along rushing brooks that flow through hilly or mountainous terrain, always in wooded habitats. The nest is usually built in a hole in the stream bank, hidden by tree roots, weeds, or grass. Louisiana waterthrushes nest throughout the East; they winter in streamside forests in Mexico, Central America, and the West Indies.

Common Yellowthroat *(Geothlypis trichas).* *Witchity, witchity, witchity* sings this bird with the gray back, black mask, yellow throat, and whitish belly. (Females lack the black mask.) In Pennsylvania, yellowthroats nest in cattail marshes, alder swamps, shrubby bogs, wet meadows, forest edges and openings, and old fields. They like thick, briery cover and often take advantage of small habitat patches. Nests are bulky, made of dry leaves and coarse grasses and lined with finer plant matter; they're built on or near the ground, hidden in tussocks, weed stalks, and shrubs. Yellowthroats eat insects (grasshoppers, dragonflies, mayflies, beetles, moths, ants, aphids, and others), spiders, and seeds. They nest statewide across Pennsylvania and are probably our most common wood-warbler. The species winters in the southern United States, Mexico, and Central America. The draining and filling of wetlands—even very small ones—harms yellowthroats and many other forms of wildlife.

Other wood-warblers breeding in Pennsylvania include the blue-winged warbler *(Vermivora pinus),* golden-winged warbler *(Vermivora chrysoptera),* Nashville warbler *(Vermivora ruficapilla),* northern parula *(Parula americana),* magnolia warbler *(Dendroica magnolia),* yellow-rumped warbler *(Dendroica coronata),* black-throated green warbler *(Dendroica virens),* Blackburnian warbler *(Dendroica fusca),* yellow-throated warbler *(Dendroica dominica),* pine warbler *(Dendroica pinus),* prairie warbler *(Dendroica discolor),* cerulean warbler *(Dendroica cerulea),* blackpoll warbler *(Dendroica striata),* prothonotary warbler *(Protonotaria citrea),* worm-eating warbler *(Helmitheros vermivorus),* Swainson's warbler *(Limnothlypis swainsonii),* northern waterthrush *(Seiurus noveboracensis),* Kentucky warbler *(Oporornis formosus),* mourning warbler *(Oporornis philadelphia),* hooded warbler *(Wilsonia citrina),* Canada warbler *(Wilsonia canadensis),* and yellow-breasted chat *(Icteria virens).*

The following warblers migrate through Pennsylvania: Tennessee warbler *(Vermivora peregrina),* orange-crowned warbler *(Vermivora celata),* Cape May warbler *(Dendroica tigrina),* bay-breasted warbler *(Dendroica castanea),* palm warbler *(Dendroica palmarum),* Connecticut warbler *(Oporornis agilis),* and Wilson's warbler *(Wilsonia pusilla).*

Curson, J., D. Quinn, and D. Beadle. *Warblers of the Americas: An Identification Guide*. Boston: Houghton Mifflin, 1994.

Dunn, J. L., and K. L. Garrett. *A Field Guide to Warblers of North America*. Boston: Houghton Mifflin, 1997.

Griscom, L., and A. Sprunt, Jr. *The Warblers of America*. Garden City, NY: Doubleday, 1979.

Morse, D. M. *American Warblers: An Ecological and Behavioral Perspective*. Boston: Harvard University Press, 1989.

TANAGERS

Two tanager species travel north from the Neotropics to breed in eastern North America—a small percentage of the more than two hundred species of tanagers, many of whom have dazzling colors, including red, yellow, green, blue, and purple. The word *tanager* comes from a South American Indian word denoting a small, brightly colored bird. In tropical forests, mixed feeding flocks may include over a dozen kinds of tanagers.

Scarlet Tanager *(Piranga olivacea).* The brightest red I've ever seen met my vision when I focused binoculars on a male scarlet tanager singing in a treetop: against a backdrop of dark storm clouds, and lit by the last rays of the evening sun, he looked positively fluorescent. Males arrive on the breeding range—eastern North America from southern Canada to the Carolinas—in late April and early May, just as the trees are leafing out. Their bodies are red, and their wings and tails are jet black. Females, which show up a few days later, are a greenish yellow color that blends with the leaves in which they rest and feed. Adults are about 7 inches long.

Scarlet tanagers favor dry oak woods in uplands. They also inhabit mixed and coniferous forest and tree plantings in suburbs and parks. Males claim 2- to 6-acre territories by singing almost constantly from prominent perches and driving away competing males. The song sounds like *jeeyeet jeeay jeeeoo jeeyeer jeeyeet,* five to nine slightly hoarse notes, "like a Robin with a sore throat," according to Roger Tory Peterson. Males whose territories adjoin sometimes perch along shared boundaries and "countersing," with one male singing a phrase several times and the other male matching it; this behavior helps settle border disputes. Males return to the previous year's territories, but it's thought that females lack this strong homing instinct, so they rarely end up mating with the same male in succeeding years.

Insects and fruits form the bulk of the diet. Females forage higher in the tree canopy than males. Both sexes work slowly and methodically, inspecting leaves, twigs, and branches and picking at leaf clusters near the ends of twigs. Sometimes they make short flights to catch flying insects, particularly bees and wasps. They eat caterpillars, moths, adult and larval beetles, dragonflies, aphids, snails, spiders, worms, and millipedes. During cold snaps, they land on the ground and hunt for beetles, earthworms, and other prey. They also eat tender buds, wild fruits and berries, and cultivated fruits such as cherries.

Scarlet tanagers nest from late May to mid-June. To rear a brood, a pair needs 4 to 8 wooded acres. The courting male flies to a perch below the female; he droops his wings and spreads his tail to show off his brilliant back. If the female strays outside his territory, he chases her back into it. Tanagers mate frequently, with the female crouching and calling to entice the male. She chooses the nest site and builds the nest herself, over three to seven days, while the male sings from perches at the midforest level. Tanagers nest lower than they forage; nests are 8 to 75 feet up (usually 18 to 50 feet), often near the end of a horizontal branch in an oak, with a view of the ground and with clear flyways from nearby trees. The nest is flattish and rather flimsy, made of twigs and rootlets and lined with grasses and stems; some nests are so loosely woven that the eggs can be seen from beneath. The female lays two to five eggs, usually four. They are a pale blue-green marked with

In the Northeast, scarlet tanagers favor dry oak woods in uplands.
They winter in remote forests in South America.

brown. The female incubates the clutch for about two weeks, with the male bringing food to her. Both parents feed insects and fruit to the young, which leave the nest after nine to fifteen days; their parents go on feeding them for two more weeks. Only one brood is produced each summer.

Fledglings are brown, with slight streaking. In late summer, the adults molt, and for a while, the male is a patchwork of red, yellow, and green; he ends up looking like the female but retains his black wings and tail. Scarlet tanagers leave Pennsylvania in September and early October. They migrate mainly through the Caribbean lowlands of Central America and spend the remainder of the year east of the Andes in remote forests in Colombia, Ecuador, Peru, and Bolivia. There they join mixed-species flocks and feed in the canopy, along with other tanagers, and in fruiting trees.

One scarlet tanager that had been banded lived for ten years; most probably don't survive for half that long. They are preyed on by accipiter hawks, falcons, and owls. Tanagers attack squirrels and blue jays, which nevertheless manage to rifle many nests. Crows also eat eggs and fledglings. Brown-headed cowbirds parasitize more than half of all tanager nests in some areas, particularly where the forest has been fragmented by logging or home development. Scarlet tanagers nest statewide in Pennsylvania and are more common than many people think. The highest populations occur in mature, extensive forests. Scarlet tanagers are absent from treeless urban areas and intensively farmed lands.

Summer Tanager *(Piranga rubra)*. This all-red tanager breeds mainly in the Southeast, where it is called the "summer redbird." Its range extends into southwestern Pennsylvania, where it nested during the 1980s in Greene, Washington, and Beaver counties. Summer tanagers inhabit dry upland forests, with a preference for slightly open oak woods. In summer, they eat mainly insects: caterpillars, moths, beetles, cicadas, grasshoppers, flies, and others; often they tear open wasp nests to feed on larvae, apparently without being stung. The summer tanager's breeding and nesting habits are similar to those of the scarlet tanager. Individuals seen in springtime in Pennsylvania may have overshot their normal range and may then turn around and move back south to find mates. Summer tanagers spend the rest of the year on a large range extending from central Mexico to Bolivia and Brazil.

Isler, M. I., and P. R. Isler. *The Tanagers, Natural History, Distribution, and Identification.* Washington, DC: Smithsonian Institution Press, 1987.

TOWHEE AND SPARROWS

At first glance, sparrows may seem to be drab, ordinary birds. Because of their apparent sameness—as well as the dense or grassy cover in which most are found—beginning and casual birders find it tough to identify the different species. In fact, the plumage of each is a distinctive, complex blending of shades and streakings of brown, and the birds' habits and adaptations work in fascinating ways to let them take advantage of many habitats. *Sparrow* comes from *spearwa,* an Anglo-Saxon word meaning "flutterer"; English settlers applied the name to New World sparrows. (In England today, birds we would call sparrows are referred to as buntings.) More than thirty species are native to North America. Eleven breed in Pennsylvania, and five more pass through the state when migrating.

Sparrows have short, thick bills for cracking the hard seeds of grasses, weeds, and trees. Most forage on the ground, scratching with their feet to expose food in dense grass and weeds and beneath shrubby growth. They keep in contact with mates or flock members by using short calls, often *chip* or *seep* sounds, which vary between the species. Sometimes sparrows make short flights to catch flying insects that they've flushed from the ground. Adults eat insects in summer and nourish their young with this high-protein fare. In late summer and fall, sparrows eat berries and fruits. And they eat many seeds, especially those of grasses and weeds.

Males defend territories mainly by singing from exposed perches, and their songs are often complicated and mellifluous. The males of some grassland sparrows perform flight-and-song displays. Males also chase away rivals. In most species, pairs nest in isolation or in loose colonies brought together less by social tendencies than by an attraction to a certain habitat. Sparrows usually nest in low bushes or on the ground. The typical nest is an open cup woven out of grass, weeds, and twigs, built mostly or entirely by the female. The eggs of the various northeastern sparrows are spotted or blotched with brown. In most species, the female incubates the eggs; the male may bring food to her. Both parents share in feeding the young. Should a female begin a second brood, her mate may assume the care of first-brood young that have fledged from the nest.

Ornithologists believe that most sparrow pairs are monogamous, but the breeding biology of many species hasn't been studied carefully enough to allow definite conclusions. In the savannah sparrow, males may have two mates whose broods are staggered, so that the male can help first with one brood, then with the second. Some male swamp sparrows also have two mates.

Sparrows do not make long migrations. Most species winter in the southern United States and northern Mexico, and none go as far as the tropics. In winter, sparrows are often gregarious and travel in flocks when searching for food. In open country, flocks often contain individuals of only one species, but in brushy areas or along woods edges, which offer a more diverse suite of foods, mixed-species flocks are the rule. The greatest threat to our native sparrows is the degrading or destruction of their habitats. Draining marshes and converting fields to housing developments relentlessly cuts into the size and diversity of sparrow populations—as well as harming many other kinds of wildlife.

A closer look at five common Pennsylvania sparrows follows.

Eastern Towhee *(Pipilo erythrophthalmus).* Formerly called the rufous-sided towhee, this large (7 to 8 inches), long-tailed sparrow breeds statewide in Pennsylvania. Adults have rusty sides, white bellies, and solid-colored backs and heads that are black in the male and brown in the female. The eyes are red. Males sing a distinctive *drink your tea,* with the middle syllable low and the last syllable drawn out and quavering. Both sexes frequently give an emphatic *che-wink* or *tow-hee* call. One way to locate the birds is to listen for the rustling they make while scratching for food in the leaf litter. The eastern towhee is sometimes called the chewink, for its call, and the ground robin, for its foraging habits.

Eastern towhees are found mainly in second-growth forests, overgrown fields, woods edges, clearcuts, hedgerows, thickets, dense brush, and the understory of open deciduous woods. Rarely do they inhabit suburban yards, cities, or intensively farmed areas. When seeking food, towhees energetically turn up leaves by hopping backward and scratching with both feet. They pick up beetles, ants, bugs, spiders, millipedes, and snails; they eat caterpillars (including late-stage gypsy moth larvae) and moths (adult gypsy moths and others); and they dine on small fruits, berries, acorns, and seeds.

In April, males arrive in the North in small bands; they disperse and, singing from high perches, claim individual territories of 0.5 to 2 acres. Females show up about a week later. Males and females spread their wings and tails to each other, exhibiting their white patches. The female gathers materials for the nest, while the male sings nearby. She scuffs out a shallow depression in the ground and builds a bulky but well-camouflaged nest of leaves, bark strips, and other plant matter, lined with fine grasses and pine needles. Occasionally the nest is placed in a bush as high as 5 feet above the ground.

The female lays three or four eggs, creamy white with brown spotting. She incubates them for twelve to thirteen days; during the day, she sneaks off to feed about once every half hour. After the eggs hatch, the male brings food for the brooding female and the young. In about a week, the female begins leaving the nest to help the male forage and feed the brood. The young leave the nest after ten to twelve days, and their parents feed them for another month. Most females build a second nest, and most pairs produce two broods. Towhees nest from late April into August in Pennsylvania. After fledging, the young birds flock together; adults do not defend their territories against juveniles.

In winter, towhees shift southward into the southern states, where they forage in loose flocks averaging fifteen to twenty-five members. Females go farther south than males. The estimated life span is four to six years. The clearing of the eastern deciduous forests around the turn of the century helped towhee populations to expand. More recently, as old fields have matured into woods, the population of this species has declined.

Chipping Sparrow *(Spizella passerina)*. This small, slim sparrow is about 5 inches long and marked with a rusty-colored cap and a line of white above each eye. Volunteer surveyors for the *Pennsylvania Breeding Bird Atlas* found the chipping sparrow to be the fourth most widespread bird in the state; only the song sparrow, crow, and robin were observed more frequently. Chipping sparrows feed and breed in suburbs, urban parks, gardens, clearings around rural homes, pastures, orchards, shrubby fields, open woodland, woods edges, and along roads through deep woods. On a continental scale, they breed from Alaska to Nova Scotia and south to Nicaragua. They are not very shy of humans. The song is a rattling or buzzing trill: a series of chips in one pitch.

Chipping sparrows forage in trees and on the ground. Their diet in early summer may be 90 percent insects, including grasshoppers, caterpillars, beetles, and moths. They eat many seeds, especially in fall and winter, of chickweed, pigweed, ragweed, foxtail, and other grasses. Males arrive on the breeding range in April, ahead of females, and claim territories of 0.5 to 1.5 acres. In early May, the females build nests, often in conifers (including suburban plantings), 3 to 10 or more feet above the ground. A female usually lines her nest with fine grasses or animal fur, including horse hair. The three or four eggs are a pale bluish green, marked with brown spots. The female incubates the eggs for eleven to fourteen days; the young fledge from the nest eight to twelve days after hatching. Chip-

*The chipping sparrow inhabits orchards, roadsides, woods edges,
and yards around houses in towns, suburbs, and rural areas.*

ping sparrows are believed to be monogamous breeders. Most pairs raise
two broods per summer.

In August and September, family flocks desert their home territories
and wander while searching for food. In late September and October, most
chipping sparrows leave the Northeast for wintering grounds in the Gulf
states. In the 1800s, the chipping sparrow was *the* common sparrow of
American towns and cities, but the introduced house sparrow largely took
over that role. Chipping sparrows are preyed on by blue jays, snakes, domestic cats, and the smaller hawks and owls; brown-headed cowbirds often parasitize first broods, but chipping sparrows raise their second broods after the
cowbirds' annual breeding period has ended.

Field Sparrow *(Spizella pusilla)*. Like the chipping sparrow, the field sparrow has a chestnut-colored cap; however, it lacks a white facial stripe and
has a noticeably pink or rust-colored bill. The song is a series of notes
speeding up into a trill *(swee-swee-swee-swee-wee-wee-wee-wee)*. Field sparrows
live in thickets, fencerows, Christmas tree plantations, and old fields with
scattered brush, brambles, and sumac clumps; they avoid open meadows,
cropland, urban areas, and deep woods. The species breeds in every Pennsylvania county but is absent from heavily developed areas and from cities.

The field sparrow ranges across the East and winters from southern Pennsylvania southward.

Field sparrows arrive in their breeding habitats in mid-April. Males' territories average 2 to 3 acres. Females build their nests on or near the ground for the season's first brood, then often select a thick shrub, such as a hawthorn, for a second-brood nest. The three to four eggs hatch after about eleven days of incubation. Unlike chipping sparrows, field sparrows rarely nest near humans' dwellings; like chipping sparrows, field sparrows permit people to come quite close. Field sparrows migrate south in September and October. *Spizella pusilla* was first described and named by ornithologist Alexander Wilson on the basis of specimens collected around Philadelphia.

Song Sparrow *(Melospiza melodia)*. An accomplished songster, this shy sparrow has a heavily streaked breast with a dark central spot. When in the species' preferred habitat of overgrown weedy areas, thickets, or abandoned pasture land, listen for the melodious song: three or four repeated notes, *sweet sweet sweet sweet,* followed by a number of shorter variable notes and a trilled ending. Song sparrows breed across North America and winter in the lower forty-eight states. They breed statewide and abundantly in Pennsylvania. More song sparrows winter in the southern half of the state than in the northern half; corn stubble and brushy thickets are prime wintering areas.

Song sparrows nest mainly on the ground in grasses, sedges, and cattails, with later nests often located in trees or bushes up to 12 feet high; on rare occasions, song sparrows nest in tree cavities. Prolific breeders, they may raise two, three, or even four broods per season, sometimes all in the same nest. The normal clutch is four eggs. The eggs of brown-headed cowbirds look very much like song sparrows' eggs (greenish white, heavily dotted and blotched with reddish brown), and, except for the yellow warbler, the song sparrow is the most frequently

The song sparrow has a heavily streaked breast with a dark central spot.

reported host for the parasitic cowbird. The song sparrow population in Pennsylvania seems to be stable or increasing slightly.

Dark-Eyed Junco *(Junco hyemalis)*. Juncos are familiar winter visitors; many people are surprised to learn that juncos also breed in Pennsylvania. These birds have slate-gray backs and heads, white bellies, pink bills, and white outer tail feathers. The springtime song is a slow, musical trill similar to that of the chipping sparrow; what's usual in winter is a string of twittering notes. Ground-loving birds, juncos scratch for food in the leaf duff, soil, and snow. In summer, insects make up about half of the diet. Seeds of ragweed, foxtail, crabgrass, smartweed, pigweed, and other grasses and weeds predominate in fall and winter. Juncos also eat springtails, the tiny "snow fleas" that pepper the snow on warm winter days.

Juncos breed across northern North America and south in the Appalachians to Georgia. In Pennsylvania, they nest on wooded ridgetops and in hemlock ravines across the forested northern third of the state. In spring, males stake out breeding territories of 2 to 3 acres, singing from tall trees—about the only time these birds ascend very far from the ground. Breeding runs from April into August. Females build nests on the ground: on vegetated cutbanks of logging roads, stream banks, and hillsides or tucked beneath exposed tree roots overhung by dirt or plants. The three to six eggs are pale blue, profusely dotted with brown. Some pairs raise two broods.

Juncos move south in flocks, mainly in October. The individuals we see wintering in Pennsylvania probably bred or were hatched farther to the north. Winter flocks tend to have same-age, same-sex members, who forage together on an area of 10 to 12 acres. Most flocks number around fifteen to thirty birds. In winter, juncos favor hedgerows, brush piles, thickets, weedy fields, and shrubbery around houses; they often forage along roadsides when snow covers other feeding habitats. At night, flock members roost together in a habitual site, usually in the dense boughs of a conifer.

Six other sparrows breed in Pennsylvania. The vesper sparrow *(Pooecetes gramineus)* is a grassland species that breeds in scattered locales across the state, including revegetated strip mines; its numbers have declined over the last thirty years. The shy, inconspicuous savannah sparrow *(Passerculus sandwichensis)* nests on the ground in open grassy areas such as meadows, hayfields, and reclaimed surface mines, mainly in the northern and western counties. Another species inhabiting grasslands and meadows is the grasshopper sparrow *(Ammodramus savannarum)*. Henslow's sparrow *(Ammo-*

dramus henslowii) breeds mainly in western Pennsylvania, in abandoned weedy fields, damp meadows, and reclaimed strip mines. The swamp sparrow *(Melospiza georgiana)* is found in Delaware River tidal marshes, in freshwater marshes in the state's northeastern and northwestern quadrants, and elsewhere in bogs, swamps, and rank growth around ponds and sluggish streams. The white-throated sparrow *(Zonotrichia albicollis)* breeds mainly in the North, often in or near forested wetlands, and its range extends south into Pennsylvania's northern tier; this chunky, colorful sparrow is frequently seen during its migration, and in some years, it winters in the southeastern counties.

As well as the above-mentioned species, other sparrows move through Pennsylvania in spring and fall. The American tree sparrow *(Spizella arborea)* is a common migrant and a winter resident. The fox sparrow *(Passerella iliaca)* and white-crowned sparrow *(Zonotrichia leucophrys)* also may winter in Pennsylvania. Nelson's sharp-tailed sparrow *(Ammodramus nelsoni)* and Lincoln's sparrow *(Melospiza lincolnii)* are rare migrants.

Rising, J. *A Guide to the Identification and Natural History of the Sparrows of the United States and Canada.* San Diego: Academic Press, 1996.

CARDINAL, GROSBEAKS, INDIGO BUNTING, AND DICKCISSEL

Cardinals, grosbeaks, and indigo buntings are equipped with stout, strong bills to crush seeds. In addition to seeds and fruit, which are important fall, winter, and spring foods, these birds eat protein-rich insects in summer and feed them to their young. They live in thick cover, including forests, woods edges, brushland, swamps, and ornamental plantings in suburbs and cities. The dickcissel is a related species that breeds mainly in the Midwest but also nests in grassy habitats in Pennsylvania.

Northern Cardinal *(Cardinalis cardinalis).* Adults are 8 to 9 inches long, slightly smaller than a robin. Both sexes have an orange-red bill and a prominent head crest. The male's plumage is an overall bright red; the female is yellowish brown with red tints on her wings, tail, and crest. The cardinal is a common bird in the southeastern United States. Before 1900, the species was rare in Pennsylvania, but over the last century, cardinals have spread as far north as Maine and southern Canada. They now inhabit all of the Keystone State, except for areas of unbroken forest on the Allegheny High Plateau. Cardinals also breed across the Midwest and in Central America from Mexico to Guatemala. They are year-round residents throughout their range.

Cardinals live in thickets, hedgerows, brushy fields, swamps, and gardens, and in towns and cities. They need dense shrubs for nesting; these can range from multiflora rose tangles between woodlots and fields, to hedges of privet and honeysuckle on shady streets. Hawthorns, lilac, gray dogwood, and dense conifers also provide nesting cover. Mated pairs of cardinals use territories of 3

The northern cardinal's thick, sturdy bill is well suited to cracking hard seeds.

to 10 acres. They eat caterpillars, grasshoppers, beetles, bugs, ants, flies, and many other insects; fruits of dogwood, mulberry, and wild grape; seeds of smartweeds and sedges; grains scattered by harvesting equipment; and sunflower seeds at bird feeders. Cardinals are not particularly fearful of humans. Once one landed on a log about 3 feet from where I was sitting. It furiously crushed a black beetle between its mandibles, discarded with a shake of its head the beetle's wing sheaths and spiny legs, swallowed the beetle, defecated, and flew off: not just a flash of pretty color, I found myself thinking, but a fearsome predator in its own right.

Cardinals begin calling in February and March, signaling the onset of the breeding season. Males and females sing equally well. The song is a series of clear whistled notes, *whoit whoit whoit* (like a child learning to whistle) or *wacheer wacheer*. Cardinals often countersing: one individual sings a song and another individual responds by matching it. Males on neighboring territories countersing to settle boundary disputes; when males and females countersing, the behavior strengthens the pair bond. As another part of courtship, a male will pick up a bit of food (such as a sunflower kernel at a feeder) in his bill and sidle up to his mate; the two touch beaks as she accepts the morsel.

It takes the female three to nine days to build the nest, a loose cup woven of twigs, vines, leaves, bark strips, and rootlets, lined with fine grasses or animal hair. Nests, rarely higher than 6 feet, are often placed in the thickest, thorniest shrub on the pair's territory.

The female lays two to five eggs (commonly three or four), which are whitish and marked with brown, lavender, and gray. She does most of the incubating, and the male brings her food. The young hatch after about twelve days. Their parents feed them regurgitated insects at first, then whole insects. The young fledge after ten days; the male may continue to feed them for a few days while the female builds another nest and begins a second clutch. Cardinals can produce up to four broods per year. Nest predators include snakes, crows, blue jays, house wrens, squirrels, chipmunks, and domestic cats. Cardinals compete with gray catbirds for food and nest sites; catbirds usually dominate in these interactions and may force cardinals to the fringe of a usable habitat.

In fall, the pair bond weakens between male and female. They stay together, however, and may join with other cardinals to form feeding groups that usually number six to twenty birds. In winter, white-footed mice sometimes move into old cardinal nests, stuff the cups with plant matter, and set up housekeeping. Cardinals are preyed on by hawks, owls, and foxes and other ground predators. The longevity record is fifteen years.

Cardinal populations rose steadily in Pennsylvania through the twentieth century. Several factors may have helped *Cardinalis cardinalis* overspread the state during that period: an increase in edge habitats caused by expanding farms and rural development; a period of warm winters in the early 1900s; a similar warming trend in the 1980s and 1990s; and an increase in backyard feeding stations dispensing high-energy seeds that help cardinals and other birds survive frigid weather.

Rose-Breasted Grosbeak *(Pheucticus ludovicianus).* My old friend and neighbor Wayne Harpster, an excellent self-taught naturalist, was a staunch fan of the rose-breasted grosbeak: no thrush could hold a candle to that rich singing, he said, and anyway, the rose-breasted was the handsomest bird in the woods. The male of the species has a black head, a massive ivory-colored bill *(grosbeak* means "big beak"), white patches on black wings that flash like semaphore signals when the bird flies, and a triangular red patch on the white breast. (The patch varies in size and shape from one individual to the next.) Adults are about 8 inches long. The female looks like a big brown sparrow. The song, given by both sexes, is somewhat like a robin's, but quicker, mellower, and sweeter.

Rose-breasted grosbeaks breed from Nova Scotia to western Canada and south in the Appalachians to Georgia. The species is statewide in Pennsylvania: scarce in the developed and agricultural southeast, abundant across the northern tier. Grosbeaks favor second-growth deciduous or mixed woods and can also be found in old orchards, parks, and suburban plantings. They eat insects (about half of the diet in summer), seeds (easily crushed by that formidable bill), tree buds and flowers, and fruits.

Males arrive on the breeding grounds in April and May, about a week ahead of the females. Males sing to proclaim a 2- to 3-acre breeding territory and may attack other males who intrude. When courting a female, the male takes a low perch or lands on the ground, then droops his wings and quivers them, spreads and lowers his tail, and slowly rotates his body from side to side while singing. Rose-breasted grosbeaks often nest in thickets along the edges of roads, streams, or swamps. The nest, built mostly by the female, is loose, bulky, and made almost entirely of twigs. It is usually 10 to 15 feet above the ground in a small tree or shrub. Since the pair do much calling (a short, metallic *chink* is often given) and singing in the vicinity, the nest is fairly easy to find.

The three to five eggs (typically four) are pale greenish blue, blotched with browns and purples. Both parents share in incubating them, and the eggs hatch after about two weeks. Both parents feed the young, which leave the nest nine to twelve days after hatching. A female may desert her first-brood young while they are still in the nestling phase; the male assumes care of the brood, while the female starts building a second nest, often less than 30 feet away from the first. Adults molt in August, and the male's new plumage includes brown and black streaks on the back, neck, and head. In September, rose-breasted grosbeaks start migrating to wintering grounds in Central and South America.

Blue Grosbeak *(Guiraca caerulea)*. Like the cardinal, this is a southern species that expanded northward during the last century. In the 1980s, blue grosbeaks were found nesting in southern Fulton, Lancaster, and Chester counties and along the border of Delaware and Philadelphia counties near the John Heinz National Wildlife Refuge at Tinicum. Males are a deep dusky blue; females are brown and sparrowlike. Blue grosbeaks inhabit open areas with scattered trees, fencerows, roadside thickets, reverting fields, brush, and forest edges. They often feed on the ground and eat many insects as well as the seeds of weeds, grasses, and other plants. Breeding males sing from treetops and utility wires. The female builds the nest, a compact cup, 3 to 10 feet above the ground in a shrub. The usual brood is four. Blue grosbeaks winter mainly in Mexico and Central America.

Indigo Bunting *(Passerina cyanea).* The indigo bunting breeds throughout the East and in parts of the Midwest and Southwest. The species is statewide and common in Pennsylvania. Adults are about 5.5 inches long, slightly smaller than a house sparrow. The male is bright blue, although he may look almost black in deep shade; the female is drab like a sparrow. Indigo buntings feed on the ground and in low bushes. They eat many insects, including beetles, caterpillars, and grasshoppers, along with grass and weed seeds, grains, and wild fruits.

Males arrive in the north in late April and May, with older males preceding younger ones and returning to their territories of past years. The 2- to 6-acre territories are in brushy fields, woods clearings, woods edges, and along roadsides and powerline rights-of-way. Males make display flights along the territorial boundaries, flying slowly with their wings fanned and tail and head held up, using rapid, shallow wingbeats while sounding a bubbly song. They also perch and broadcast a more complicated song: a series of high, whistled notes described as *sweet-sweet-chew-chew-seer-seer-sweet.* Females, by contrast, are so shy and retiring that it's often hard to determine when they've arrived on the breeding range.

The male spends much time singing from prominent places and little time helping with brood rearing. The female builds a neat, cup-shaped nest out of leaves, dry grasses, bark strips, and other plant materials, 1.5 to 10 feet up (usually around 3 feet) in a dense shrub or a low tree, often an aspen. She lays three to four eggs, which are white or bluish white and unmarked. She incubates the clutch until the eggs hatch after twelve or thirteen days. Some observers report that the male helps to feed nestlings, while others say that he does not help or that he gives food to the female, who then carries it to the nest. Sometimes a male will have more than one mate nesting in his territory.

Young indigo buntings leave the nest ten to twelve days after hatching. In some cases, males take over the feeding of newly fledged young while females start a second brood. Males keep singing well into August. Most pairs raise two broods. Brown-headed cowbirds may parasitize the nests, and various predators—particularly blue jays—eat eggs and nestlings.

Some researchers believe that only 30 to 50 percent of indigo bunting nests are successful.

The adults molt in August. The male in his winter plumage looks much like the female, but he retains blue streaks in his wings and tail. Buntings migrate south from late August through October. Many individuals cross the Gulf of Mexico, reversing their spring passage. Indigo buntings winter in loose flocks in southern Florida, Central America, and northern South America. The longevity record is ten years.

Dickcissel *(Spiza americana)*. The dickcissel is a bird of the prairies and a common resident of the Midwest. The male dickcissel looks a bit like a small meadowlark, with a yellow breast and a black bib; the female looks like a pale sparrow. A rare breeding species in Pennsylvania, *Spiza americana* has nested in Clarion, Westmoreland, Somerset, Fayette, Franklin, Cumberland, and York counties, mainly on reclaimed strip-mine sites and also on cut hay-fields, especially in years when drought stunts the regrowth of grasses. Nests are on or near the ground, hidden in dense grass, weeds, or a shrub. Dick-cissels forage on the ground and in low vegetation, picking up insects and seeds. The dickcissel is classified as a threatened species in Pennsylvania.

Osborne, J. *The Cardinal*. Austin, TX: University of Texas Press, 1992.

BLACKBIRDS, COWBIRD, ORIOLES, AND STARLING

Except for the European starling, the birds described in this chapter are all blackbirds, a group found only in the Americas. The introduced starling is covered here because starlings often join feeding flocks containing several kinds of blackbirds. In the Northeast, blackbirds live mainly in open areas such as marshes, fields, and woods edges. Some blackbirds are drab; others are brightly colored. Most species are social, living in flocks outside of the nesting season.

Blackbirds eat mainly insects in summer and seeds in winter. Orioles prefer to eat berries; grackles consume a range of foods, including the eggs and nestlings of other birds. Many blackbirds employ a feeding technique called gaping, in which an individual sticks its bill into a crevice or vegetation or beneath a rock or a stick, then suddenly opens its mandibles to push aside or pry away the screening object, in hopes of exposing something to eat, such as an insect, spider, or seed. Blackbirds have a wide range of nesting habits: some species place their nests on the ground, while others build them in marsh vegetation or trees. The starling nests in cavities. The brown-headed cowbird does not build a nest at all but lays its eggs in the nests of other birds.

Bobolink *(Dolichonyx oryzivorus)*. Bobolinks breed across southern Canada and the northern United States. The males are black, with white on the back and yellow on the nape of the neck; the females look like large sparrows. Bobolinks feed on beetles, grasshoppers, caterpillars, ants, other insects, millipedes, spiders, seeds of weeds and grasses, and grain. They nest on the ground in moist meadows and fields of hay, clover, alfalfa, or weeds. The adults land away from the hidden nest and walk to it. Most clutches contain five or six eggs. In Pennsylvania, bobolinks nest most successfully in the northwest and northeast on farmland at high elevations, where cool spring and early-summer temperatures retard hay growth and delay cutting until after broods have fledged. Bobolinks start their southward migration in August and September; en route, flocks may damage rice fields in the South. Most bobolinks cross the Caribbean and winter in South America.

Red-Winged Blackbird *(Agelaius phoeniceus)*. Many ornithologists believe the red-winged blackbird is the most populous bird in North America. The species breeds across the continent and as far south as Costa Rica and the Caribbean islands. Adults are 7 to 9 inches long. The jet black male has on each shoulder a vivid red patch, or epaulet, bordered below by a stripe of yellow; females and juveniles lack the epaulets and are drab brown with darker streaks. The male's song is a bubbling *ook-a-leee,* and both sexes sound a harsh *chack* as an alarm note.

Red-winged blackbirds arrive on the breeding grounds in late February and early March, with the males preceding the females by a week or two. They inhabit cattail marshes, swamps, wet meadows, pastures, and hayfields; individuals may temporarily leave their home territories to feed in nearby fields. In summer, red-winged blackbirds eat dragonflies, mayflies, caddisflies, midges, mosquitoes, caterpillars, beetles, grasshoppers, cicadas, and many other insects. In fall and winter, they turn to seeds, which make up about three-quarters of the annual diet. They consume seeds of grasses and weeds, and grains dropped by farm machinery. Flocks of red-winged blackbirds may damage corn, wheat, oats, barley, rice, and sunflower crops.

Male red-winged blackbird, above, and female below.

Adults usually breed within 30 miles of where they were hatched. In spring, the males perch prominently, displaying their epaulets and calling to attract females and to intimidate other males. When venturing across or into other territories to feed, males hide their epaulets by covering the red with adjoining black feathers, making it less likely that they will be attacked by resident males. Each male guards a breeding territory of up to 0.25 acre; within this area, one to several females will nest. A male may mate with several females, and a female may mate with more than one male. Females first breed when they're one year old. Yearling males often do not breed, although they continually try to take over older males' territories; sometimes yearlings displace reigning males, but more often they must wander about until a territory opens up after its owner is killed by a predator.

Red-winged blackbirds nest in loose colonies. They aggressively attack crows and hawks and drive them out of the area. Males do not help with nest building. Females attach their open-cup nests to cattail stalks or other marsh vegetation, or place them in low trees near or over the water; in hayfields and upland sites, females hide their nests in grass, weeds, or shrubs. Each female lays three or four pale bluish eggs, which are blotched with browns and purples. Incubation takes ten days to two weeks. Both parents feed insects to the hatchlings, and the young leave the nest after about two weeks. In the Northeast, most females raise one brood per year, renesting if a predator destroys an early clutch. Nest predators include crows, marsh wrens, raccoons, and minks.

In winter, red-winged blackbirds often feed alongside grackles, cowbirds, starlings, and robins. Red-winged blackbirds usually fly between food sources in long, strung-out flocks. At night, they roost communally, the males grouped separately from the females. Most individuals winter in the southeastern United States, with huge concentrations in the lower Mississippi Valley. In times past, red-winged blackbirds were more limited to wetlands; the population increased after the species began nesting in agricultural areas. About 40 percent of adults perish each year. A typical life span is two to four years.

Eastern Meadowlark *(Sturnella magna)*. Males and females have a brown-streaked back and a bright yellow breast with a prominent black V; the outer tail feathers are white. Meadowlarks live in pastures, hayfields, fallow fields, and strip mines that have been replanted to grass. In summer, they eat grasshoppers, crickets, beetles, ants, caterpillars, and many other insects; they also eat seeds and waste grains. Males arrive in the spring two to four weeks before the females and stake out territories, which average 7 acres. The males perch on phone poles, trees, and fence posts, singing a sweet, slurred, whistling song. Sixty to 80 percent of males have two or three mates. The female builds a ground nest in grass or weeds 10 to 20 inches high; the nest, usually hidden in a slight depression, is made of dry grasses with a woven, dome-shaped roof and a side entry.

Females lay eggs from late May through June. The early mowing of hayfields destroys many nests. The three to five eggs are white, heavily blotched with brown. The female incubates her clutch for about two weeks, and after the young hatch, both parents feed them insects. Fledglings leave the nest after ten to twelve days and are fed by their parents for another two to four weeks. Some females raise two broods over the summer. In August, meadowlarks abandon their breeding territories and forage in small flocks. In September and October, most shift southward, migrating at night and feeding during the day. Some meadowlarks winter in the southern half of Pennsylvania (where they forage in stubble fields over which manure has been spread), but most individuals go farther south. The population has declined in the Northeast as development has wiped out agricultural land and formerly farmed areas have grown up into brush and woods.

Common Grackle *(Quiscalus quiscula).* Grackles are sleek, black birds with purple, green, and bronze highlights in their plumage. Adults are about 1 foot in length and have long, wedge-shaped tails. Grackles live in suburbs, towns, farming areas, and streamside groves statewide in Pennsylvania. They forage mainly on the ground and eat insects (beetles, grubs, grasshoppers, caterpillars, and others), millipedes, spiders, earthworms, crayfish, minnows, frogs, the eggs and young of other birds, and even small rodents. In spring, males display in front of females by raising their bills, fluffing out their feathers, spreading their tails, and sounding a loud, ascending *reedeleek.*

Unlike most other songbirds, grackles remain social throughout the year. Most nest in colonies of ten to thirty pairs, usually in evergreen trees, where mated pairs defend only a small area right around their nest. Grackles breed from April into July. The female builds a cup-shaped nest out of grasses and mud. The typical clutch has four or five eggs. Only the female incubates, and the eggs hatch after twelve to fourteen days. Both parents feed the young, which fledge after sixteen to twenty days. In the fall, grackles roost in large flocks along with starlings, red-winged blackbirds, and cowbirds. Most grackles winter to the south of Pennsylvania, but some stay on in the state.

Brown-Headed Cowbird *(Molothrus ater)*. The brown-headed cowbird is a bird of farms, fields, and woods edges. Males have black bodies and brown heads; females are brownish gray. Seeds of grasses and weeds, plus waste grains, make up about half of the birds' diet in summer and more than 90 percent in winter. Cowbirds also eat insects, particularly grasshoppers, beetles, and caterpillars. I have watched our own small resident flock walking along behind my wife's Icelandic horses in our 3-acre pasture, darting this way and that to nab insects kicked up by the grazing equines. In the past, cowbirds followed bison herds on the Great Plains, where they were known as "buffalo birds."

In the spring, the male cowbird displays for females by fluffing up his body feathers, spreading his wings and tail, and singing a bubbly *glug-glug-gleee*. The species builds no nest. The cowbird is a brood parasite: the female lays eggs in the nests of other birds, which, guided by their instincts, raise the young cowbirds as their own. Ornithologists believe that cowbirds did not live in forested Pennsylvania before European settlement, a theory bolstered by the fact that few of our native songbirds have evolved defensive behaviors against its parasitism. Today cowbirds are common breeders statewide in Pennsylvania, mainly in farmland and in areas where development has fragmented the forest, giving cowbirds access to the nests of woodland birds. *Molothrus ater* has been reported to parasitize over 220 different species. In the Northeast, cowbirds particularly plague warblers, vireos, flycatchers, finches, thrushes, and sparrows.

A female cowbird will sneak in to a nest that is temporarily unoccupied, quickly lay an egg, and fly off, sometimes after removing or eating one of the host's eggs. Cowbird eggs are whitish with brown and gray spots. Young cowbirds, hatched and fed by the host parents, grow rapidly; they monopolize food and may crowd the other young out of the nest. Juvenile cowbirds fledge ten to twelve days after hatching. Nests with two or more cowbird eggs often do not fledge any host young. In one study, a successfully raised cowbird caused a reduction in the brood of a host pair by approximately one fledgling. Other ornithologists cite cowbird predation as a major factor—along with habitat loss—in declines of many species, including the wood thrush. A female cowbird may lay up to forty eggs in one season; of these, two or three will yield young that ultimately mature to adulthood.

Cowbirds migrate in large flocks in spring and fall. They winter mainly in the southern states and in Central America, often sharing huge winter roosts with starlings and with other blackbirds.

Orchard Oriole *(Icterus spurius)*. The adult male is chestnut and black, and the female is olive and yellow. This robin-size oriole inhabits open areas with scattered large trees, including parks, old orchards, and shade groves; it avoids deep woods. The species breeds most commonly across the southern part of Pennsylvania. Orchard orioles feed on insects, berries, nectar, and flowers. Pairs are thought to be monogamous. The female builds a hanging, basket-like nest among dense leaves in a tree, usually 10 to 20 feet above the ground. The three to seven eggs are incubated for twelve to fifteen days. Both parents feed the young, which leave the nest about two weeks after hatching. Brown-headed cowbirds often parasitize orchard oriole nests. Long-distance migrants, orchard orioles winter in Mexico and Central America.

Baltimore Oriole *(Icterus galbula)*. The male Baltimore oriole has a brilliant orange body and a black head: the species' name arose because orange and black were the heraldic colors of Lord Baltimore, an English colonist and a founder of Maryland. The female Baltimore oriole is yellow-orange. Baltimore orioles breed throughout eastern North America (and statewide in Pennsylvania) in open woods, residential areas, parks, fencerows, and tall trees along streams, often sycamores or willows; formerly elms were a favorite before a disease epidemic killed most American elms. Adults feed on insects, particularly caterpillars; spiders; snails; berries, including mulberries, serviceberries, and blackberries; cultivated fruits; and flowers. Baltimore orioles visit feeding stations for sugar water and pieces of fruit.

The species is best known for its sacklike hanging nest, intricately woven by the female out of plant fibers, pieces of string, grapevine bark, and grasses. A central chamber is lined with hair, fine grasses, and cottony plant matter. Nests are usually hung at the ends of pliant branches, probably to deter predators, which include snakes, blue jays, and crows. Females lay three to six eggs that hatch after twelve to fourteen days. Both parents feed the nestlings, which leave the nest after two weeks. Flocks depart from the breeding range quite early, in July and August. The species winters in southern Mexico, Central America, and northern South America, where the birds feed on insects and nectar.

European Starling *(Sturnus vulgaris)*. From a hundred birds released in the 1890s in New York City's Central Park have descended the more than two hundred million starlings populating North America today. Starlings are chunky birds with short tails and long, straight bills; airborne, they show a distinctly triangular body shape. The plumage is black with iridescent highlights. Starlings are adaptable, hardy, and wary. They thrive in farmland, suburbs, cities, and woods edges and are absent from marshes and extensive forests. Starlings eat almost equal amounts of animal and plant food, including earthworms, beetles, grasshoppers, ants, flies, caterpillars (gypsy moth and tent caterpillars are frequent prey), seeds, grains, berries, and wild and cultivated fruit. When foraging on lawns, starlings often probe their bills into the soil and pry apart grass roots to get at beetle larvae.

Starlings begin to defend nest cavities in late winter, preempting them before native cavity nesters start breeding. Starlings nest in woodpecker holes, crevices in trees and buildings, and birdhouses. In April, males perch outside the cavities; when they see other starlings, they sing and windmill their wings to attract a mate. The male's song includes shrill squeals, squawks, and imitations of other birds. The female fills the nest cavity with grasses, weed stems, twigs, old cloth, and dry leaves, and lines a central cup with fine grasses and feathers. She lays four to six unmarked pale bluish green eggs. Both parents incubate the eggs, which hatch after about twelve days. The nestlings are fed by both parents and leave the nest three weeks after hatching. By now their droppings have so fouled the cavity that the adults must search out (or take over) another nest hole in which to rear a second brood: often they drive native birds from their nests, including woodpeckers, nuthatches, great crested flycatchers, tree swallows, house wrens, and bluebirds. Harassment by starlings may be causing declines in populations of the northern flicker and red-headed woodpecker.

Starlings feed in flocks and roost together at night. In late summer and fall, their roosts may contain thousands of birds. Some individuals shift southward for the winter, while others remain in the Northeast; many roost in cities, where buildings give off heat, and then fly out into the surrounding agricultural land to feed during the day. Winter roosts can be huge and noisome. Of the starlings that are alive in January, about half die in the coming year, with one-third of the deaths happening in January and February. The average adult lives for one and a half years.

Skutch, A. F. *Orioles, Blackbirds, and Their Kin.* Tucson, AZ: University of Arizona Press, 1996.
Jaramillo, A., and P. Burke. *New World Blackbirds.* Princeton, NJ: Princeton University Press, 1999.
Ortega, C. P. *Cowbirds and Other Brood Parasites.* Tucson, AZ: University of Arizona Press, 1998.

FINCHES AND HOUSE SPARROW

Finches are small to medium-size songbirds whose sturdy bills let them crack open the tough hulls of seeds, their main food. The house sparrow was introduced from Europe, and the house finch is a western species liberated in the Northeast whose population has exploded in recent years. Finches are sociable birds, and outside of the breeding season they gather in flocks. They feed on the ground and in tall weeds, shrubs, and trees; many of these birds visit bird feeders. Even during summer, when insect populations burgeon, many finches continue to eat seeds and nourish their young with a pulp composed of regurgitated seeds.

Male finches sing to attract females and to maintain the pair bond. In most species, the female builds a cup-shaped nest hidden in the thick foliage of a tree. The house sparrow nests in cavities. Female finches do most or all of the incubating of eggs, and males and females team up to feed the young. In addition to our breeding species, several other finches breed farther to the north and sometimes winter in the Northeast.

Purple Finch *(Carpodacus purpureus)*. Don't look for a purple bird when trying to pick out this species: the male purple finch is a burgundy or raspberry color, and the female is brown with darker streaks. The species breeds across Canada and in the Northeast south to West Virginia. In Pennsylvania, purple finches nest mainly in the northern tier, and in winter, individuals from farther north overspread the state. Purple finches inhabit conifer plantations (including Christmas tree farms), spruce bogs, hillside pastures, woods edges, and mixed and open woods. In winter, they eat weed, grass, and tree seeds (including those of elm, ash, sycamore, and tuliptree); in early spring, they consume buds and flowers of trees and shrubs; they take some insects in late spring; and they concentrate on fruits in summer.

The male has a melodious warbling song. The female builds a nest 15 to 20 feet above the ground on a horizontal branch, usually in a conifer; she weaves a compact open cup out of twigs, weeds, rootlets, and strips of bark, lining it with fine grasses or animal hair. The three to five eggs are a pale greenish blue, dotted with black and brown. The female incubates them for around thirteen days. Both parents feed the nestlings, mainly with seeds, and they fledge about two weeks after hatching. In the East, only one brood is raised per year. In winter, purple finches, American goldfinches, and pine siskins may join together in mixed-species foraging flocks. At feeding stations, house finches and house sparrows dominate purple finches and often drive them away. Purple finches winter as far south as Florida.

House Finch *(Carpodacus mexicanus).* House finches in the eastern United States descend from birds released in New York City in 1940. The species is native to the U.S. Southwest; today *Carpodacus mexicanus* breeds from coast to coast. Females are sparrowlike, and males range in color from pale yellow to bright red, with streaking on the breast. The red coloring in both the house finch and purple finch comes from beta-carotene, a pigment found in many plants, particularly in red fruits; the red blush to the plumage intensifies as the males age. House finches live in cities, suburbs, and farms. They feed on seeds, flowers, buds, berries, small fruits, and insects.

Pairs often form within flocks during the winter. Males do not stake out territories but instead defend areas around their mates. House finches begin nesting as early as March and produce two or more broods per year, each with four or five young. Females nest in a variety of sites, including conifers, ivy on building walls, abandoned nests of other birds, and above porch lamps and in hanging flower baskets. The population of this western species is expanding in the East, with winter survival possibly aided by backyard bird feeders. House finches seem to outcompete house sparrows for food, habitat, and nest sites, contributing to a decline in the house sparrow population since the 1960s.

Pine Siskin *(Carduelis pinus).* With their brown colors and streaked breasts, pine siskins look like sparrows; patches of yellow in the wings and tails are good field identifiers. Pine siskins nest in New England and Canada and in scattered sites southward in the Appalachian Mountains. In Pennsylvania, they breed irregularly, mainly in the northern tier and the high mountains, nesting in stands of hemlocks, pines, spruces, and larches and in ornamental conifers in backyards. These rather tame birds become much more visible when they flock to feeding stations in winter. As well as eating seeds put out by people, siskins consume the seeds of trees (alder, birch, spruce, and others), weeds, and grasses. They also eat buds, flower parts, and some insects. They usually forage in flocks, even during the nesting season; in winter, they often keep company with goldfinches. In some years, many siskins winter in the Keystone State; in other years, few appear.

American Goldfinch *(Carduelis tristis).* The male goldfinch in summer is one of our most conspicuous birds: bright yellow, with black wings and a black forehead. The female is a dull olive-gray. In winter, both sexes look like the summer female. Gregarious birds, goldfinches often fly in groups, showing a characteristic bouncing or undulating flight pattern: bursts of wingbeats followed by short glides during which the birds lose a few feet of height. While airborne, flock members sound a *per-chickoree* call. American goldfinches nest across North America and statewide in Pennsylvania. They forage in a variety of habitats, including brushy areas, roadsides, open woods, woods edges, and suburbs.

In the spring, goldfinches eat seeds, insects, and insect eggs. In summer, they turn mainly to the seeds of thistles, dandelions, rag-

The breeding plumage of the male American goldfinch is an eye-catching mix of black, white, and bright yellow.

weeds, sunflowers, and grasses. They eat elm seeds, birch and alder catkins, flower buds, and berries. They clamber around in weeds and shrubs, picking out seeds. In winter, a flock may seem to roll across a field as birds in the rear leapfrog over other flock members on the group's leading edge: this strategy gives each individual access to fresh foraging areas while requiring only a short flight to get there.

In April and May, goldfinches move in from the south, returning to breed in the areas where they hatched; they remain in flocks and do not set up territories until late June or early July. Goldfinches start nesting later in the season than any other bird in the Northeast; perhaps breeding occurs late so that the young hatch when seeds are maturing on favorite food plants, particularly thistles. Males claim territories that may be small (100 feet in diameter where goldfinches nest in loose colonies) or as large as a quarter of an acre. The male sings from a perch, voicing a clear, canarylike song, and makes high, circling flights. Goldfinches often nest in thornapples, shrub willows, and gray dogwood. The female builds a neat cup lined with thistle or cattail down, 4 to 14 feet up in a horizontal or upright fork of a small tree or shrub. The nest is woven so tightly that it will hold water; it is flexible and expands as the young increase in size. The female lays four to six pale bluish eggs. She incubates the clutch, with the male bringing her food, and the young hatch after twelve to fourteen days. Their parents feed them mainly on seeds. The young fledge from the nest after another eleven to seventeen days. Some pairs raise a second brood, and fledglings have been found as late as September. Cowbirds sometimes lay eggs in goldfinch nests, but the young cowbirds often die because they do not get enough protein from the regurgitated seeds that goldfinch parents feed to their nestlings.

House Sparrow *(Passer domesticus).* Although we call it a sparrow, this common bird is actually a species of weaver. House sparrows have spread out from Eurasia to live with humankind around the globe. People introduced them in North America between 1850 and 1886 in an attempt to control insect pests, particularly the elm spanworm caterpillar. At first, the bird was called the English sparrow, because most imports were brought from England. Male house sparrows have black chin and breast patches (the amount of black varies among individuals), white cheeks, and a chestnut nape. Females are a dingy brown.

House sparrows live year-round on most of the species' continent-wide range. Never far from humanity, they inhabit cities, suburbs, towns, rural areas, and farms. They eat weed and grass seeds, waste grain, chicken feed, insects and spiders (about 10 percent of the diet), fruit tree buds, flowers, and garbage. They nest in protected places such as holes in trees and buildings, on porch and barn rafters, behind shutters and awnings, in bluebird houses, and in ivy growing thickly on the sides of buildings. Sometimes they destroy the eggs and young of native cavity nesters. House sparrows use their nests for shelter during most of the year. Both sexes work at lining the cavity with grass, weeds, feathers, and trash. Pairs are monogamous and mate for life; prolific breeders, they produce two or three broods of three to seven young annually. Recently fledged juveniles form flocks in the summer and are joined by adults after the breeding season ends in August and September. In late fall, pairs return to their nest cavities.

When house sparrows overran the United States in the late 1800s—ousting native breeders, fouling buildings with their droppings, and offending people with their aggressive, noisy habits—those who had championed the species' introduction were roundly castigated. The population of *Passer domesticus* peaked in the early twentieth century, and it has fallen since then. Several factors may be involved. Tractors and automobiles have replaced horses, and farming operations have been sanitized, so that grain is no longer widely available in winter. The house finch, a bird of western America accidentally introduced in the East in 1940, competes with the house sparrow for food and territory.

Winter Finches. Four finches breed in the far north and visit the Northeast in winter, when they descend on feeding stations in people's yards. In some years, many finches invade our area; in other years, they stay to the north. Ornithologists believe that finches come south when key food sources, particularly the seeds of conifers, fail in their boreal homes.

Red crossbills *(Loxia curvirostra)* and white-winged crossbills *(Loxia leucoptera)* have oddly shaped bills, the tips of whose mandibles cross. A bird will stick its bill between the scales of a spruce cone, then open the mandibles, prying apart the scales; the bird lifts out the exposed seed with its tongue. The male red crossbill is brick red in color, and the female is a mix of olive-gray and yellow. The white-winged crossbill has white wingbars in both sexes; the male is a rosy pink, and the female is colored much like the red crossbill female. Both types of crossbills eat the seeds of various conifers, and they also feed on buds and weed seeds. In the years when they winter in Pennsylvania, they may arrive with cold fronts in late October and November. Red crossbills are rare nesters in Pennsylvania.

The common redpoll *(Carduelis flammea)* has a red forehead and a black chin. It is the size of a goldfinch. Redpolls feed actively in brushy and weedy fields and along woods edges, picking up seeds of trees, weeds, and grasses. Often they forage in mixed flocks with pine siskins and goldfinches.

The evening grosbeak *(Coccothraustes vespertinus)* is a big, husky bird. The male is dull yellow with prominent white wing patches, and the female is yellowish gray; the massive bill is white in both sexes. Wintering flocks wander widely in search of food, although a feeding station frequently restocked with sunflower seeds will hold them in one area. Evening grosbeaks forage in mixed woodlands, coniferous forests, towns, and suburbs. At bird feeders, they often displace one another, as well as the local birds, giving strident chirping calls and putting on aggressive displays while competing for food.

AMPHIBIANS
AND REPTILES

Like birds and mammals, the amphibians and reptiles are vertebrates—animals with backbones. Unlike birds and mammals, amphibians and reptiles cannot internally control their body temperature and are said to be cold-blooded (biologists prefer the term "ectothermic"), meaning that an individual's temperature depends on that of its surrounding environment. To remain active, an amphibian or reptile may need to move about, seeking cool or warm places so that it can keep its temperature within a range in which the animal can function. Most amphibians and reptiles hibernate in winter, and some estivate (become dormant) during the heat of summer.

The amphibians include salamanders, frogs, and toads. Most amphibian species have a two-stage life cycle that begins in the water, where adults lay eggs covered with a protective jelly. The eggs hatch into aquatic larvae, or tadpoles, that obtain oxygen from the water by breathing through gills, like fish. Later the larvae metamorphose: they transform into air-breathing juveniles that either remain in the water or emerge to live on land. Although the adults of many species have lungs, they retain an ability to breathe underwater by drawing in oxygen through their skin. In most species, the larvae eat microscopic plants, and the juveniles and adults feed on insects and other invertebrates. Unlike reptiles, amphibians do not have claws on their feet or skins that are protected by scales. Thirty-seven species of amphibians live in Pennsylvania, including twenty-one salamanders and sixteen frogs and toads.

Snakes, lizards, and turtles are reptiles. Reptiles have adapted to life on dry land. Their skin has scales or plates to protect the body and slow the evaporation of fluids. Adults lay eggs with tough protective shells or give birth to live young. All of the reptiles—including juveniles and species that spend much of their time in the water—breathe through lungs. Although some of the turtles eat vegetable foods, most reptiles are meat eaters, feeding on insects, small mammals, birds, fish, and carrion. Pennsylvania has thirty-nine reptile species: twenty-one snakes, four lizards, and fourteen turtles.

Biologists believe that amphibians arose some thirty million years ago, and reptiles appeared about fifty million years later. Birds and mammals both evolved from early reptilian lines. The study of amphibians and reptiles is known as herpetology. The word derives from the Greek *herpein,* to creep—which, of course, is the way most amphibians and reptiles move through the world. Amphibians and reptiles are sometimes referred to as herpetofauna or simply as "herps."

Today many reptile and amphibian species seem to be dwindling, probably because of human-caused changes to the environment. Acid rain and the use of pesticides and herbicides may interfere with reproduction in some species. New highways, suburban sprawl, pollution of waterways, and the draining of wetlands all modify or destroy natural habitats. Herpetologists do not know how much habitat degradation can occur, or how low a population can fall, before a species is wiped out in a given area.

Currently, Pennsylvania has six state endangered species of reptiles and amphibians, three threatened species, and three species under consideration for threatened or endangered status. Three others—the eastern mud turtle *(Kinosternon subrubrum)*, the midland smooth softshell turtle *(Trionyx muticus)*, and the eastern tiger salamander *(Ambystoma tigrinum)*—apparently have died out in Pennsylvania since European settlement.

Green, N. B., and T. K. Pauley. *Amphibians and Reptiles in West Virginia*. Pittsburgh: University of Pittsburgh Press, 1987.

Shaffer, L. L. *Pennsylvania Amphibians and Reptiles*. Harrisburg, PA: Pennsylvania Fish and Boat Commission, 1995.

Hulse, A. C., C. J. McCoy, and E. J. Censky. *Amphibians and Reptiles of Pennsylvania and the Northeast*. Ithaca, NY: Cornell University Press, 2001.

Mitchell, J. C. *The Reptiles of Virginia*. Washington, DC: Smithsonian Institution Press, 1994.

Ernst, C. H., and R. W. Barbour. *Snakes of Eastern North America*. Fairfax, VA: George Mason University Press, 1989.

Ernst, C. H., and R. W. Barbour. *Turtles of the United States*. Lexington, KY: University Press of Kentucky, 1972.

Pfingsten, R. A., and F. L. Downs. *Salamanders of Ohio*. Columbus, OH: Ohio Biological Survey, 1989.

Tyning, T. F. *A Guide to Amphibians and Reptiles*. Boston: Little, Brown, 1990.

HELLBENDER AND MUDPUPPY

These huge, grotesque salamanders look like they just crawled out of the primordial slime—or somebody's screaming nightmare. The eastern hellbender and the mudpuppy spend their lives in the water. They are classified in separate families, in the group known as the giant salamanders. Hellbenders and mudpuppies are more common than is generally assumed, but because they're nocturnal and live in places out of view of most people, they are seldom seen. The hellbender inhabits cold running water of creeks and rivers. The mudpuppy is at home in slightly warmer waters of lakes, ponds, rivers, streams, and canals.

Eastern Hellbender *(Cryptobranchus alleganiensis).* I was canoeing on Tionesta Creek in northwestern Pennsylvania. The river flowed through a wooded valley, its stretches of deep water punctuated by riffles. At the end of a gentle rapids, I looked down and saw a creature hugging the bottom next to a large rock. I backpaddled to hold the canoe in place. What appeared to be a monstrous, blunt-headed, dun-colored salamander half crawled and half swam beneath the rock, and vanished. What had I seen? My mind flashed back to an illustration of a hellbender, in a guide to reptiles and amphibians, that had both fascinated and revolted me as a child. And I understood that I'd glimpsed the largest salamander found in Pennsylvania.

Hellbenders are olive gray to almost black, sometimes with dark blotches; the belly is paler than the back. Adults are 11 to 20 inches, although individuals as long as 29 inches have been found. A hellbender has a head that is flattened and appears rounded when seen from above and a rudderlike tail. The creature's antediluvian appearance is furthered by tiny, lidless eyes and prominent gill slits in the neck. Along each side of the body, between the fore and hind legs, is a crinkled, fleshy fringe: in this fold of skin are papillae, small projections through which gas exchange takes place. The hellbender is thought to rely more on these external respiratory structures than on its lungs for getting rid of carbon dioxide and obtaining oxygen from the water.

Hellbenders inhabit the Susquehanna and Ohio river watersheds in Pennsylvania; reports of hellbenders in the Delaware River drainage have

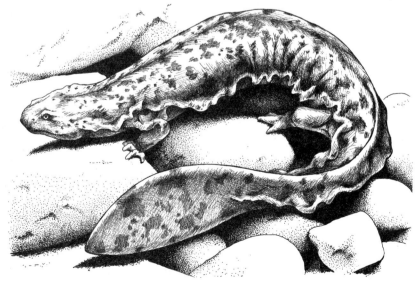

Despite its fearsome appearance, the hellbender is a docile creature.

not yet been confirmed. The species' range extends south and west through West Virginia and Ohio into Kentucky and Tennessee. Hellbenders favor large, quick-running streams and, in rivers, channels with strong, well-oxygenated flows. Their flattened body shape helps them stay immobile on the bottom, even in a strong current. A study conducted in the 1960s on French Creek in Crawford County found that hellbenders spent the day in crevices beneath large, slab-shaped rocks, in water generally 5 to 18 inches deep. The median activity radius for an individual was about 18 feet, and the home range was estimated at 120 square yards. Individuals defended the area beneath their home rocks against other hellbenders.

At night, hellbenders leave their dens and forage on the bottom among boulders, stones, and submerged logs, feeding on crayfish, worms, snails, small fish, fish eggs, and aquatic insects. Crayfish are the main food item. Sometimes a fisherman will catch a hellbender; when this happens, the creature should simply be freed back into the stream. Despite their ferocious appearance, hellbenders do not bite people. They are thought to be rather docile creatures.

Hellbenders breed in August and September. The male digs a shallow nest in the stream bottom, in which the female lays a string of two hundred to five hundred marble-size eggs; the male covers the eggs with seminal fluid. The young hatch after eight to ten weeks. The hatchlings are 0.5 inch long, shaped like their parents, and have external gills. The gills are absorbed after two years, when the hellbenders are about 5 inches long. Individuals mature sexually after three to four years, when they are about 13 inches long.

Dams, water pollution, and the discharge of silt can ruin streams for hellbenders. The species has declined in parts of its range, especially where acid mine drainage or industrial effluents have killed off crayfish populations. Individuals may live for up to thirty years in the wild. *Cryptobranchus* bones found in Native American middens excavated along the Ohio River imply that this salamander was eaten by the Woodland Indians. The origin of the name "hellbender" is obscure; I find the moniker almost as unsettling as the appearance of the beast itself. Another quirky name is the "Allegheny alligator."

Mudpuppy *(Necturus maculosus)*. Adults are 8 to 13 inches long, with a few ancient specimens reaching 17 inches; among salamanders, only the hellbender is larger. Seen by day, a mudpuppy is muddy gray to rusty brown, usually with small, scattered blue-black spots on the back; the belly is grayish, also with dark spots. By night, the creature may become pale, almost white. The body is long, the legs short, and the tail long. Most striking are the external gills, red and plume-shaped, fanning out on either side of the neck. Mudpuppies have these gills at all ages: in essence, they remain as larvae throughout their lives.

Mudpuppies live in clear, cool streams, backwaters of creeks and rivers, ponds, canals, reservoirs, and drainage ditches. Although basically nocturnal, they sometimes feed during the day among thick weeds or in muddy water. They hide under rocks and debris and beneath overhanging banks. In fall, they may gather in leafy beds in slack water. Mudpuppies remain active year-round. They crawl on the bottom, feeding on insects, crayfish, small fish, worms, and snails. When alarmed, mudpuppies swim rapidly, propelled by their bodies and tails. Young mudpuppies are eaten by predatory fish.

Mudpuppies mature sexually after five or six years. They mate in September and October. The male stimulates the female by slithering around, over, and under her. Following this courtship, the male deposits a spermatophore (a packet of sperm) on the stream bottom, which the female picks up inside her cloaca. The following spring, the female excavates a nest under a stone, board, or piece of debris submerged in running water. She attaches the eggs singly to the underside of the projecting object; clutches range from 30 to 125 eggs. The female may guard her nest. The eggs hatch after six to eight weeks.

In Pennsylvania, mudpuppies are widely distributed in the tributaries and the main stream of the Allegheny River; they have also been collected in the Shenango River. In North America, *Necturus maculosus* ranges from New York and southern Canada south to Tennessee. In the South, mudpuppies are often called "water dogs"; both names stem from an erroneous belief that these salamanders make barking noises. Sometimes fishermen hook mudpuppies. People think the mudpuppy is venomous; it is not. Like the hellbender, it is harmless to humans.

MOLE SALAMANDERS

These sturdy, medium-size salamanders are like moles in the way that they spend their lives under the soil or burrowed beneath rocks, logs, boards, or piles of damp leaves. Mole salamanders have prominent eyes, which are protected by lids, and five toes on each hind foot and four on each front foot. They belong to the genus *Ambystoma,* a North American group with thirteen species in the eastern United States, three of which are fairly common in Pennsylvania.

On humid or rainy nights, mole salamanders leave their subterranean hideaways to wander about searching for insects, spiders, and other invertebrate prey. In the spring, males and females migrate to water, often to vernal or ephemeral ponds, depressions that fill up with rain and melted snow in the spring and may be dry again by July or August. Sometimes the salamanders crawl across ice and snow to reach the ponds. The adult salamanders enter the water to court, mate, and lay eggs—activities that may be crowded into just a few nights or may stretch out over several weeks. An exception is the marbled salamander, which lays its eggs in autumn; the eggs hatch in the fall, and the larvae, which may remain active during winter, are ready and waiting to eat the young of other amphibians come spring.

The eggs of mole salamanders are covered with a thick protective gelatinous coating and are attached in various ways to underwater twigs and stems. After they hatch, the larvae resemble streamlined tadpoles. Unlike frog and toad tadpoles, which eat plant matter, salamander larvae prey on smaller creatures living in the breeding pools. Salamander tadpoles swim well. A high tail fin extends forward onto the back, and long, feathery gills project from the sides of the neck. Over a period of months, forelimbs sprout from the body, hind limbs emerge, and finally the gills vanish. This changeover to an adult form, known as metamorphosis, can be speeded up if drought starts drying out the home pool prematurely. After they have metamorphosed, juvenile mole salamanders crawl out of the pool and disperse onto the land.

Many mole salamanders are killed on roads when moving to and from their breeding sites. Salamander populations are particularly vulnerable to pollution and to acid rain, which lowers the pH levels in vernal ponds, leaching out heavy metals in the surrounding soil. The heavy metals interfere with the normal growth of larvae in the egg masses.

Jefferson Salamander *(Ambystoma jeffersonianum).* Four to 7 inches long, Jefferson salamanders are dark brown or gray. The belly is pale, and sometimes bluish flecks stand out on the limbs and the lower sides. The species is named after Jefferson College in Washington County, Pennsylvania, near which the first specimen of this type of salamander was collected in the early 1800s. The Jefferson salamander ranges from New England south to Virginia and Kentucky. In Pennsylvania, it is statewide in suitable habitats.

Jefferson salamanders live in moist deciduous forests, mainly in the uplands, near temporary or permanent ponds. A classic habitat would include steep, rocky slopes with rotting logs and deep layers of leaf litter. Adults spend most of the year underground in the burrows of small mammals or in masses of fallen leaves, where several individuals may cluster together. Jefferson salamanders feed on snails, worms, insects, spiders, and centipedes. The salamander snatches its prey with a quick snap of the jaws, further securing it with the sticky tongue; before swallowing, it may shred its victim by rubbing it against the teeth or the roof of the mouth.

Jefferson salamanders are among our first salamanders to breed in spring. When rains fall in March, these salamanders home in on a woodland pond, which may lie 250 to 650 yards from their summertime territories. Courtship takes place at night. The male grasps the female from behind; after letting go, he deposits a spermatophore—a gelatinous structure capped with sperm—on debris in the water. The female nips off the spermatophore with the lips of her cloaca, the cavity into which the genital, urinary, and intestinal tracts open, and fertilization takes place inside her body. Females lay up to two hundred eggs, in sausage-shaped masses of fifteen to thirty eggs each, attached to twigs and stems in the pond. After breeding, adults leave the pool and return via an essentially straight line to their summer retreats. Individuals use the same breeding ponds and terrestrial habitats year after year.

The eggs hatch after two to four weeks. The larvae feed on toad and frog tadpoles, larvae of other mole salamanders, and small fish, if present. The larval stage lasts for about three months, with newly transformed juveniles leaving the pond in July or August. Jefferson salamanders often share breeding sites with wood frogs, spring peepers, chorus frogs, and other amphibians.

Spotted Salamander *(Ambystoma maculatum).* On a rainy March night, standing knee-deep in a pond on our neighbors' land, I directed my flashlight downward and watched five big salamanders swimming in the cold, clear water. They were black, with prominent yellow spots on their backs, and their long tails waved back and forth as they writhed in a silent dance, swimming past and over and around one another, bumping noses into their fellows' sides. It was a revelation to see such beautiful, mysterious creatures, which heretofore I had not known lived so close at hand.

Spotted salamanders can reach 8 inches in length from the nose to the tip of the tail. They favor hardwood forests and also may live in mosaics of wooded, open, and lightly developed land. They feed on earthworms, snails, slugs, pill bugs, and various insects. The species inhabits the East from Canada to the Gulf coast states and occurs throughout Pennsylvania.

Ambystoma maculatum breeds in ephemeral ponds, floodplain swamps, marshes, bogs, beaver ponds, farm ponds, and backwaters of streams and small rivers. Males often precede females by several days to the breeding sites, and some studies imply that only about a third of the adult females

Spotted salamanders are best seen in early spring, when they gather to breed at woodland pools, swamps, farm ponds, and the backwaters of creeks and streams.

breed in any given year. The male deposits a spermatophore, which the female picks up in her cloaca. A female will lay up to two hundred eggs in a large, globular mass—sometimes clear, sometimes with a cloudy aspect—attached to a submerged twig. I've found over a dozen such egg masses in the 30-foot-diameter pond near my home. The eggs hatch after forty-five to fifty-five days, and the tadpoles feed on small crustaceans, such as water fleas; aquatic insect larvae; and other larval amphibians. Spotted salamander larvae are themselves taken by diving beetles, dragonfly naiads, larger salamander larvae, snakes, turtles, birds, and mammals.

Spotted salamanders metamorphose after about three months, when they're 1.5 to 3 inches long, although they can transform earlier and at a smaller size if the pond threatens to dry up. They cannot, however, adjust to acid rain, and reproductive success falls as vernal ponds become more acidic. Writes Floyd Downs in *Salamanders of Ohio:* "It seems a real possibility that continued acidification will entirely eliminate the spotted salamander from many of its breeding ponds."

Marbled Salamander *(Ambystoma opacum).* This 3.5- to 4.5-inch salamander is shiny black, with a series of pale silvery bands across its back. It ranges from New England to Florida and west to Illinois and Texas. Two populations occur in Pennsylvania: one group lives in the west, from Westmoreland and Indiana counties to Crawford County; and a second, larger concentration occupies the southeastern third of the state. The marbled salamander seems to be absent from the Allegheny Mountains. *Ambystoma opacum* lives in swamp forests and dry, deciduous woodlands, often near sand and gravel deposits and on well-drained rocky slopes. Marbled salamanders are more tolerant of drier habitats than are spotted and Jefferson salamanders. Adults hide under surface objects and burrow into the soil. They eat earthworms, snails, slugs, and insects.

Unlike the other mole salamanders, marbled salamanders gather to breed in the fall, mainly in September and early October. Males precede females to breeding areas: dried-up ponds, stream oxbows, wetlands, or depressions in the forest floor. After mating, the females lay 50 to 150 eggs. The eggs are unattached to each other and are deposited under leaves or debris or in the ground. The female may guard her nest until autumn rains fill the area with water. During dry winters, unhatched embryos may lie dormant until spring. But most eggs hatch in autumn or early winter, within about two weeks of immersion. The larvae are about 0.75 inch long. They may remain active through the winter, hiding in submerged vegetation at the edge of the pond by day and swimming to the middle of the pond and feeding on the bottom at night. They prey on fairy shrimp, other invertebrates, and, in spring, on the larvae of other mole salamanders and frogs, especially spring peepers and wood frogs. Metamorphosis takes place in the summer, and the juveniles are 2 to 3 inches in length when they leave the pond.

Eastern Tiger Salamander *(Ambystoma tigrinum tigrinum).* Mature tiger salamanders are 7 to 11 inches long. These big, chunky amphibians have dark brown bodies marked with tan or yellow spots in an irregular pattern. In the East, the tiger salamander lives along the Atlantic coastal plain from northern Florida to Long Island. One Pennsylvania record exists, from the mid-1800s in Londongrove, Chester County. In 1973, a herpetologist searched there for tiger salamanders and found none, but concluded that "the area is at least potentially suitable" for the species.

The life history of *Ambystoma tigrinum* is similar to that of the other mole salamanders, except that the tiger salamander tends to use deeper, more permanent ponds for its reproduction. Tiger salamanders are adept at digging and spend much of their lives in underground burrows.

RED-SPOTTED NEWT

The red-spotted newt, *Notophthalmus viridescens,* is one of the most common salamanders in Pennsylvania and the Northeast. The four currently recognized races of *Notophthalmus viridescens* range from southern Canada to Florida and Texas. All sixty-seven Pennsylvania counties have newt populations. Newts are unique among our amphibians in exhibiting three distinct life stages. They hatch from eggs into tadpolelike larvae in ponds, lakes, and streams. Within one season, the larvae grow rapidly and change into a land-dwelling form, which has its own name: red eft. Finally, after several years on land, the efts metamorphose into adult newts, which complete the species' life cycle by breeding in the water, where they spend the rest of their lives.

The adult stage is the common "salamander" that children carry home from state parks and lakes. It's hard to recommend that people make pets out of reptiles and amphibians, so many of which are losing habitat and dwindling in number. But the abundant, adaptable newt offers a good opportunity for nature observation. Efts can be kept in a terrarium and fed on small pieces of dog food, then later returned to the woods. *A Field Guide to Reptiles and Amphibians of Eastern and Central North America,* by Roger Conant, includes a pertinent chapter, "Care in Captivity."

Biology. Adult newts are around 4 inches long. The back and sides are a light to deep olive green, peppered with black. The back is also marked with large, bright red dots, which are ringed with black; these dots vary in number and positioning from one individual to the next. The belly is pale yellow speckled with black. The skin is fairly smooth, although not as slippery or slimy as that of most other salamanders.

Adults swim in shallow water, crawl along on the bottom, or float a few inches below the surface, ascending now and then to gulp in air. They eat fairy shrimp and other small crustaceans and mollusks; the larvae of aquatic insects, including mayflies, caddisflies, and mosquitoes; fish eggs; worms; and the eggs and young of many amphibians, including their own kind. They may be active year-round, swimming below the ice. If the home pond is so shallow that it freezes solid, newts may migrate to a nearby deeper pond or hibernate beneath leaves or logs on the pond bottom, on land underneath logs, rocks, or leaf litter, or in mammal burrows. Should the home pond dry up in summer, the adults burrow into mud or hide beneath debris in the pond basin, emerging after rains restore their habitat.

Most adults probably return to breed in the pond where they hatched and spent their larval stage. Breeding commences in March or April and may last into June. The male's tail fin enlarges, and black, horny structures grow on the undersurfaces of his hind legs. A male will court a female by nudging her with his snout and swimming near her in an undulating fashion. He may clasp her around the neck with his hind limbs and vibrate his tail, creating currents in the water that carry a stimulating pheromone, produced in the male's cloaca, to the female's nostrils. The male advances in front of the female and deposits a spermatophore, a gelatinous structure containing sperm cells. The female picks up the spermatophore with her cloaca (the cavity in which her sexual organs are located), and when she lays eggs several days later, they are fertilized by the male's sperm. A female may mate twenty to thirty times, often with different males, during the drawn-out breeding season. Theoretically, each batch of eggs that she lays may be fertilized by a different mate.

Females lay eggs singly, attaching them to underwater plants and submerged, decaying leaves. Each batch numbers six to ten eggs. The eggs of the red-spotted newt are toxic to some degree and repel predators. The eggs hatch after three to five weeks, depending on the water temperature. Hatchlings are 0.25 inch long and light green in color, lack forelimbs, and breathe through bushy gills. They feed on small aquatic animals and are preyed on by other salamander larvae (including the mole salamanders of genus *Ambystoma*), larger newt larvae, and adult newts.

Some of the larvae reach the next life stage: the land-based eft, a sexually immature juvenile that crawls out of the pond in late summer or early fall. When they emerge, efts are about 1.8 inches long and are marked with black-bordered dots on their backs. They are an overall brilliant orange or orange-red, a color advertising them as toxic creatures and warning off most predators. (Other amphibians, including the northern red salamander, have evolved to look like red efts, to persuade predators that they, too, taste terrible, even if they don't.) Efts have a dry, granular skin studded with glands that produce a neurotoxin that irritates the mucous membranes lining predators' mouths. Efts are ten times as toxic as the aquatic adult newts. They seem aware of their favored status, boldly marching across the forest floor by day or by night. They shelter in the leaf litter or under logs and brush and emerge when it rains and when temperatures top 50 degrees Fahrenheit. A few predators—probably skunks and raccoons—have figured out that efts' bellies are not as toxic as their backs; the predators turn the amphibians over and eat their internal organs.

The red eft is the terrestrial, or land-dwelling,
phase of the aquatic red-spotted newt.

Efts remain on land for three to seven years. They may move a quarter
mile from the home pond and range over 300 to 600 square yards, mainly
in forested areas. They are not territorial and may feed side by side with
other efts. In wet years, they forage more often and grow faster. They prey
on whatever they can find and subdue, mainly worms, grubs, spiders, mites,
and small insects and snails; in New York, one scientist found up to two
thousand springtails (small insects) in the stomachs of efts. Efts hibernate
underground. As they mature, they undergo a second metamorphosis, turning yellow and then green before returning to the water—usually in August
and September—as sexually mature adults.

In some coastal areas, where conditions on land are harsh and unstable,
red-spotted newts may skip the land or eft stage. In other regions, the
aquatic adult stage is eliminated, and terrestrial efts migrate to vernal breeding ponds much like the *Ambystoma* or mole salamanders.

Habitat. Adults live in quiet pools (both semipermanent and permanent),
farm ponds, swamps, lake margins, and backwaters of streams and rivers.
They thrive in sunny, weedy shallows and prefer still or slow-moving water.
Lower numbers occur in cool, shady habitats such as vernal pools, woods
ponds, mountain brooks, and springs.

Efts inhabit moist hardwood and pine forests and may also use bordering pastures and meadows. If they find ample food, they may not venture far from the pond or lake from which they emerged.

Population. The red-spotted newt is common to abundant in much of the Northeast, including Pennsylvania. Most newts are four to five years of age at their first reproduction, and they generally survive one or two breeding seasons. Fish avoid adult newts because of their toxicity, but toads, bullfrogs, snapping turtles, painted turtles, and garter snakes may feed on them. In some areas, amphibian blood leeches kill many newts.

LAND SALAMANDERS

The lungless salamanders belong to the family Plethodontidae. Fifteen species inhabit Pennsylvania: "lungless," because these amphibians lack lungs, having evolved to exchange gases through their thin skins and the capillary-rich membranes lining the mouth and throat. These membranes must remain moist to take in oxygen, and so the lungless salamanders must live in damp places. Seven species are found on land, under rotting logs and in spaces beneath stones, in the humid, dim to pitch black world where leaves become humus and tiny invertebrates abound. Another eight species are tied more directly to the water, relying on springs, streams, or ponds during all or part of their lives (see the following chapter, "Streamside Salamanders," for information on these semiaquatic types).

Of the land salamanders, many are drab and nondescript, and others show bright colors and intricate patterns. The lungless salamanders have prominent tails; in some species, the tail is markedly longer than the body. Related species look very similar, and in some cases herpetologists must use genetic analysis to tell these small woodland salamanders apart.

We know little about the reproduction of many species. In general, the male deposits a spermatophore, a gelatinous structure containing his sperm; the female picks up the spermatophore in her cloaca, the chamber where the intestinal, urinary, and reproductive systems exit from the body. Fertilization is internal—less chancy, perhaps, than the frogs' and toads' strategy of spreading milt (sperm) in the water in an attempt to fertilize eggs. After a female lungless salamander lays eggs, she typically guards her clutch. Sometimes several females lay their eggs in the same area, often a chamber in a rotting log or stump. In land salamanders, the larval stage occurs within the egg, so that hatchlings emerge as already metamorphosed juveniles.

Salamanders are mainly "sit-and-wait" predators that feed on insects, slugs, spiders, worms, immature amphibians—in short, anything happening past that can be seized, subdued, and worked down the gullet. During rainstorms, salamanders leave their ground-level abodes to search for prey on the forest floor. Habitats can be densely populated with these seldom-seen amphibians. In a New Hampshire study, biologists estimated that the total biomass of five species of salamanders was greater than the biomass of birds and approximately equal to the biomass of mice and shrews in a given watershed. Salamanders are a key component of many ecosystems, keeping small invertebrates in check and providing meals for snakes, birds, shrews, and other predators.

Mountain Dusky Salamander (*Desmognathus ochrophaeus*). Writes the herpetologist Roger Conant regarding dusky salamanders: "Identifying these salamanders is like working with fall warblers—only worse!" Individuals have a dark background color (gray, brown, olive, darkish yellow, or dull orange) with a lighter stripe (orange, yellow, gray, tan, or reddish), bordered by darker pigmentation, running down the back. Mountain dusky salamanders are found from New York south to Alabama. The species inhabits about two-thirds of Pennsylvania and is absent from the southeastern counties.

Mountain dusky salamanders are slender and up to 4 inches in length. Nocturnal, and among the most terrestrial of the land salamanders, they often forage far from water, preying mainly on earthworms and small insects. They inhabit the leaf litter, damp zones beneath bark and stones, and cracks and crevices in rock formations. They can be plentiful in hemlock ravines, a favorite habitat. Agile for an amphibian, a mountain dusky salamander will leap, wriggle, writhe, and run to avoid being captured. It may also shed its tail, drawing a predator's attention while the salamander escapes.

Desmognathus ochrophaeus apparently breeds in both spring and fall. The female lays twelve to nineteen or more eggs in a grapelike cluster, attached by a single stalk to the underside of a rock in a damp area, in a crevice through which water trickles or under moss in a seepage. Coiling her body around the eggs, she protects them until they hatch. In winter, mountain dusky salamanders cluster around springs, seeps, and bogs, hibernating between the ground litter and the saturated soil. Herpetologists estimate the life span for a typical individual at fifteen years.

Redback Salamander *(Plethodon cinereus)*. This small, slender salamander may reach 4 inches in length. It is dark brown and usually has a straight-edged red stripe running down the back. Many herpetologists believe it is the most common salamander throughout its range, which extends from southern Canada to North Carolina, Indiana, and Minnesota. Statewide in Pennsylvania, the redback salamander inhabits hardwood and coniferous forests, spending its days beneath sloughed-off bark, rocks, logs, leaves, and debris, and emerging to feed at night. I have found many redback salamanders on my land; when I lift a stone, the small salamander beneath it will remain coiled and frozen in place until I touch it, whereupon it becomes galvanized into motion, leaping and rushing off into the leaves.

Redback salamanders eat snails, earthworms, spiders, mites, millipedes, and various insects, particularly ants. They mate in the fall and lay eggs in spring and fall. Females lay grapelike clusters of three to fourteen eggs in rotting logs and stumps, beneath stones, under loose bark, and under moss. After the eggs hatch in August or September, the young remain in the nest with the female for one to three weeks. By the following spring, the young are about 0.75 inch long from snout to vent. They grow rapidly during their first summer and are mature two to three years after hatching. In *Salamanders of Ohio*, Ralph Pfingsten notes that redback salamanders are extremely adaptable in their use of breeding and feeding habitats, and he characterizes the species as "one of the most successful animals in all of eastern North America."

Ravine Salamander *(Plethodon richmondi)*. Almost snake- or wormlike with its long body, long tail, slender head, and tiny legs, the ravine salamander may reach 5 inches in length. It is brown to black, with silver or bronze speckling. It inhabits the wooded slopes of valleys and mountain ravines, where it hides by day under leaves and rocks. Ravine salamanders live in western and southwestern Pennsylvania, on the eastern edge of a rather small range that includes parts of Ohio, Indiana, West Virginia, and some adjoining states.

Not much is known about the ravine salamander's life history. Ants and sow bugs are important prey items, along with small beetles, earthworms, snails, crustaceans, and spiders. Ravine salamanders seem to go underground from June to September to avoid summer's heat, emerging again in the fall. Females apparently lay their eggs in obscure places in and under the ground. Ravine salamanders have been found in winter when people removed snow and exposed the masses of leaves covering the ground.

Valley and Ridge Salamander *(Plethodon hoffmani).* This species ranges from south-central Pennsylvania south in the Ridge and Valley physiographic province through Maryland and into Virginia and West Virginia. It is long (3 to 5.5 inches), slender, and short-legged, dark brown to black with whitish speckling. It lives beneath logs and rocks on the slopes of wooded hills and preys on insects, centipedes, and mites. Courtship and reproduction probably occur in spring, with females laying four or five eggs. Until 1972, the valley and ridge salamander was considered to be the same species as the ravine salamander, which it closely resembles.

Slimy Salamander *(Plethodon glutinosus).* Named for a whitish secretion produced by its skin, the slimy salamander is large (up to 8 inches) and colored black with white or creamy flecks. Found in much of the Northeast and the South, it is statewide in Pennsylvania. The slimy salamander lives in shady ravines, shale banks, talus slopes, and cave entrances, in piles of leaves, and under rocks and logs. According to herpetologist Ralph Pfingsten, in Ohio, the slimy salamander appears in late spring, around May, and vanishes by mid-October with the onset of cold weather. During drought, it burrows down from the surface or slips through cracks and crevices in rocks to find damp areas.

Slimy salamanders are active at night and especially on rainy summer evenings. They eat ants, beetles, centipedes, earthworms, flies, various insect larvae, mites, roaches, slugs, snails, sowbugs, and spiders. Slimy salamanders become sexually mature after four or five years. Females lay their eggs from late spring through midsummer in rocky crevices, rotting logs, and underground burrows. A study conducted in Pennsylvania and Maryland found that about half of the females laid eggs each year, and that individuals reproduce every other year. The slime on the skin of *Plethodon glutinosus* is a defensive mechanism that irritates the mucous membranes in the mouths of predators.

Wehrle's Salamander *(Plethodon wehrlei).* Specimens collected by R. W. Wehrle of Indiana, Pennsylvania, let taxonomists classify this salamander as a species in 1917. Individuals are usually dark gray or brown with white, bluish white, or cream-colored markings on the sides. Length is 5 to 6 inches. *Plethodon wehrlei* ranges from southern New York south through Pennsylvania to North Carolina in the Appalachian Mountains. At higher elevations, Wehrle's salamander inhabits red spruce forest, and at lower elevations it lives among mature mixed hardwoods. It seems to prefer a rockier habitat than do redback and slimy salamanders and has been found in rock crevices and in the entrances, or "twilight zones," of caves.

Wehrle's salamanders are nocturnal feeders that take prey typical for land salamanders: ants, beetles, mites, spiders, springtails, crickets, and insect larvae. They are believed to mate in March and April and to lay eggs in underground burrows and crevices in late spring and summer; reproduction may also take place in fall and winter. Females produce clutches of up to sixteen eggs.

Green Salamander *(Aneides aeneus)*. The green salamander's body is 3 to 5 inches long. It is brown to blackish and camouflaged with green markings, making this amphibian look like a lichen-studded rock and blending it in with its habitat: shaded cliffs and outcroppings on slopes in humid forests. The species ranges from southwestern Pennsylvania (Fayette, Westmoreland, and Somerset counties) through the Appalachians to northern Alabama. In Pennsylvania, the green salamander is listed as a threatened species.

Green salamanders live in moist crevices on rock faces. On cloudy days and at night, they emerge to take small insects, snails, slugs, and spiders. Males bite and shove one another when contending for territories. Green salamanders mate during the spring and apparently also in the fall. In late spring or early summer, females lay about seventeen eggs, which are held together in a cluster and cemented to the roof of a rock crevice with strands of mucus. A female will guard her clutch until the eggs hatch, twelve to thirteen weeks later. Young green salamanders are about an inch long. They seek out crevices of their own or remain in the crevice where they hatched, perhaps moving deeper into the rocks as temperatures fall during winter. Adults hibernate, sometimes in groups, below the frost line in crevices.

Herpetologists have found around a dozen sites in Pennsylvania inhabited by green salamanders. Land managers can protect habitats on public land by preserving the tree cover that helps keep humidity levels high.

STREAMSIDE SALAMANDERS

With over 230 species, the lungless salamanders, family Plethodontidae, represent the largest family of salamanders in the world. The highest population numbers and the greatest diversity of species occur in three regions: the Appalachian Mountain and Pacific coastal areas of North America, and Central America.

As adults, lungless salamanders breathe through their skin and through membranes lining the throat and mouth. They must keep their bodies moist for this gas exchange to take place. Some species live on land, beneath rocks and logs and in other damp places. And some species—eight in Pennsylvania—never move far from standing water, spending the larval and the adult stages of their lives in and on the edges of streams, brooks, and ponds, mainly in wooded areas.

The streamside salamanders, as I have chosen to call them, depend on actual water more completely than do the land salamanders described in the preceding chapter (although some of the land salamanders are quite at home in standing or moving water as well). Not only do streamside salamanders produce larvae that develop in the water, but the adults may enter it to find food, escape from predators, or lay eggs.

The Greek word *amphibios,* the root for our modern term *amphibian,* describes a creature that has two lives—in the case of salamanders, a larval and an adult existence. Streamside salamander larvae look like tadpoles; they breathe through bushy external gills and have fins to help them swim. After some time in the water (usually several months to a year), they metamorphose into an adult form. Adults have legs and tails; in adults of some species, a swimming fin, or caudal fin, adorns the tail. During both the larval and adult stages, salamanders are completely carnivorous, eating insects, other invertebrates, and other amphibians. In turn, streamside salamanders are preyed on by raccoons, skunks, shrews, birds, snakes, and larger amphibians.

Northern Dusky Salamander *(Desmognathus fuscus fuscus).* The northern dusky salamander is stout-bodied and 3 to 5 inches in length. It is variable in color, basically brown or gray, usually with a rust-brown stripe down the back, edged with a darker color. Northern dusky salamanders are common to abundant from southern New Brunswick south through New England and all of Pennsylvania to the Carolinas. They live in woodlands, on the rocky edges of streams and brooks where they shelter beneath flat rocks, logs, leaves, and moss. Mainly nocturnal, they may also become active on cloudy or rainy days. They burrow into muddy stream banks and climb onto the sides of rocks and into low vegetation. Northern dusky salamanders feed on worms, grubs, snails, slugs, mites, spiders, and other salamanders and their larvae.

Courtship and mating take place on the land. Around July, a female will lay a cluster of ten to as many as forty eggs beneath a stone, under the loose bark of a log, in wet leaves, or in moss at the edge of a small stream. The female guards her clutch. After five to ten weeks, in late summer or early fall, the larvae hatch: they drop directly into the stream or squirm through tunnels or cracks in the ground to reach the water. They metamorphose the following summer. During winter, larvae remain active. Adults hibernate under logs and rocks in deeper water, or they stay active on stream bottoms or in springs and seeps.

Appalachian Seal Salamander *(Desmognathus monticola monticola)*. This robust salamander is 3.25 to 5 inches long. Its tan or brown back is scattered with darker wormlike or netlike markings. The species ranges through the Appalachians from southwestern Pennsylvania to Alabama. The Appalachian seal salamander inhabits the banks of cool, well-oxygenated mountain streams, as well as seepages and springs. It looks like a seal as it perches on a rock near the water's edge, bathed in spray or mist. Seal salamanders take refuge under rocks and in burrows in the bank. They are agile and slippery. In summer, females lay around thirty-six eggs beneath leaves and in rotting wood and on the undersides of rocks along streams; they remain with the clutch until the eggs hatch in September. The aquatic larvae are believed to metamorphose after one year.

Four-Toed Salamander *(Hemidactylium scutatum)*. This small (2- to 3.5-inch) salamander has four toes on each hind foot, rather than five, as in most other salamanders. The four-toed salamander has a reddish or grayish brown back and a white belly with scattered black dots. It also has a noticeable constriction at the base of its tail, where the tail will easily break off— or be cast off by its owner, as a diversionary tactic—should a predator attack. The species ranges throughout much of the East and is statewide in Pennsylvania. The four-toed salamander lives in damp forests with spring seeps, bogs, and permanent or temporary ponds studded with mossy logs, roots, and grass clumps.

Adults mate in September and October. They hibernate in rotting logs, subterranean burrows, decayed root channels, or piles of rotting leaves; many individuals may congregate in one place. In the spring, the females migrate to streams, springs, woods ponds, or bogs, where they nest in damp cover at the water's edge. A favorite nesting habitat is a hummock of sphagnum moss in a bog; several females may nest communally in one hummock. The female lays fifteen to sixty eggs and guards them against predators. The larvae hatch after one or two months and then wriggle or fall into the water. After three to eight weeks, the larvae metamorphose into the terrestrial form.

Northern Spring Salamander *(Gyrinophilus porphyriticus porphyriticus).* There's an old Appalachian folk song that goes: "Wish I was a lizard in that spring." No doubt the line describes the spring salamander, whose presence in a spring would not be surprising but would certainly be eye-catching, since this "lizard" is big (up to 8 inches), chubby, and colored orange, salmon, or pinkish, mottled or freckled with darker pigment. Young spring salamanders are brighter in color; probably the species evolved to look like the red eft, which is toxic to predators. The northern spring salamander is found from Maine to Alabama. Spring salamanders live throughout Pennsylvania in or near cool woodland streams, springs, and caves. During rainy weather, they may wander about at night hunting for insects, crustaceans, spiders, worms, snails, and other salamanders, including two-lined and dusky salamanders and their larvae.

In winter, individuals live beneath rocks in the streambed, in springs, or burrowed into saturated, unfrozen soil; they may feed all winter, since their habitats don't freeze. Mating probably takes place from fall through spring. Females lay their eggs in the summer, attaching them individually to the undersides of flat rocks wedged into the banks of cool, flowing streams or springs. The egg mass, guarded by the female, may contain forty to a hundred eggs. Newly hatched larvae are just under 1 inch in length. For three to four years, they grow in their watery habitat. They are thought to metamorphose in summer.

I once flipped over a rock along Lushbaugh Run in Cameron County and grabbed for the showy specimen lying in the water beneath: I ended up with a wet fist, and watched as an orange-red spring salamander went writhing into the stream's swift flow. Bayard Green and Thomas Pauley, in *Amphibians and Reptiles in West Virginia,* recommend coating the palms of your hands with sand before trying to catch this slippery amphibian.

Northern Red Salamander *(Pseudotriton ruber ruber).* The northern red salamander has evolved to look like the red eft, which is toxic and apparently tastes horrible to predators; *Pseudotriton ruber* is also toxic, but less so, and its resemblance to the eft may reinforce the predators' recognition and boost the chance that they will also leave the northern red salamander alone. Scientists term this form of deceit "Batesian mimicry," after the British naturalist H. W. Bates, who first described the phenomenon in 1862.

Adult northern red salamanders are 4 to 7 inches long. Their stocky bodies are colored red to dark orange, with many black spots. The eyes are yellow. The New York–Pennsylvania line is the approximate northern border of the species' range, which blankets Pennsylvania and also covers most of the southern states. Look for red salamanders in springs and small streams, usually in woods but sometimes in open meadows. They shelter beneath rocks and logs, and in the winter move into deeper springs where temperatures hold steady. In autumn, females attach their eggs (up to seventy of them) to the undersides of rocks sunk in the soil next to springs and small streams. The eggs hatch after eight to ten weeks. Northern red salamanders spend three to four years as aquatic larvae before transforming, by which time they are 3 or 4 inches long.

Eastern Mud Salamander *(Pseudotriton montanus montanus)*. The eastern mud salamander looks like its close relative, the northern red salamander, except that its eyes are brown and its body has fewer black spots. Mud salamanders burrow into the muck and mud around spring seeps and along the banks of streams. The species ranges from New Jersey southward in the Coastal Plain and Piedmont regions. The type, or first, specimen of the eastern mud salamander was found near Carlisle in Cumberland County. Today, herpetologists know of only one site in the state—in Adams County— where eastern mud salamanders live, and the species is considered endangered in Pennsylvania.

Observations made in the Carolinas and Virginia indicate that mud salamanders breed in the fall. Females lay eggs in December, with the average female depositing around 127 eggs every other year. The larvae hatch in February and metamorphose after about a year and a half.

Northern Two-Lined Salamander *(Eurycea bislineata bislineata)*. This common, widespread salamander ranges from Canada to Virginia east of the Great Lakes. It is statewide in Pennsylvania. The body is yellowish to brown; a pale band runs down the back, bordered on each side by a dark line starting at the eye and running to the end of the tail. Slender, and a maximum of only 4 inches long, the two-lined salamander sometimes falls prey to the larger streamside salamanders, which it tries to escape by wriggling away rapidly or jumping, a feat it accomplishes by curving the body, then suddenly straightening it out. The two-lined salamander can also shed its tail.

Two-lined salamanders stay close to streams, hiding under rocks, logs, and beds of fallen leaves. In winter, they hibernate beneath the leaf litter or remain active in spring seeps, streams, and waterlogged soil. They eat small insects and other invertebrates. Courtship takes place on land. The adults mate during fall, winter, and early spring, and the females lay their eggs from March to May. A female will turn herself upside down to cement her clutch of fifteen to a hundred eggs (forty on average) to the underside of a rock beneath the flowing water of a spring or brook. The larvae hatch after one to two months and then spend up to two years in the water, where they eat insect larvae and tiny crustaceans. People often use adult two-lined salamanders as fish bait.

Longtail Salamander *(Eurycea longicauda longicauda).* Longtail salamanders range from southern New York to the Gulf coast and west to Missouri and Arkansas. They are statewide in Pennsylvania, except for lands bordering Lake Erie. Adults are yellow to deep orange with black spots densely marking the sides, back, and tail. This showy creature lives up to its name by having a tail that constitutes almost two-thirds of its total length, which can exceed 7 inches. The tail breaks off easily under a predator's assault and may provide enough of a distraction to let the salamander escape.

Longtail salamanders live near streams, although they rarely enter the water. It is said that when a two-lined salamander and a longtail salamander are found beneath the same rock near a brook, the two-lined will race for the water while the longtail runs toward land. Longtail salamanders move about on land mainly at night, eating small insects, mites, spiders, worms, centipedes, and snails. Adults often spend the winter in caves. Females lay their eggs from late fall into winter, in clutches of fifty to sixty in underground crevices feeding into springs, temporary pools, and streams. The eggs hatch from March through May. The larvae are aquatic. It is believed that they transform into adults after three and a half to seven months.

TOADS

Frogs and toads are known as anurans, a scientific term denoting the fact that these closely related amphibians lack tails. Toads are less agile and more terrestrial than frogs; their earthy colors and nubby skin help camouflage their squat, plump bodies. Since toads' hind legs are shorter than those of frogs, they cannot leap as strongly as frogs can. But toads do use their hind legs to hop away from predators and toward prey. When a toad gets within a foot or two of a prospective meal, it raises up on all fours and closes in by stalking, moving both feet forward on one side of the body, then the other. Toads have long, sticky tongues, attached to the front of the mouth rather than the rear, that they flip out to catch prey up to 2 inches away. They use their front feet to cram their victims into their toothless mouths. Toads feed on a wide range of small animals, including insects and their larvae, earthworms, spiders, snails, slugs, centipedes, and millipedes—whatever is available and easy to catch.

Toads are "explosive breeders," which means the adults all congregate in one place at the same time to mate and lay eggs. In the spring, toads migrate to temporary ponds, stream backwaters, puddles, water-filled ditches, and the shallows of lakes and rivers. Males attract females by calling, producing trills, bleats, or squawks, depending on the species. Breeding males inflate a large, bubblelike throat sac and then release air through the larynx, with the vocal sac acting as a resonating chamber to amplify the sound. The males usually wait along the water's edge. They clasp the females from behind in a mating posture known as amplexus; during the melee in the breeding pond, males may accidentally clasp other males or nonbreeding females, which respond with a squeaking release call, causing the clasping male to let go. Some females drown when grabbed by two or more competing males. While in amplexus, the female swims about laying long strings of eggs, and the male, holding on to her, releases clouds of sperm to fertilize the eggs. One female can lay thousands of eggs over a few days or, in some species, in just a few hours.

The eggs hatch within several days, and the larvae, or tadpoles, begin a race to grow and develop into land-dwelling toadlets before the pond dries up or predators nab them. Tadpoles—also called pollywogs—lack limbs; they propel themselves through the water by lashing their strong tails back and forth. They eat algae and bacteria that they filter out of the water, and they also feed on carrion. When you see tadpoles nosing into the pond bottom or against rocks, they are working to dislodge algae for feeding. As tadpoles mature, their bodies reabsorb their tails and send forth limbs. Many changes take place internally as well.

Toad tadpoles are preyed on by insects, including predaceous diving beetles, and by fish: one advantage to toads' breeding in temporary ponds is that fish aren't present there. Of the hundreds of thousands of tadpoles that may hatch from toad eggs laid in a pond, a small minority survive to become adults.

Toads are active in spring, summer, and early fall, mainly in the evening and at night. During the heat of the day, they shelter in leaf litter; in damp soil beneath logs, boards, and stones; and in burrows they dig themselves. When the weather turns cold, toads bury themselves in the earth or crawl into abandoned mammals' tunnels or natural crevices.

The skin of toads feels dry to the touch. The rough, gland-studded skin secretes bad-tasting toxins that cause many predators to spit toads out and permanently put them on their mental list of inedible critters; some predators die after biting toads. The skin secretions irritate mucous membranes, so if you handle a toad, wash your hands before touching your eyes or mouth. People cannot get warts from toads.

Three species of toads are found in Pennsylvania.

Eastern American Toad *(Bufo americanus americanus).* The American toad ranges throughout eastern North America from coastal Labrador south to Texas. The subspecies known as the eastern American toad is statewide in Pennsylvania. Toads are abundant and can be found in almost any setting—woods, meadows, marshes, gardens, suburban yards, city parks, and farmland. Individuals are 2 to 4 inches long. They come in many colors, generally brown and gray, sometimes olive or brick red. They may have patches of yellow and tan, and warts that are accented with yellow, orange, red, or brown.

It's easy to overlook specific aspects of a toad's appearance, what with all of the warts and the complex color pattern. But behind each eye is a large oblong or kidney-shaped bump, the parotid gland, which secretes a powerful steroid that can affect a predator's heart functioning and blood pressure. On a toad's back are around half a dozen smaller black patches, each islanded with one or two warts. (Fowler's toad, found in the southern two-thirds of Pennsylvania, has from three to seven warts in its black spots.) Male American toads have dark, almost black throats; females' throats are pale. Toads possess some ability to adjust their color to match their surroundings.

Toads sit near outdoor lights, waiting for flying insects. They hunt in the evening and on humid or rainy nights. They consume a great many earth-

The large parotid glands, located on the back behind the eye bumps
of the eastern American toad, release a powerful steroid that can
affect the functioning of a predator's heart.

worms. A female toad (judging from her pale throat, as well as her great size; females are often a fifth larger than males) lived for several years under our front stoop. She was an attractive maroon color, similar to the local soil. I once rescued her from a window well—or perhaps only displaced her, since it was a fine, damp spot with plenty of worms, spiders, and bugs.

Confronted by a predator, a toad may lower its head to appear larger and to make its poison glands the first thing its enemy will contact. Toads urinate to deter predators, and they puff themselves up, making it harder for their enemies to swallow them. Despite toads' toxicity, some creatures take them regularly, particularly hognose and garter snakes. Hawks, herons, and some waterfowl eat toads; raccoons feed on toads' undersides and avoid the glands on the back.

Toads dig themselves into the ground by backing in and shuffling to one side and then the other, using their hind feet to push the dirt out of the way. In September or October, American toads burrow deeply into the soil and become dormant.

Toads emerge from hibernation in March and April. The males enter vernal ponds, marshes, flooded meadows, and lake and river shallows. Their calling, a high, sweet tremolo, "can easily be imitated by whistling and humming a high note at the same time," writes Massachusetts naturalist Thomas Tyning. A toad will call for up to thirty seconds, pause for a few seconds, and begin again. Early breeders often quit for a while during cold snaps. At the peak of the breeding season, usually in April in Pennsylvania, toads call during the day and at night. There is no sweeter chorus than a pond filled with singing toads, some voices pitched slightly higher than the others—now loud, as several toads sing at the same time, now softer—the sound a clear, compelling announcement of spring.

Female toads lay twisted double strands of small eggs that can be 3 feet long and can contain two thousand to twenty thousand eggs. After three to twelve days (on average, four days), the eggs hatch into tiny tadpoles. Soon the tadpoles, black in color, begin foraging on the bottom of the pond. At night, they rest individually, on the bottom in deeper water; as the sun heats the pond by day, they move into the shallows and form schools. Research has shown that schooled tadpoles are more efficient than solitary individuals at filtering food, evading predators, and maintaining an optimum body temperature. Some predators avoid toad tadpoles, apparently because they taste bad.

The tadpoles grow for five to ten weeks. In some breeding habitats, such as roadside ditches and tire tracks, toad tadpoles perish when the water dries up before they can transform and emerge. Newly metamorphosed toadlets are about half an inch long. In mid- to late summer, hundreds of them may be found hiding in plant growth around the pond.

After breeding, adults may wander far from standing water. Toads reach breeding maturity after two or three years and can live for ten or more years in the wild. Captives have survived for over thirty-five years.

Fowler's Toad *(Bufo woodhouseii fowleri)*. Fowler's toad is named for two nineteenth-century American naturalists. It is slightly smaller than the American toad, with adults measuring 2 to slightly over 3 inches in length. Fowler's toad is gray or greenish gray, with a pale stripe down the center of the back. It lacks the rusty colors found on many American toads. A key identifier is the presence of three or more warts in the large dark spots on the back. The species ranges from southern New England to the Carolinas in the East. In Pennsylvania, Fowler's toad occurs in the southern two-thirds of the state and along Lake Erie.

Fowler's toads eat many ground-dwelling insects, especially beetles and ants. They show a preference for sandy soil of floodplains, river bottoms, lake edges, beaches, and roadsides. They also inhabit open meadows, suburban gardens, farmland, brushy areas, and woods edges. In most years, Fowler's toads do not emerge from hibernation until after American toads have finished breeding. Fowler's toads will hybridize with American toads, and the two species are kept separate by slightly different habitats, breeding periods, and mating vocalizations.

Woods and farm ponds, lake edges, stream backwaters, marshes, and pools between dunes and on beaches are all breeding sites for *Bufo fowleri*. The male's call is a nasal, low-pitched bleat, described as *w-a-a-a-h*. It lasts for two to seven seconds and carries over a long distance. In Pennsylvania, breeding peaks in May. Each female lays around eight thousand eggs. Within a week, tadpoles hatch. They are black and look like the pollywogs of American toads. The tadpoles transform into half-inch toadlets by midsummer. Fowler's toads begin hibernating earlier in autumn than do American toads and spend the winter burrowed deeply into sandy soil.

Eastern Spadefoot Toad *(Scaphiopus holbrookii holbrookii)*. Spadefoot toads are found from Massachusetts to Florida. In Pennsylvania, the species inhabits two ranges: the Susquehanna River Valley from Maryland to the northcentral region; and the Delaware River Valley, along Pennsylvania's eastern border, as far north as Monroe County. At 2.5 inches, spadefoots are smaller than American and Fowler's toads. They are also smoother and have less prominent warts. Most spadefoots are gray or brown, with a wavy yellowish line beginning at each prominently bulging eye and extending down the back. Unlike other Pennsylvania toads, which have round pupils, spadefoot toads have vertically elliptical pupils. The inner surface of each hind foot has a dark, horny projection used in digging. A squatting toad can work its way down below the surface of loose soil simply by rocking or twisting back and forth, shifting the dirt aside with the "spades" on its hind legs. Spadefoot toads live in sandy lowland habitats in forested, brushy, and farming country.

Heavy, sustained downpours any time from March to October can trigger a mass migration to a breeding pool. Spadefoot toads may mate, lay eggs, and depart from a pool in a single night. In years without hard rains, spadefoots may not reproduce in local areas; in rainy years, local populations may congregate and breed twice. The male's call is a loud, abrupt squawk that has been compared to the complaints of young crows, the honking of geese, and the noise made by rubbing a balloon. The call carries for half a mile or farther. Unlike our other toads, a male spadefoot will not clasp a female until she approaches and touches him. Each female lays up to twenty-five hundred eggs, which hatch after one to seven days. The tadpole stage lasts from two weeks to two months.

Spadefoots are active at night, when they move about in a home range of approximately 70 square feet, which includes one or several burrows. In dry weather, spadefoots dig themselves into the ground. Writes Thomas Tyning: "During periods of extended drought, spadefoot toads excrete a fluid, curl into a tight ball, and lie dormant. The fluid hardens the earth around the toad and forms a compact chamber that will hold whatever moisture is there." After rains soak the ground again, the toads emerge. Individuals have survived after losing water equaling 40 percent of their body weight.

TREEFROGS

Few outdoor people have seen these small, retiring, well-camouflaged amphibians. The treefrogs of the family Hylidae (from the Greek *hyle*, "belonging to the forest") are slim and have longish limbs; the feet of most species end in long digits tipped with suction cups, which help these frogs cling to bark and climb in brush and trees. Treefrogs are agile and are excellent jumpers. They eat small invertebrates, mainly insects and spiders. Females are larger than males.

You will know that these frogs inhabit your area by hearing them singing in springtime. Treefrogs gather to breed in temporary and permanent bodies of water: woodland pools, cattail swamps, flooded meadows, spring seepages, river backwaters, roadside ditches, artificial ponds, and sometimes even in swimming pools. As with birds, the males do the calling, and each species has a distinctive song. Females lay eggs in the water. The eggs hatch into tadpoles within a few days. The tadpoles feed on algae and transform into froglets (immature frogs) during summer. In winter, treefrogs become dormant, hibernating beneath leaf litter, logs, and loose bark, and in crevices extending below the ground. Some treefrogs build up sugar compounds in their blood that retard dehydration and act as a kind of antifreeze, keeping cells from freezing.

Seven species of treefrogs live in Pennsylvania: the six detailed in this chapter, plus the well-known spring peeper, *Pseudacris crucifer*, covered in the next section.

Northern Cricket Frog *(Acris crepitans crepitans)*. The northern cricket frog looks like a little toad with its rough, wart-studded skin. The cricket frog is an anomaly among treefrogs, because it does not climb; instead it pursues a semiaquatic existence, living on the open, sunny edges of ponds, marshes, bogs, and lakes, and on sand and gravel bars of sluggish rivers and streams. Cricket frogs often rest on mats of floating vegetation or on the leaves of emergent water plants such as spatterdock. Adults are 1.5 inches long, tan or gray with green, brown, or yellow patches. The northern cricket frog is found in southeastern and south-central Pennsylvania.

Cricket frogs commence breeding in April and May, and the males' singing continues off and on into summer; *Acris crepitans* is the latest breeding of our treefrogs. The call is a string of sharp clicking noises, like pebbles being struck together, starting off slowly, speeding up, and lasting for twenty to thirty seconds. The easiest way to spot these frogs is to visit a breeding pond at night: the calling males' vocal sacs reflect a flashlight's beam. Cricket frogs are powerful leapers; they can also escape from danger by diving into the water and burrowing into a muddy bottom. Females lay eggs singly or in small clumps that become attached to aquatic plants.

Eastern Gray Treefrog *(Hyla versicolor versicolor)*. This woodland species has a large range that spreads across eastern North America and includes all of Pennsylvania. Adult gray treefrogs are 1.25 to almost 2.5 inches long. The warty skin is an overall gray color, often tinged with green or brown and marked with dark-colored blotches; an individual's color will gradually brighten or darken to match its surroundings. The inner surfaces of the hind legs are bright yellow-orange, although the color is concealed when the frog is sitting. The toes end in large round pads, climbing aids made more effective by sticky mucus. Gray treefrogs can climb vertical tree trunks and are sometimes found clinging to window glass on houses.

In late spring, males begin calling after nighttime temperatures top about 50 degrees Fahrenheit. At first, they call individually, before migrating to the breeding ponds. At the ponds, they stake out perches on branches of nearby trees and on boughs that arch over or have fallen into the water. Males space themselves at least 30 inches apart. The call, a prolonged musical trill, carries over long distances; the sound at a pond filled with calling males can seem almost deafening. Females arrive throughout the breeding season, while the males remain on station at the ponds. A female will pick a mate by approaching closely, then touching or jumping onto him. With the chosen male clasping her from behind, the female descends the tree, enters the water, and lays her eggs: up to two thousand of them, in groups of ten to forty scattered throughout the pond. As the eggs are laid, the male sheds his sperm to fertilize them. The eggs hatch after two to five days. Tadpoles are 0.25 inch long at hatching; after thirty to sixty days, they transform into brilliant green froglets 1.5 inches long.

It's hard to find a gray treefrog outside of the breeding season. When resting, this creature pulls in the long digits of its front limbs, tucks its fists beneath its chin, and presses its body tightly against the tree bark that it so closely resembles.

Chorus Frogs (*Pseudacris* species). Four species or subspecies of chorus frogs are found in different parts of Pennsylvania. Chorus frogs are quite small, at around 1 to 1.5 inches in length. They are brownish, with stripes on their backs. The stripes may be thin, broad, or broken, depending on the species or race. These frogs are lively jumpers. They do not climb often, although they will clamber into weeds and low shrubs to take prey. They breed in shallow water from late winter to early summer. In Oak Pond, the vernal pond near my home, I have heard chorus frogs trying to insert their songs into the cacophony raised by the much more numerous wood frogs and spring peepers.

The western chorus frog *(Pseudacris triseriata)* calls with an ascending, fluttering *prreep*, which Roger Conant, in *A Field Guide to Reptiles and Amphibians of Eastern and Central North America,* says can be "roughly imitated by running a finger over approximately the last 20 of the *small* teeth of a good-quality pocket comb." Western chorus frogs are early breeders, and in some years they start singing before the ice has melted from their ponds. They are found on the Allegheny High Plateau.

The New Jersey chorus frog *(Pseudacris feriarum kalmi)* lives along the Atlantic seaboard and is at the inland limit of its range in extreme southeastern Pennsylvania. This frog has dwindled as the city of Philadelphia and its suburbs have sprawled across its habitat. The New Jersey chorus frog is found in tidal wetlands of the Delaware River at the John Heinz National Wildlife Refuge at Tinicum. It is classified as an endangered species in Pennsylvania.

The upland chorus frog *(Pseudacris feriarum feriarum)* occurs in south-central Pennsylvania north to Lycoming County. Its *crreek* or *prreep* call is similar to that of the western chorus frog. It often inhabits upland areas, mating and laying its eggs in heavily vegetated ponds. Perhaps because it breeds during both daylight and dark, the upland chorus frog has a shorter reproductive season than the mountain chorus frog, which is a nocturnal breeder.

The mountain chorus frog *(Pseudacris brachyphona)* inhabits forested mountains and plateau country in the southwestern corner of the state. Its call, described as a rasping *rake, rake,* increases in rate and intensity as air and water temperatures rise. Males stay close to the breeding pond through the spring, returning to meet receptive females after each rainfall. Each female visits the pond on only one night, mating and laying three hundred to nine hundred eggs over several hours.

SPRING PEEPER

I played the flashlight's beam over the brown grass edging the pond. He clung between two stems, his rump barely touching the water. He was not quite an inch long, and his reddish tan body was traversed with brown streaks that met in a wavery X in the center of his back. He was a northern spring peeper, of the species *Pseudacris crucifer* (formerly *Hyla crucifer*), a singing male broadcasting his *prreeep* call again and again into the March night. At his throat, his vocal sac swelled like a pale bubble. His calls joined those of scores of other peepers, in shrill counterpoint to the clacking calls of wood frogs also gathered to breed at Oak Pond.

The spring peeper, among the best known and most beloved of the harbingers of spring, is found from Atlantic Canada south to the Gulf coast in eastern North America. This diminutive treefrog inhabits moist, forested terrain in wilderness areas, farmland, suburbs, and cities. Although quite obvious in spring, *Pseudacris crucifer* becomes secretive and unobtrusive for the rest of the natural year.

Biology. Adult males are about 0.75 inch long, and females are 1.5 inches long. Peepers have cream-colored bellies. Their backs show a range of colors, including pale cinnamon, tan, various shades of darker brown, gray, and olive. An individual can change colors within fifteen minutes to take on the hue of its surroundings. The X marking on the back may be complete or broken. Peepers have long limbs ending in digits tipped with round, mucus-moistened pads, which help the frogs climb trees and shrubs. The hind feet are strongly webbed.

Peepers eat spiders, ants, beetles, flies, caterpillars, springtails, pill bugs, mites, ticks, snails, and other invertebrates. They feed on the ground and 2

The northern spring peeper often forages several feet up in vegetation.

to 3 feet up in ferns, high grass, shrubs, and small trees. They do not take aquatic prey, even when they are clustered at breeding ponds for days or weeks on end. Peepers are preyed on by fish, larger frogs, salamanders, snakes, birds, and mammals; some blunder into the webs of large spiders, with fatal results. Although active mainly at night, peepers sometimes move about by day. Hikers walking through the woods, particularly during or just after rains, may flush out these small frogs, which hop strongly away.

After leaving their winter dormancy, males call tentatively from hillsides above breeding areas. The males move to the breeding ponds before the females arrive. Peepers congregate in temporary ponds, flooded meadows, bogs, swamps, the marshy edges of lakes, and other shallow waters having aquatic vegetation or debris. The males stake out tiny territories, usually 4 to 16 inches in diameter, depending on how densely an area becomes crowded with males.

In Pennsylvania, peepers begin calling in March. Only the males call, and they have two main vocalizations. The one that earns the species its common name is the "advertisement call," a single, high-pitched *prreeep* that slurs upward at the end. The advertisement call is given about once per second. The calling becomes especially intense on warm, rainy nights, when the combined chorus of hundreds of peepers reaches far into the surrounding habitat to alert and summon females. From a distance, calling peepers sound like sleigh bells jingling. Closer in, the noise seems deafening, getting inside a listener's head until, like a grating, swelling pressure, it is felt more than heard. Males call from the ground at the water's edge or from perches on twigs or vegetation. The males in a pond will call for fifteen minutes to half an hour, then rest for about five minutes, with only a few males calling sporadically. During the height of the breeding season, males will also call during the day. An individual may sound his *prreeep* call some forty-five hundred times nightly during the breeding season.

The second type of call, the lower-pitched "aggressive call," is used for close-range rather than long-distance communication. When another male ventures onto or near a male's territory, the owner of the territory sounds a warning trill that slurs upward and lasts for a second or longer. Thomas Tyning, in *A Guide to Amphibians and Reptiles,* explains the two calls in detail: "Early in the season, when nighttime temperatures are relatively cool and few peepers are singing, you can easily hear aggressive calls," which become harder to pick out when peepers are in full chorus. As the season progresses and more male frogs enter the breeding pools, males become more tolerant of crowding and give aggressive calls less often.

Field research has shown that females often choose older, larger males (presumably fitter and more genetically desirable), who call at a faster rate than smaller, younger males. The female seeks out the mate of her choice and touches him, whereupon he clasps her from behind. Swimming below the surface of the water, she releases eight hundred or more eggs, singly or in small masses, while the male discharges sperm to fertilize them. The pair stay linked, in a position known as amplexus, for up to four hours. "Satellite males" lurk in the pond without calling; they may temporarily take over and sing from the territories of males that are off with females in amplexus. Satellite males also try to intercept females lured in by other males' calling.

Over the breeding season, members of both sexes arrive at the ponds in staggered numbers. The chorus becomes most intense in March and fades in April, although it may resume in May or June if rains replenish the ponds. The males' calling probably requires tremendous amounts of energy; early in the season, intent on breeding, peepers do not feed but rely on stored body fat. They are fortunate that few of their predators are active so early in the year. After breeding, adults move out into woodlands, old fields, and shrubby areas. Isolated males may call away from the breeding ponds, especially in rainy weather.

Peepers' eggs hatch within a week. The tadpoles are brown or green and have gold-colored flecks on their upper surfaces. In shallow, open-water areas of the pond, they graze on algae and other tiny plants. During their two-month larval stage, they are fair game for dragonfly larvae, predaceous diving beetles, larval salamanders, snakes, frogs, turtles, and fish. Some peeper froglets still have tails when they metamorphose and emerge from the water. Froglets hide in vegetation near the pond before moving off into forested or brushy habitats.

In a study in southeastern Michigan, researchers found home range diameters of 4 to 18 feet for adult peepers, and they documented individual daily movements of 20 to 130 feet. Peepers spend most of the daylight hours hidden in leaf litter on the forest floor. They enter hibernation in November, becoming dormant on the surface of the soil, covered by leaf litter, overarching tree roots, logs, moss, and bark; sometimes several individuals hibernate in clumps. The bodies of hibernating peepers may freeze, but a glucose-based compound causes the ice to form in the spaces between cells, so that the cells themselves are not damaged.

Most spring peepers are sexually mature in the spring following their metamorphosis. The life span of a typical adult is about three years.

Habitat. Deciduous woodland is the major habitat, but peepers can also live in coniferous woods. They inhabit forested areas with a dense understory, old fields, floodplains, gardens, and even urban areas, so long as unpolluted breeding habitats are nearby. Peepers breed on the edges of marshes, swamps, wet meadows, and sphagnum bogs; on lake and reservoir margins; in temporary and permanent ponds; and (probably futilely) in extremely impermanent spots such as water-filled ruts on dirt roads.

Population. Spring peepers are abundant in Pennsylvania, with populations in every county. *Pseudacris crucifer* becomes especially plentiful in brushy second-growth or cutover woods. Even though local populations seem secure, they can be reduced or wiped out by the draining or filling of small, seemingly insignificant wetlands. If such areas are destroyed, or are cut off from adjacent habitat by development or roads, even common species like the spring peeper will disappear.

TRUE FROGS

The true frogs are pond and stream dwellers, the amphibians most people think of when they hear the word *frog*. They have smooth skin, long legs, and webbing between the toes of the hind feet and are medium to large in size. They feed on a wide range of prey, with insects the mainstay for most species. In turn, frogs are eaten by wading birds, fish, snakes, turtles, raccoons, otters, minks, and other predators. In spring and early summer, frogs mate and lay eggs in the water: a female may lay from one thousand to as many as forty thousand eggs, depending on her species and size. The eggs hatch into tadpoles, which feed by filtering algae, bacteria, and plant detritus out of the water. Tadpoles are preyed on by snakes, turtles, fish, birds, predaceous insects, and salamander larvae.

The adults of some species venture away from the water and feed in grassy areas during summer; one type, the wood frog, is decidedly terrestrial and visits the water only for its annual springtime mating. In winter, frogs hibernate in saturated mud, beneath submerged branches or debris at the bottoms of ponds and streams, and under leaves, logs, and rocks on land. When underwater, frogs take in oxygen through their skin.

The true frogs of North America belong to the large genus *Rana,* with about 250 species worldwide, most of them in the Old World. Six species or subspecies live in Pennsylvania, including our largest frog, the bullfrog, which is treated in a separate chapter.

Northern Green Frog *(Rana clamitans melanota).* Green frogs look like small bullfrogs: individuals are green, yellowish green, olive, brown, or a variable mix of these colors, with a creamy belly. Unlike bullfrogs, green frogs have dorsolateral ridges that run the length of their backs. Adults are 2.25 to 3.5 inches long. Common over much of eastern North America, green frogs are statewide and are probably the most abundant frog species in Pennsylvania. They live in ponds, streams, brooks, marshes, bogs, lake margins, and springs, in open and wooded terrain. Green frogs seldom get more than a few yards away from the water. Often they rest, hidden among vegetation on the bank, where they can intercept prey and where their aquatic habitat is but one leap away: when springing to safety, green frogs utter a loud yelp or scream.

More active at night than during the day, green frogs wait for prey to happen by, or they walk slowly through the habitat, sometimes closing in on their victims in a series of short hops. A study in New York found that beetles, flies, grasshoppers, caterpillars, and spiders were favored prey. Green frogs also eat snails, slugs, crayfish, smaller frogs, and fish.

From October to March, green frogs hibernate in the muddy bottom of a stream or pond. They breed in April. Males stake out territories, about 20 to 30 feet across, in shallow water; they defend these zones by sounding growling calls and wrestling with trespassers. Males try to attract females using an "advertisement call," three or four resonant notes often described as sounding like a plucked banjo string. Females swim slowly through the breeding pond before selecting a mate, probably basing their choice on the quality of the habitat he is defending. Mating takes place at night. As in other frogs, the male clasps the female from behind and releases sperm to fertilize the three thousand to five thousand eggs she extrudes. Tadpoles that hatch early in summer transform into froglets by mid- to late summer; those that hatch later remain as tadpoles over winter and transform the following spring. Wading birds, fish, mammals, and the larvae of predaceous insects prey on the pollywogs.

Wood Frog *(Rana sylvatica)*. The wood frog is about 3 inches long. It is easily identified by the dark "robber's mask" outlining its eyes and the contrasting white line along its upper lip. Wood frogs are the most terrestrial of the true frogs; adults live on land, in both hardwood and coniferous forests. A good place to look for these brownish amphibians is along abandoned logging roads, where sunlight strikes the forest floor and stimulates the growth of low plants. Sometimes wood frogs move from forested areas into adjoining pastures and grasslands. The species ranges farther north than any other reptile or amphibian, reaching Alaska and northern Quebec. *Rana sylvatica* lives in suitable habitat throughout Pennsylvania.

From October to March, wood frogs hibernate on land, beneath leaf litter and in rotting wood. To survive freezing temperatures, they produce large amounts of glucose, which acts as a natural antifreeze to keep ice from rupturing body cells. Wood frogs eat insects, spiders, snails, slugs, and earthworms. In March, they may not even wait until all the ice has melted from a pond before entering the water to breed. During spring, males are very dark, even black, so that their "masks" are not always discernible. The males call both day and night, repeating loud quacking calls; they sound like ducks, or like round stones clacked together. Each year, wood frogs breed in a temporary rain-fed pond about 300 yards from our house. The earliest date I have heard them commence calling was March 9, 1998; the latest was April 6, 1993. It can be hard to observe these shy amphibians, since approaching the pond may send them diving to the bottom.

The breeding season is short, lasting only about two weeks. Males do not defend territories, and as soon as a female enters the pond, she is seized by one or several males. (Some females are drowned by their overly aggressive suitors.) Pairs may stay clasped for two or three days, until the female lays all her eggs: up to two thousand of them, in a globular mass. After about three weeks, the eggs hatch, and for the next sixty to seventy days, the tadpoles feed and grow. In a dry year, with the pond shrinking rapidly, the tadpoles transform into froglets and leave the water as quickly as they can. In a year when rains replenish the pond, the larval stage is more drawn out: apparently the pond is a safer, more comfortable place for a juvenile wood frog than the land.

Northern Leopard Frog *(Rana pipiens)*. Like its feline namesake, the leopard frog has spots: dark, rounded ones irregularly dispersed over a green, greenish brown, or brown back. Adults are 2 to 4 inches long. The northern leopard frog's range includes much of northern North America. In Pennsylvania, it is absent only from the Pocono region, the extreme southeast corner, and the Allegheny Mountains in the north-central part of the state. Leopard frogs prefer open, sunny habitats, including ponds, marshes, swamps, bogs, and river and stream edges. In spring, they congregate in shallow waters for breeding; in summer, adults venture away from the water into meadows, fields, and pastures; and in early fall, they shift back to permanent waters, such as larger ponds, streams, and lakes, where they hibernate underwater, buried in mud.

Leopard frogs concentrate on insects—beetles, caterpillars, wasps, bugs, crickets, grasshoppers, and ants—and also take spiders, small crayfish, and snails. The male's breeding call is a deep, rumbling snore punctuated with hoarse croaks. The female lays four thousand to sixty-five hundred eggs in shallow water. The eggs hatch after two to three weeks, and the tadpole stage lasts for nine to twelve weeks, with transformation coming in July and August. *Rana pipiens* is the frog commonly sold for use in experiments and for dissection in science classes. Since the late 1960s, field researchers and naturalists have reported declining leopard frog populations. It is not known whether the declines come from illegal capture for the biological supply and fish bait industries, an ongoing loss of habitat, acid rain changing the pH of breeding waters, the overuse of pesticides, or a combination of these factors.

Coastal Plain Leopard Frog *(Rana utricularia).* This southern version of the northern leopard frog barely ranges into Bucks and Delaware counties in southeastern Pennsylvania, where it is found in tidal wetlands of the Delaware River. Because its habitat is threatened by pollution and development, the coastal plain leopard frog is listed as an endangered species in the Keystone State. This frog is more common in New Jersey, Delaware, Maryland, Virginia, and south to Florida and west to Texas. Its life history is similar to that of the northern leopard frog.

Pickerel Frog *(Rana palustris).* Common and statewide in Pennsylvania, pickerel frogs prefer the cool, clear waters of streams (often in shady mountain ravines), springs, bogs, ponds, and lakes. On summer days, they may wander into grassy fields or weedy areas. I once tried to catch a pickerel frog in a damp meadow, and it was several minutes before I could tire out that agile jumper enough to cover him with my hand. Pickerel frogs are light brown to olive, overlaid with dark squarish markings on the back. Adults are 1.75 to 3 inches in length.

In spring, pickerel frogs congregate at woodland ponds, bogs, creeks, and water-filled ditches. Breeding takes place from March to May. Males attract females with a low-pitched snoring call lasting one to three seconds. Females lay two thousand to three thousand eggs in a globular mass attached to underwater plants. The eggs hatch after eleven to twenty-one days; the tadpole stage lasts for about three months, with the pollywogs transforming into froglets from July to September. Pickerel frogs winter in mud on the bottoms of ponds, beneath rocks in damp ravines, and occasionally in caves. Adults do not breed until their second or third year. The skin of pickerel frogs produces secretions that repel some predators.

BULLFROG

The bullfrog, *Rana catesbeiana*, is the largest frog native to North America. The call of the breeding male, a low, rumbling *jug-o-rum*, is a familiar sound of summer. Bullfrogs live across the eastern two-thirds of the United States and in adjoining parts of Canada, and populations have been introduced in many western states as well. Bullfrogs favor larger bodies of water than do most other frog species. Some people hunt bullfrogs for their meaty hind legs, and the harvest of these amphibians is regulated by the Pennsylvania Fish and Boat Commission.

Biology. Bullfrogs, like all amphibians, grow throughout their lives, and old adults can be as long as 7 inches from nose to tail. Most bullfrogs are 3.5 to 6 inches long, colored olive green, brown, or yellowish on the back; the belly is off-white. Bullfrogs have smooth backs with no ridges, which distinguishes them from green frogs and other true frogs. Male bullfrogs have yellow throats, which bulge visibly when the frog is calling, and females have white throats. Another way to tell the sexes apart is by the size of the tympanum, the round, external eardrum visible on the side of the head: in males, the tympanum is about one and a half times as large as the eye, and in females, it is the same size as the eye or a bit smaller.

Bullfrogs are solitary and rarely move far from water. They rest and hunt among dense plant growth along the edges of rivers, lakes, and ponds. They swim powerfully, using their long, muscular hind legs and their webbed feet. Aggressively they will lunge forward, open-mouthed, to take creatures as large as ducklings, small turtles, fish, snakes, newts, smaller frogs (including bullfrogs), and mice; sometimes they attack fishing lures. Herpetologists have seen bullfrogs stalking through the shallows of temporary ponds, seizing male spring peepers as they called. But mostly bullfrogs eat insects, spiders, crayfish, snails, and other small invertebrates. Extremely wary, bullfrogs will leap into the water to escape from danger, sounding a single yelp as they jump. Minks, otters, raccoons, herons, water snakes, snapping turtles, bass, and pike feed on bullfrogs. When seized by a predator, a bullfrog may utter a loud, wailing cry.

Bullfrogs spend the winter underwater, hibernating on the bottom beneath dead leaves and muck. When the water reaches 55 to 60 degrees Fahrenheit, usually in April, bullfrogs become active. Listen for the first calls of males in mid-May; the vibrant bass notes will continue into July as the males try to attract females. Males choose breeding territories in the water away from the bank and where plants grow thickly. The territories range from 6 to about 20 feet in diameter. Should one male try to take over another's territory, the rivals will shove, gouge, and grip each other, churning up the water. Males guard their territories during the day and even more actively at night.

The oldest, largest males appropriate the best territories. They call while partly submerged or perched on a log or a floating mass of plants. Slightly smaller males may swim about in the habitat, calling periodically but retreating from a fight. Even smaller males swim without calling, trying to intercept females as they home in on the territorial males; these "satellite males" are thought to make up 14 to 43 percent of breeding populations.

About a week after the males start calling, the females—singly or a few at a time—enter the breeding areas. A female will pick a mate, nudging him with her leg or nose. The male then clasps the female from behind. As she swims, she extrudes her eggs in bunches of fifty to sixty, and the male sheds sperm into the water to fertilize them.

Bullfrog eggs are tiny: 0.05 inch in diameter. As the female releases them, they rise to the surface and spread out in a thin, frothy mass. The egg masses, up to 2 feet in diameter, lodge against plants; they don't last long, with leeches and other predators eating the eggs, and tadpoles hatching out after just three or four days. Most females lay around six thousand to seven thousand eggs, and really big individuals can deposit almost twenty thousand. After laying her eggs, the female returns to a life along the shore, and the male goes back to his territory and tries to attract another mate.

The tadpoles are greenish brown on top, matching the silty pond bottom, and pale on their bellies. They eat algae, other aquatic plant matter, and carrion. Some bullfrog tadpoles become huge: 5 to 6 inches, including the tail, by the time they transform into adults. During their first winter, tadpoles hibernate by digging themselves into mud or leaves on the lake or pond bottom. Most tadpoles transform after twelve to fourteen months, at the end of their second summer, but many go through another year in the larval stage. As adults, bullfrogs reach breeding size when they are 4 to 4.5 inches long. Bullfrogs hibernate between mid-October and April or May in

Male bullfrogs stake out territories, from which they call to attract mates.

Pennsylvania. Males do not live as long as females, since their calling and breeding behaviors expose them to more predators. Bullfrogs have lived for fifteen years in captivity.

Habitat. Bullfrogs prosper in many different habitats. They prefer larger bodies of water, including lakes, reservoirs, stream and river shallows and backwaters, marshes, floodplain swamps, and bogs. Bullfrogs also reside, in smaller numbers, in ponds on farms and golf courses, as well as quarry ponds and ornamental ponds. They need permanent water with plenty of submerged and emergent water plants. Juvenile bullfrogs are sometimes found in temporary ponds and small streams, sanctuaries they may resort to when dispersing, moving between habitats during wet weather.

Population. These large, adaptable frogs are common in Pennsylvania. The largest populations center on habitats where underwater and shoreline plants grow thickly. Herpetologists have noticed declining numbers of bullfrogs in some eastern states and suggest that habitat destruction, overharvesting for food or the biological supply trade, the use of pesticides and herbicides, and drought may all cut down on the number of bullfrogs.

SNAPPING TURTLE

On a sunny April day, while canoeing at Black Moshannon State Park in Centre County, my family and I came upon one snapping turtle after another. All were in the shallows fringing the lake, and all were dug down into the muck on the bottom. Because the water plants hadn't yet begun pushing up, the turtles were fully exposed. Was that why each had buried itself out of view? Or were the turtles still hibernating? I prodded one of the mucky mounds with my paddle. Instead of displaying the vaunted temper of *Chelydra serpentina,* the snapper simply swam a few feet ahead, then commenced squirming and digging itself back into the lake bottom—although not before we'd gotten a good look at its massive head, formidable jaws, rough-ridged carapace, and long tail topped with saw-tooth projections.

The common snapping turtle resides in many watery habitats, including ponds, lakes, streams, and rivers. It is statewide in Pennsylvania and is our largest turtle species. *Chelydra serpentina* lives across the eastern two-thirds of North America from Nova Scotia to Florida and from

Saskatchewan to Texas; three other subspecies extend the species' range into northern South America.

Biology. The shell, or carapace, of a mature snapping turtle can surpass 17 inches in length, and the turtle may weigh over 40 pounds. The shell is dark brown, usually covered with algae and mud, with a serrated rear margin and with three parallel ridges running front to back. The plastron (the part of the shell that covers the belly) is cross-shaped and relatively small, so that much of the fleshy underparts are exposed. A snapping turtle has a pointed tail about as long as its shell. The feet have webbing between the claw-tipped toes. The eyes are beady, and the upper jaw ends in a prominent hook. Loose, warty skin drapes the neck.

During the day, a snapping turtle will lie on the bottom of a deep pool or sink itself into the mud in shallow water, periodically stretching out its neck and raising its nostrils to take in air; or it may float near the surface with only its eyes and nostrils exposed. Snappers don't bask out of the water as frequently as other turtles do, because their bodies lose moisture rapidly and they tolerate heat poorly. Instead, they soak up warmth by moving into shallows that get a lot of sunlight.

In the evening and at night, snappers become active, creeping along on the bottoms of streams, lakes, and ponds. If disturbed, they swim away speedily. On land, they are fierce when threatened, hissing and lunging forward, and a bite from one can cause severe injury. Always keep at least one shell's length away from an angry snapper. To hold a snapping turtle, grip it by a hind leg, not the tail, which might injure the creature's spine. Carry it well away from your body with its belly toward you; it can extend its head back over the shell much farther than it can reach under the shell. Snappers are not nearly so feisty when in their natural element. Writes Joseph Mitchell in *The Reptiles of Virginia*, "I have stepped on and bumped into snappers [underwater] without their becoming aggressive."

Snapping turtles eat a tremendous variety of foods: insects, crayfish, clams, snails, worms, fish (particularly sunfish, suckers, bullheads, and perch), fish eggs, frogs and toads, tadpoles, salamanders, snakes, small turtles, waterfowl, and carrion. They also eat algae, pondweeds, knotweeds, water lilies, duckweeds, cattails, wild celery, cow lilies, and naiads; plant matter may make up 35 to 70 percent of an individual's diet. Snapping turtles sometimes catch swimming waterfowl, drag them below the surface, and drown them. People have watched snappers, resting on the water's edge, lunge out to seize cottontail rabbits. Young snappers tend to forage more actively than older individuals, which generally hide and wait. Snappers feed underwater; they swallow small items whole, and they hold large prey in their jaws, then rip it to shreds with their strong front claws.

Both males and females become sexually mature when their shells are about 8 inches long. Mating, which takes place in the water, can occur from April to November. Some zoologists believe that the male's sperm remains viable in the female for several years. Females nest from late May through September, with a peak in June. A gravid female will leave the water and travel as far as several hundred yards to find a well-drained, sunny area in which to deposit her eggs. Using her hind claws, she digs out a flask-shaped nest with a narrow opening expanding to a large chamber 4 to 7 inches below the ground's surface. In the chamber, she lays twenty to forty eggs (a large female can deposit as many as eighty eggs) and buries them. The eggs are spherical, about 1 inch in diameter, and wrapped in a tough white shell. Some females lay more than one clutch each season.

The eggs hatch after 55 to 125 days. Hatchlings usually emerge from late August to early October, but they may remain in the nest until the following spring. They head for the water straightaway. The hatchlings are

Common snapping turtle.

about 1 inch long and weigh 2 ounces. They fall prey to herons, hawks, crows, bullfrogs, large fish, and snakes; raccoons, skunks, foxes, minks, and crows often dig up and eat the eggs.

Should a habitat dry out in midsummer, snapping turtles may estivate by digging themselves into wet mud—or they may strike out overland to find water. I once encountered a snapping turtle in the flower garden where I worked, a mile and a half from the nearest stream or pond. People have seen snapping turtles in winter, crawling along on the bottoms of ponds beneath the ice. Most, however, hibernate: in late October, they burrow into the soft bottom, squirm under logs or debris, or hide in muskrat burrows and lodges. Sometimes large numbers of snappers hibernate together. In Virginia, herpetologists found where raccoons had dug adult snappers out of shallow mud and then eaten the cold, lethargic reptiles, wisely starting from the rear.

Snapping turtles have lived for over thirty years in captivity. A study in Canada found an annual survival rate among adult females of over 96 percent.

Habitat. *Chelydra serpentina* is found in many aquatic habitats, including rivers, lakes, reservoirs, ponds, beaver marshes, creeks, bogs, and even salt marshes and tidal creeks. The preferred habitat is a freshwater stream or river with a muddy bottom and plenty of vegetation along the banks. Snapping turtles are more tolerant of polluted water than are many other reptiles and amphibians and may live in urban rivers and wetlands near industrial sites.

Population. The snapper is one of our most common turtles, which, through sheer size—and not only strength or pugnaciousness—may dominate a given habitat. According to Joseph Mitchell in *The Reptiles of Virginia,* "Because of their large size, the biomass of snapping turtles may exceed that of all other species in the freshwater turtle community." He cited studies in South Carolina and Michigan that projected a biomass ranging from 10.5 pounds to over 30 pounds of snapping turtles per acre of habitat.

Many snappers, particularly nesting females and young, are killed by automobiles each year. Humans are the main predators of snappers, legally taking them with baited set lines and in traps. Ponds or local waterways can be trapped so heavily that a decade may pass before these areas support snappers again. The flesh of *Chelydra serpentina* has an excellent flavor and is often used in soup. However, turtles gotten from some habitats may not be safe to eat. Scientists have found toxic chemicals, including organochlorides (from pesticides) and heavy metals, concentrated in the tissues of snapping turtles.

STINKPOT AND MUD TURTLE

The twenty-two species of musk and mud turtles occur only in the New World; they are found in freshwater habitats from Canada to Argentina. These relatively small turtles have smooth, high-domed shells. As a means of self-defense, they emit a foul-smelling musk from glands in the skin under the carapace, or upper shell. Today only one species in the group inhabits Pennsylvania: the stinkpot turtle, living in lowland areas in the northwestern and southeastern corners of the state. The eastern mud turtle once inhabited the southeast, but herpetologists believe it has been extirpated.

Musk and mud turtles walk along on the bottoms of shallow streams and ponds while searching for food. They rise to the surface to breathe, and they often bask in shallow water, exposing the upper parts of their shells to sunlight. They do not welcome being handled, and their sharp jaws can deliver a painful bite.

Stinkpot Turtle *(Sternotherus odoratus)*. Sometimes called the common musk turtle, this reptile is 3 to 5 inches long. The shell is olive-brown to black, often with an irregular pattern of dark streaks or spots and usually patched or covered with algae; it looks much like a rounded stone. The large head has distinct yellowish stripes on each side, one above and one below the eye. The species ranges from New England and southern Canada south to Florida and Texas. In Pennsylvania, the stinkpot inhabits wetlands in the Shenango River watershed and Lake Erie in the northwest; it is also found across much of south-central and southeastern Pennsylvania.

Stinkpots thrive in the slow-moving or still waters of rivers, streams, lakes, ponds, ditches, and swamps. They emerge from hibernation in March or April. Active mainly at night, they trundle along on the bottoms of waterways feeding on insects, snails, mollusks, crayfish, tadpoles, fish eggs, small fish, carrion, the seeds of aquatic plants, and algae; they may venture onto land to feed on slugs and earthworms. In turn, stinkpots—especially the juveniles—are preyed on by raccoons, skunks, herons, foxes, large fish, bullfrogs, snapping turtles, and water snakes. Stinkpots spend much of their time underwater; studies have shown that submerged individuals need only one-eighth the oxygen that they require when breathing atmospheric air. Although they do most of their basking in shallow water, stinkpots sometimes climb out onto banks and crawl up into the limbs of leaning or fallen trees.

Most male stinkpots mature sexually in their third or fourth year; females may take somewhat longer. Stinkpots breed sporadically throughout the year, with peaks in the spring and fall. Stinkpots court and mate in the water. Females lay their eggs from May through August. They leave the water and deposit their eggs in shallow nests dug in sand, humus, or rotting stumps; under logs, debris, or mats of vegetation; and even in the walls of muskrat houses. Normal clutches range from two to five eggs, although some females lay up to nine. Several females may lay so close together that their clutches become intermingled. In most cases, the eggs hatch in September and October. Hatchling stinkpots are about the diameter of twenty-five-cent pieces.

Stinkpots spend the winter burrowed in soft muck on the bottoms of their watery homes. Some take shelter in muskrat burrows and beneath debris in the water, and sometimes they hibernate in large groups. Stinkpots do not move very far within their habitats. Of twenty-three Pennsylvania stinkpots monitored by herpetologist Carl Ernst, the average distance between recaptures was about 150 yards. Stinkpots sometimes annoy fishermen by stealing the bait off their hooks. One stinkpot lived for more than fifty years in the Philadelphia Zoo.

Eastern Mud Turtle (*Kinosternon subrubum subrubum*). Habitat loss in the urbanized, densely populated Delaware Valley of southeastern Pennsylvania sounded the death knell for this small, drab turtle, which was recently removed from the state endangered species list and reclassified as "extirpated." The eastern mud turtle was near the northern limit of its range in southeastern Pennsylvania. It still exists in the Piedmont and Atlantic coastal plain areas of New Jersey, Delaware, Maryland, and Virginia, and in states south to Florida.

The eastern mud turtle has a dark brown carapace, usually without markings, and reaches a length of 3 to 4 inches. Prime habitats are shallow, sluggish waters of ponds, lakes, creeks, marshes, swamps, and bogs, with plenty of aquatic and emergent plants and a soft, mucky bottom. Mud turtles sometimes live in brackish water. Active mainly at night, they eat insects, crustaceans, amphibians, algae, plant parts and seeds, carrion, and other items. Mud turtles mate from mid-March to May. In May and June, females lay one to six eggs in nest cavities dug into mounds of plant debris or sandy or loamy soil. The young hatch in August and September. Mud turtles often venture onto land to look for food, particularly after rainstorms.

POND AND MARSH TURTLES

These mostly medium-size turtles live in and near permanent bodies of water, with the different species showing varying degrees of preference for an aquatic habitat. As well as the eight species described in this chapter, the box turtle, *Terrapene carolina,* is also grouped with the pond and marsh turtles. (This familiar species is covered in a separate chapter.) The pond and marsh turtles belong to the family Emydidae, which includes a total of twenty-five species across the United States.

Turtles are reptiles that, over millions of years, have evolved a protective shell. When danger threatens, a turtle can pull its legs and head back into its shell; the head retracts when the neck bends vertically into an S shape. The shell is made up of two parts: an upper portion, called the carapace, knit together out of about fifty bones; and the plastron, whose nine interlocking bones protect the turtle's belly. Pigmented horny plates, called scutes, cover the bones of the shell on the outside.

Pond and marsh turtles are active mainly during the day. They emerge from hibernation in late March or early April, when air and water temperatures reach 60 degrees Fahrenheit and above. These turtles often bask in open areas; by soaking up the sun's rays (and, alternatively, by moving into shade or water), they can maintain their blood temperature at a level that lets them stay active. For a habitat to be suitable to a pond or marsh turtle, it must include basking sites (rocks, logs, or ledges) near the escape cover provided by deep water.

Omnivorous feeders, pond and marsh turtles eat small animals, including many insects and spiders, and the stems, leaves, and fruits of plants. Typically, these turtles mate in the spring. In June and July, females lay eggs that hatch during August and September. The female uses her hind feet to dig a nest in soil, humus, or sawdust. She deposits her eggs in the nest, buries the clutch, and then abandons it to be incubated by the sun. Mammals and birds may break into turtle nests and eat the eggs; many predators, including large fish, birds, and mammals, catch and eat young turtles. Around November, turtles go into hibernation after digging into the soft bottom of a waterway or into leaf litter or soil, or by creeping into a muskrat house or a crevice in a stream bank.

Spotted Turtle *(Clemmys guttata).* This small turtle is 4 to 6 inches long, with bright yellow-orange spots on the head, neck, and scattered across the black or bluish black carapace. The spots fade with age, and very old individuals may not have them. *Clemmys guttata* ranges from Maine south and west to Georgia and Illinois. Spotted turtles live in wet meadows, swampy woodlands, marshes, bogs, ponds, and meandering creeks. In Pennsylvania, they are found at lower elevations, both east and west of the Alleghenies, in about two-thirds of the state.

Spotted turtles feed in the water and on land. They eat insects, worms, slugs, snails, crayfish, spiders, millipedes, tadpoles, and carrion, plus assorted vegetation, including aquatic grasses and filamentous algae. Spotted turtles are most active in the spring; during the heat of summer, they become inactive, estivating in mud and muskrat burrows and under creek banks (sites that are also used for hibernating in winter). Spotted turtles often bask on logs, stumps, tree roots sticking out of the water, and mats and tussocks of grass.

From May to July, females dig their nests in soil among grass tussocks and sphagnum moss, often in marshy pastures. They lay their eggs in clutches of three to eight; the eggs hatch after about ten weeks. A prime habitat may hold sixteen to thirty-two individuals per acre. Spotted turtles have been known to live for twenty-six years in the wild, and captives have survived for more than forty years.

Bog Turtle *(Clemmys muhlenbergii)*. A German naturalist first described this turtle in the late 1700s from a specimen sent to him by Heinrich Mühlenberg, the first president of Franklin College in Lancaster. The bog turtle is our smallest turtle; adults are 3.5 to 4.5 inches long. The shell is dark brown with a pattern of faint orangish markings, and a large orange or yellow blotch decorates both sides of the head. The bog turtle has a discontinuous range, living in widely separated habitats from western Connecticut and eastern New York south through eastern Pennsylvania and New Jersey to North Carolina. *Clemmys muhlenbergii* is a state endangered species in Pennsylvania, at risk from collectors who unlawfully capture these rare turtles and from the illegal draining, filling, and polluting of wetlands. In 1997, the federal government designated the bog turtle a threatened species.

Bog turtles live in damp, grassy fields and meadows with slow-moving streams and boggy areas fed by pure water from springs. They eat mostly insects, and also worms, snails, amphibians, carrion, berries, and seeds. In a Pennsylvania study, the average home range for an adult was a little over 3 acres, and the farthest any individual traveled was 740 feet. When frightened, bog turtles dive and burrow into the mucky bottoms of shallow streams. A mature female typically lays a clutch of three to five eggs in June. Adults are preyed on by raccoons, skunks, foxes, and dogs.

Wood Turtle *(Clemmys insculpta)*. Other than the box turtle, this is our most land-loving turtle species. The wood turtle is 5.5 to 7.5 inches long, with very old adults becoming as long as 9 inches. The carapace is broad and flattened, with each scute taking a pyramidal form; the scutes are brown, with concentric growth rings, faint black and yellow lines or ridges that look like the contour lines on a topographic map. The plastron (the bottom portion of the shell) bears a striking black and yellow pattern, and the blackish skin of the forelegs and neck usually has some orange or red coloration—hence the nickname "redlegs turtle." The wood turtle ranges from southern Canada to northern Virginia. It is found throughout Pennsylvania, except along the state's western border.

From April to November, wood turtles range over the land, living in forests, bogs, wooded swamps, meadows, and old fields. Individuals are active during the day and often bask on logs; during dry summer weather, they soak themselves in mud puddles. Wood turtles eat algae, moss, grass, leaves, and berries; animals such as insects, earthworms, snails, tadpoles, and newborn mice; and carrion. The wood turtle's year also may include an aquatic phase. The winter is spent on the bottom of a creek or stream, or perhaps tucked in under a bank overhang or in a muskrat burrow.

Annual growth rings mark the pyramid-shaped scutes,
or plates, on the wood turtle's shell.

Wood turtles do not mature sexually until they are more than ten years old. Males and females mate in the water. From May to early July, the female lays a clutch of four to twelve eggs (usually seven or eight), burying them in soil or humus in direct sunlight. Should mating take place in autumn, a female will not lay her eggs until the following spring. The eggs hatch after forty to sixty-seven days. The hatchlings leave the nest from mid-August to early October.

A good habitat can support five adult wood turtles per acre. In south-central Pennsylvania, researchers found that wood turtles' home ranges followed lowland stream corridors. Wood turtles can live up to forty years in the wild.

Map Turtle *(Graptemys geographica).* Adults reach 7 to 11 inches in length. The olive-brown carapace is covered with fine yellow lines in a squiggling, branching pattern that suggests the markings on a map. Map turtles range from southern Canada west to Wisconsin and south to Arkansas and Georgia. The species inhabits three separate ranges in our state: along Lake Erie and its tributaries; in part of the Ohio River drainage in western Pennsylvania; and throughout much of the Susquehanna and lower Delaware River basins. While canoeing, I have come upon map turtles basking on logs, stacked up one on top of another. It's impossible to drift very close to these gatherings. Inevitably one of the skittish creatures slides into the water, the rest following in a series of splashes.

Map turtles feed mainly on freshwater snails and clams, crushing them with their powerful jaws. They also eat insects, crayfish, carrion, and plants. Strong swimmers, they paddle along using their large webbed feet. They also walk about on the bottoms of rivers, creeks, and lakes, searching for food among aquatic vegetation. They avoid areas of swift current. Map turtles nest from May to mid-July. The average clutch is twelve to fourteen eggs, buried in a double layer. The young hatch from mid-August to September.

Map turtles generally do not leave the water except to bask or lay eggs. They hibernate under rocks and rock ledges in deep, slow water, although people have spotted them moving sluggishly under the ice in winter. Map turtles emerge from hibernation earlier in the spring than do our other species of turtles.

Painted Turtle *(Chrysemys picta).* The most widespread turtle in North America, *Chrysemys picta* is found from coast to coast. Two subspecies, or races, occur in Pennsylvania: the midland painted turtle *(Chrysemys picta* *marginata)* and the eastern painted turtle *(Chrysemys picta picta).* The painted turtle has a dark, smooth carapace that is edged with red and yellow; individuals also have yellow neck stripes and spots, and yellow- and red-striped limbs. Adults are 4 to 7 inches long.

Painted turtles inhabit shallows of lakes and ponds, marshes, and slow-moving rivers and streams with soft bottoms and ample vegetation. They often bask on rocks and logs, where they may share space with turtles of other species. Male painted turtles sometimes wander overland in spring and early summer.

Active from March through October, painted turtles feed on a range of animals and plants. The young are mainly carnivorous, but they change to a more herbivorous diet as they grow larger. Males mature sexually at age two to five, and females mature when four to eight years old. Females lay one or two clutches per year, each having two to twenty eggs. In winter, painted turtles hibernate in water as deep as 3 feet, burying themselves up to 18 inches below the muddy bottom. In a study conducted over three years at White Oak Sanctuary, Lancaster County, painted turtles comprised three-quarters of all turtles captured, even though six other species were present. Young painted turtles are eaten by raccoons, minks, muskrats, crows, black racers, snapping turtles, bullfrogs, and large fish.

Redbelly Turtle *(Pseudemys rubriventris).* Fully grown redbelly turtles can be over a foot in length (the record is a whopping 15.75 inches). The carapace is olive-brown to black, with vertical reddish lines, and the plastron is reddish; yellow stripes mark this turtle's head and neck. A threatened species in Pennsylvania, the redbelly turtle is found from Harrisburg south along the Susquehanna River and east across the southeastern part of the state to Philadelphia; development in this populous region has destroyed much turtle habitat.

Redbelly turtles live in creeks, rivers, ponds, lakes, and marshes with dense vegetation and basking sites—rocks and logs—near deep water. They often bask in the company of painted turtles. In some areas, juvenile redbelly turtles must compete for limited basking sites with red-eared sliders, a non-native species that has become established in several parts of Pennsylvania. If the redbelly turtles cannot displace the larger adult sliders from preferred basking sites, the juvenile redbelly turtles end up sunning themselves on the exposed banks of waterways, where they are more vulnerable to raccoons, minks, and other predators.

Redbelly turtles nest from late May to July. Females lay ten to twelve eggs and bury them about 4 inches below the ground's surface, often in a cultivated field. The eggs hatch after seventy to eighty days. Both adults and young eat snails, tadpoles, crayfish, and aquatic vegetation. In the past, redbelly turtles were captured and sold for food. Although these turtles are rare in Pennsylvania, populations remain larger and more secure in Virginia and North Carolina.

Blanding's Turtle *(Emydoidea blandingii)*. Named after Dr. William Blanding, the nineteenth-century Philadelphia naturalist who first identified it, this species has a smooth carapace marked with yellowish spots, and a yellow throat and chin. The plastron has a hinge, like a box turtle's, but unlike the box turtle, Blanding's turtle cannot close its plastron completely. *Emydoidea blandingii* ranges from southern Ontario through Ohio and the Great Lakes states, and west to Nebraska. Isolated populations exist in New York, Massachusetts, New Hampshire, and in Erie County in Pennsylvania. Adult Blanding's turtles may occasionally migrate along the Lake Erie shore into Pennsylvania from Ohio; no eggs or juveniles have been found in Pennsylvania for many years. Blanding's turtle is a candidate for inclusion on the Pennsylvania endangered and threatened species list.

Blanding's turtles prefer the shallows of lakes, ponds, marshes, and creeks that are grown up with aquatic vegetation. They eat many crayfish. Blanding's turtles also feed on insects, frogs, fish, snails, earthworms, and plant matter. Most breeding takes place from March to July, and clutches contain six to eleven eggs.

Red-Eared Turtle *(Chrysemys scripta elegans)*. Also known as the red-eared slider, this species is native to the Midwest and the South, ranging from Indiana to Alabama and Texas. It is the turtle commonly sold by pet stores and scientific supply companies. The red-eared turtle has a broad red stripe behind the eye, an identifying mark usually visible through binoculars. Adults are 5 to 8 inches long. Reproducing populations of red-eared turtles have shown up in several watersheds in southeastern Pennsylvania and in Allegheny, Erie, and Crawford counties; they probably come from people releasing pets and animals used in school science classes. It is illegal, as well as unwise, to turn loose non-native species. In southeastern Pennsylvania, the red-eared turtle competes with the redbelly turtle, a threatened species, for food and basking sites.

Red-eared turtles inhabit quiet water. They bask on logs, rocks, and masses of floating vegetation, where they may pile up several layers deep. Females nest in June and July, laying clutches of ten to twelve or more eggs, sometimes as far as 100 feet from the water.

BOX TURTLE

In *Pennsylvania Amphibians and Reptiles,* the range map for the eastern box turtle, *Terrapene carolina carolina,* implies that this well-known reptile does not occur in northwestern Centre County. I would like to object, and happily, for I often come upon these beautiful animals on and near my wooded acres near the brow of the Allegheny Front. Four subspecies of the box turtle—the most terrestrial of the pond and marsh turtles—live from southern New England south and west to Florida and Texas. In Pennsylvania, box turtles are found mainly in the southern two-thirds of the state and in the Lake Erie Basin; they are absent, for the most part, from the Allegheny High Plateau.

Terrapene carolina played a prominent role in the rituals and legends of the Eastern Woodland Indians; the genus name, as well as the word *terrapin,* come from the Algonquian tongue. The "box" in this turtle's common name tells of its ability to shut itself up like a box when danger threatens.

Biology. Adults are 4.5 to 6 inches; the largest box turtle ever measured had a shell nearly 8 inches long. A box turtle has a high-domed, helmet-shaped, dark-brown carapace with yellow or orange lines, spots, or bars radiating outward from the center of each scute (the horny plates that cover the shell in checkerboard fashion). Box turtles often live along woods edges, where the blotched patterns of their shells provide camouflage in dappled sunlight. The patterns on box turtles' shells differ among individuals, which also may have varying amounts of yellow, orange, or red on the head, neck, and forelimbs. In most male box turtles, the iris of the eye is bright red; females' eyes are brown.

The box turtle has a hinge in the center of its plastron, or lower shell, that lets the turtle close its shell completely, sealing off its limbs and head. When frightened, a box turtle clamps its shell shut, giving out an audible hiss; after a while it opens its plastron a crack, and peers out; finally, if not menaced, it sticks its head and appendages back out and resumes its activity. Some box turtles I have met were quite suspicious, staying in their shells for many minutes; others barely closed their shells when I picked them up and were back in action almost the moment I replaced them on the ground.

Box turtles usually hibernate on land rather than in mud beneath the water, as is the practice of their close relatives, the pond and marsh turtles. Box turtles emerge from hibernation in late April or early May; if caught by a cold snap, they may freeze to death. During warm weather, they feed and move about early in the morning or during rain showers; on hot days, they bury themselves under leaves or in rotting logs, hunker down in cool soil, rest in a mammal's burrow, or soak in a rivulet or puddle. Box turtles are active only during the day. At night, they scoop out shallow depressions in which to sleep.

Box turtles eat snails, slugs, crayfish, spiders, millipedes, centipedes, insects, fish, frogs, toads, salamanders, small snakes, and carrion. They also relish vegetable foods: roots, stems, leaves, fruits, and seeds. Food plants include blackberry, blueberry, mulberry, strawberry, wintergreen, viburnums, mosses, grasses, and garden vegetables. There is some evidence that box turtles can eat poisonous mushrooms without harm. Young box turtles favor a high-protein meat diet, and older ones consume more vegetation. Some scientists believe that *Terrapene carolina* is an important agent in the life cycle of woodland plants: the turtles eat the fruits of jack-in-the-pulpit, mayapple, pokeweed, black cherry, and grape, whose seeds, passed through the turtles' digestive systems, are more widely disseminated and more apt to germinate than seeds that go uneaten.

A box turtle's home range is usually 750 feet in diameter or smaller; a Maryland study found the average diameter to be 330 feet for males and 370 feet for females. An individual may make a circuit of his or her range over several days or weeks. The ranges of individuals overlap, and box turtles generally do not show antagonism when they meet others of their kind. They grow at a rate of about 0.375 inch per year, become sexually mature in four or five years, and reach their full growth after about twenty years. Box turtles show some homing instincts if removed from their territories. Should you find a box turtle crossing a road, the best course is to carry the turtle across the road in the direction it was headed, rather than releasing it elsewhere.

Box turtles breed on land. They mate from May until October. When looking for mates, males sometimes fight with each other. The courtship of box turtles has three phases. First, the male circles the female, nipping and shoving at her shell before mounting her from behind; an indentation in the male's plastron fits over the rear of the female's highly domed carapace. In

*Box turtles eat the fruits of many woodland plants; the seeds,
after passing through the turtles' digestive systems, are more
apt to germinate than ones that go uneaten.*

the second phase, the male hooks his hind feet into the female's open plastron near the hinge. In the third phase, the male bites the forward part of the female's shell, and the actual mating takes place. On rare occasions, a male will die when, after copulating, he falls on his back in a place where he can't get leverage to right himself.

Females in captivity have laid fertile eggs up to four years after mating. In nature, females nest from May through July, with a peak in mid-June. The female seeks out an open patch of sandy or loamy soil. Using her hind limbs, she digs a flask-shaped cavity about 3 inches deep and in it lays three to nine eggs (four or five are most usual). The eggs are elliptical and from 1 to 1.5 inches long. The female covers her clutch with soil, tamps the earth down with her feet and plastron, and leaves the site; the whole process can take up to five hours. Often a female will start digging her nest at twilight and finish it after dark.

The eggs—if not unearthed and eaten by a fox, raccoon, skunk, or crow—hatch after about ten weeks. Most eggs hatch from early September into October. Young box turtles are 1.25 inches long and weigh 0.25 ounce. Their shells are soft, and the hinge is undeveloped. They are good at hiding, but many get caught by birds, mammals, and snakes. Adults are fairly predator-proof, but some have missing or truncated limbs that testify to attacks. Hogs have been seen crushing and eating box turtles. A great many box turtles perish when run over by automobiles, farm equipment, and lawn mowers.

In late October or early November, box turtles dig themselves into the soil beneath the leaves or a clump of grass. They may hibernate in stump holes or mammal burrows, occasionally in the mud in a pond or stream bottom. An individual may use the same spot—called a hibernaculum—several years in a row. Several box turtles may hibernate together. Box turtles enter hibernation earlier and emerge later than the other pond and marsh turtles, which are buffered from radical temperature changes by their aquatic habitat. If the ground is soft, box turtles dig themselves in deeper as winter temperatures fall. Some go as deep as 2 feet beneath the ground; others hibernate with the upper parts of their shells poking above the ground. Probably many box turtles die during cold winters.

Box turtles are long-lived—up to 138 years, according to one record. A study on an estate on Long Island, New York, documented ages of 48 to 86 among eighteen individuals. Very old box turtles have shells that are worn so smooth that they no longer show discernible annual growth rings on their scutes.

Habitat. An ideal habitat for box turtles would be a mix of old fields and deciduous woods with sandy, well-drained soil, not far from a small stream or pond. Box turtles also live in woodlots, logged-over land, powerline cuts, wet meadows, pastures, and swamps. The box turtle can accommodate itself to a wider range of habitat types than can the wood turtle, another terrestrial species. Unfortunately, much box turtle habitat has been erased by sprawling cities and suburbs.

Population. A surprising number of box turtles may live in a good habitat. In a Maryland study, a researcher collected 245 adults on 29 acres and, using recapture data, estimated the population to be 4 to 5 turtles per acre. Some of the turtles had ranges enclosed by the study tract, others had ranges that extended beyond it, and still others seemed to be transients that were moving through the area and did not have a home range. Transient box turtles may promote gene flow from one small population to another and may help the species colonize new areas.

Because they are long-lived, box turtles may survive for many years in pockets of undisturbed habitat hemmed in by suburbs and roads. But if small, increasingly isolated populations cannot exchange genes with other populations—if adults are killed by cars, collected as pets, or simply die of old age—box turtles ultimately may vanish from areas where they now live.

SOFTSHELL TURTLES

Softshell turtles are uncommon aquatic reptiles. Altogether, twenty-two species of softshell turtles inhabit parts of Africa, Asia, and North America, including four species in the United States. One species is found in Pennsylvania; another species once lived here, but biologists believe it is gone from the state.

A softshell turtle has a rounded, flattened carapace that most authorities describe as resembling a pancake; this not-quite shell is covered with a flexible, leathery skin instead of the hard, horny scales or plates of other turtles. The feet are webbed between the toes and tipped with three sharp claws. The neck is long. The head ends in a long, tapering snout that looks like a funnel's narrow end; the turtle pokes its snout above the water to breathe.

Softshell turtles are powerful, maneuverable swimmers, and on land they can crawl with agility and speed. They defend themselves fiercely. They prefer the well-oxygenated running waters of rivers, streams, and creeks, and they also inhabit the still waters of lakes and ponds. Carnivorous feeders, softshell turtles take crayfish, insects, fish, and other prey. In winter, they hibernate by digging themselves into the soft, mucky bottoms of watercourses.

Eastern Spiny Softshell Turtle *(Trionyx spiniferus spiniferus)*. Females reach a maximum size of 17 inches; males are about half the size of females. A spiny softshell turtle has a highly rounded shell (when viewed from above) that is olive-gray to yellowish brown, with dark spots or blotches for camouflage. At the front of the shell, its surface has many small spines that give it a rough, sandpapery texture. The shell margins are flexible. The species' range extends from western New York to Wisconsin and Tennessee. The eastern spiny softshell is native to the western third of Pennsylvania, where it inhabits swift rivers, marshy creeks, stream oxbows, lakes, and ponds in the Lake Erie and Ohio River watersheds. In recent years, spiny softshells have established themselves in the middle and lower Delaware River and some of its tributaries; herpetologists don't know how these non-native populations were introduced to the Delaware Basin.

Spiny softshell turtles spend most of their lives in the water, venturing onto dry land only to lay eggs. Individuals will sometimes haul out onto logs or sandbars to bask in the sun, but most would rather lie on mats of debris or float on the surface of the water. A softshell turtle will bury itself in a sandy or muddy bottom, its neck craned upward through the shallows and its nose breaking the water's surface like a snorkel. Here it reposes, sipping in breath and waiting for prey: adult and larval insects, crayfish, fish, tadpoles, and frogs. Spiny softshell turtles also eat worms, mussels, and snails. They prowl about on the bottom when hunting, pushing their snouts under stones and into vegetation and exploring among the branches of trees that have fallen into the water. Scientists have found vegetation in the stomachs of softshell turtles, but it's not known if the vegetation was eaten as food or gobbled up incidentally along with prey. Adults have few predators other than humans who relish turtle meat; juvenile spiny softshells are taken by herons, snapping turtles, and large fish.

Spiny softshell turtles emerge from hibernation in April or May. They breed from May to August. A month or more after mating, the female clambers out of the water and, using her hind feet, digs a flask-shaped cavity 4 to 10 inches deep in a stretch of sand or gravel that receives plenty of sunlight. The eggs are white, rounded, and slightly more than 1 inch in diameter. Clutch sizes range from four to thirty-two eggs. After laying, the female returns to the water, and the eggs, warmed by the sun, hatch from late August to October. The young tunnel upward through the soil and head for the water. Skunks, raccoons, and crows eat the eggs, and a wide range of predators take the juveniles.

In late fall, spiny softshell turtles dig themselves 2 to 4 inches into a mud bottom beneath deep water. They absorb oxygen from the water through membranes in the cloaca and through the skin.

Midland Smooth Softshell Turtle (*Trionyx muticus muticus*). This is a midwestern turtle whose range formerly extended into western Pennsylvania. No specimens have been found in the state since 1901. Female smooth softshell turtles are 7 to 14 inches long; males are 5 to 7 inches. In both sexes, the sandpapery carapace is colored gray, brown, or olive, with indistinct dots, dashes, and mottling. The species lives in rivers and larger creeks in the Mississippi River Basin as far west as Nebraska and Texas.

FENCE LIZARD AND SKINKS

Lizards look a little like salamanders, but there are significant differences between them. Lizards are reptiles, and salamanders are amphibians. Lizards have five toes per limb, with claws; salamanders have four toes on the front feet and usually five on the rear feet, and lack claws. Lizards have dry skin covered with scales; in salamanders, the skin is moist or slimy, and smooth. Lizards have external ear openings, which salamanders lack. Although short-legged, lizards are agile and swift and could run rings around any salamander.

Pennsylvania has one lizard and three skink species, all considered to be lizards in the suborder Sauria. The different species can be told apart by the markings on their backs. These reptiles live in trees or shrubs, on fences, on the ground, or in burrows under the ground. They are active during the day. Lizards eat insects, spiders, worms, and other small invertebrates: an individual spots the movement of its prey from a few feet away, then rushes in to secure its meal. Lizards are not venomous, but most will try to bite if captured. Should a predator grab a lizard, the lizard may shed its tail in an effort to escape or to deflect the attack away from its body. Shed tails grow back rapidly. For protection, skinks have a layer of bones, called osteoderms, just below their scales; one herpetologist calls skinks "the tanks of the lizard world" and characterizes them as "difficult prey for many predators."

Lizards and skinks hibernate all winter in the ground, in rock crevices, beneath bark, or under boards or logs. They mate in the spring and early summer. Females lay their eggs in soil, sawdust, and the rotting wood of logs and stumps. The eggs are elliptical in shape and have tough, flexible shells. When hatching, a young lizard cuts its way out of the egg using an "egg tooth" on the premaxillary bone toward the front of the head; this tooth is shed soon after hatching.

Northern Fence Lizard (*Sceloporus undulatus hyacinthinus*). The northern fence lizard is gray or brown with narrow dark crossbands on the back. Adults are 4 to 7 inches long, and about half of that length is tail. The northern fence lizard is the only spiny lizard in the Northeast. The creature's overlapping scales each end in a point or spine that feels rough to the touch. Fence lizards inhabit the southern two-thirds of Pennsylvania. The species' range extends from southeastern New York to Florida and Texas. A number of related species live in the U.S. West and Southwest and in Mexico.

Northern fence lizards bask in the sun,
stretched out on rocks, fences, and stumps.

Fence lizards favor dry pine or mixed woods, particularly sunny glades or the edges of wooded areas. They live in powerline rights-of-way, near farm buildings, and in old fences, stone outcrops, and slab piles. Individuals spend much time basking in the sun. They quickly climb trees to escape from danger. Fence lizards eat spiders, millipedes, snails, and a variety of insects, including beetles, grasshoppers, caterpillars, leafhoppers, and ants.

After emerging from hibernation in April, male fence lizards defend small territories. They bob their heads up and down and make "push-up" motions, showing off the bright blue patches on their throats and bellies. Females enter the males' territories to mate. In May or June, a female lays three to thirteen eggs (on average, six) in a nest dug in the soil or in a rotting log or stump. The eggs hatch after about forty-five days. Older females may lay several clutches each summer.

Northern Coal Skink *(Eumeces anthracinus)*. The smooth, flat scales of skinks give them a shiny or a silken appearance. *Eumeces anthracinus* was first described by the naturalist Spencer Baird in the early nineteenth century from specimens collected near Carlisle in Cumberland County. Adults are 5 to 7 inches long, including the tail; maximum head-and-body length is 2.8 inches. The brown body is marked on each side by a pair of pale stripes enclosing a dark band and stretching from neck to tail. Breeding males have reddish patches on the head, and juveniles have violet-blue tails. The coal skink ranges from New York to Virginia and Kentucky. In Pennsylvania, the species is found mainly in the north-central region and the northwest.

Coal skinks are secretive and scarce, and little is known of their life history. They inhabit rocky, wooded slopes with sunlit openings; they are often found near springs. They live under logs, rocks, and leaf litter and readily dive into water when threatened. Coal skinks mate in the spring and early summer. Females lay eight or nine eggs in late June, and the eggs hatch after four to five weeks.

Five-Lined Skink *(Eumeces fasciatus).* This lizard is 5 to 8 inches long, with the tail constituting some 60 percent of the overall length. The body is marked with five white or yellowish stripes against a dark brown background. In males, the stripes become faint or even fade away altogether as the animals age, but females keep their stripes for life. Juveniles have a bright blue tail. If menaced by a predator, a young skink will dive into the leaf litter, then lash its tail back and forth above the surface. The tail, once seized, breaks off easily and keeps twitching for a while, upping the odds that the predator will be distracted long enough for the skink to escape.

Five-lined skinks range from New England to Florida. In Pennsylvania, they occur in about two-thirds of the state, south of a diagonal line from Crawford County in the northwest to Bucks County in the east. *Eumeces fasciatus* lives in a variety of moist, wooded habitats, including pine and hardwood forests, woods edges, woodlots, and around buildings in woods and fields. Individuals hide under logs, boards, and debris. A Maryland study showed that five-lined skinks eat a wide range of insects, including grasshoppers, crickets, cockroaches, leafhoppers, beetles and beetle larvae, flies, butterflies, ants, and dragonflies. Domestic cats, black racers, milk snakes, and birds prey on five-lined skinks.

During the breeding season, males fight with each other and follow the scent trails of females. In June and early July, the females lay eggs in decaying logs or stumps. They stay with the eggs for about a month, turning them daily, but provide no care for the juveniles. Young skinks are about 2 inches long at hatching. In a good habitat, adult males range over an area radiating out about 20 yards from a central point; females' and juveniles' ranges are smaller.

Broad-Headed Skink *(Eumeces laticeps).* The largest of our lizards, the broad-headed skink reaches a maximum head-body length of 5.6 inches and a total length of 1 foot. Adult males are olive-brown with reddish heads. Females and young have stripes on their backs similar to those of five-lined skinks; juveniles also have bright blue tails. Adult males are considerably larger than females. The species occurs throughout the South, and west to Kansas and Texas; its range barely extends north into Pennsylvania, where it has been found only in Chester County in the southeast corner of the state. *Eumeces laticeps* is being considered for inclusion on Pennsylvania's endangered and threatened species list.

Broad-headed skinks spend much of their time above the ground in trees, often climbing onto high limbs. They live in wooded areas and the edges of woods in drier habitats than those used by five-lined skinks. They occupy hollows and holes in trees, brush piles, and crevices beneath loose bark. In winter, groups of broad-headed skinks have been found hibernating together underground. Individuals forage in trees and under the leaf litter, taking a wide range of insects and larvae, as well as snails, spiders, and juveniles of their own species. They often attack their prey by seizing it behind the head.

Females emit a scent from glands in the base of the tail. A male will track a female, flicking his tongue against the ground to pick up her scent. After mating, the female nests in a decaying log or stump. She lays six to sixteen eggs, usually in June, and guards her clutch until the young hatch after three to four weeks. Some females produce two clutches in a summer. Domestic cats are known to be major predators of broad-headed skinks.

NONVENOMOUS SNAKES

The nonvenomous snakes belong to the family Colubridae and are sometimes referred to as colubrid snakes. They have a head that is cylindrical to slightly broadened and flattened, and only a little larger than the body. In contrast, the pitvipers, which are venomous snakes, have a head that is triangular in shape (when seen from above) and noticeably broader than the body. The nonvenomous snakes have round pupils, and the vipers have vertically slitted, elliptical pupils.

Nonvenomous snakes live in many habitats, including streams and ponds, damp lowlands, dry uplands, and wooded, brushy, and grassy areas. Some species climb into trees and shrubs. Some are nocturnal, and others are active during the day. All are carnivorous. Their prey varies greatly, from tiny invertebrates, to amphibians and other snakes (including venomous ones), to small mammals. Some snakes immobilize their prey with a bite; others are constrictors, coiling their muscular bodies around prey and asphyxiating their victims by preventing them from breathing. All snakes swallow their prey whole, since their teeth are suited only to gripping and not to cutting or shearing flesh. A special bone arrangement in the jaws, along with ribs that lack a sternum and therefore are expandable, allow snakes to swallow prey that is much larger than their own body diameter.

Snakes do not have ear openings. Instead, they sense vibrations through the ground or whatever surface they are lying upon. Their sense of smell is different from that of the mammals. A snake picks up chemicals using its tongue, which it flicks in and out rapidly; the tongue transfers the chemicals to a special organ (called the Jacobson's organ) inside the head, where they are interpreted and from where sensory messages are sent to the brain. Snakes find prey, avoid enemies, and locate mates using their taste-based sense of smell. Folk beliefs notwithstanding, the long, forked tongues of snakes are harmless.

A snake moves in two basic modes. In rectilinear movement, the snake proceeds forward slowly and in a straight line, the belly scales moving in sections, with alternating sections gripping the ground and allowing the body to slide. Serpentine locomotion lets a snake go faster: the body contracts into a series of S shapes, with the sides of the outer curves shoving off from stones, rough ground, or other objects, advancing the snake in a wavelike manner.

Snakes often bask. Lying in the sun or on a sun-warmed surface will raise a reptile's body temperature, letting it move quickly, if needed, or speeding up its digestion or promoting the internal development of eggs. In the fall, snakes enter hibernation. They spend the winter in a dormant state, lying in rock crevices, beneath boards and rotting logs, in the leaf litter, and in the soil. They emerge in the spring, usually in April in the Northeast.

Most species mate in spring, fall, or both. Some species reproduce by laying eggs; others give birth to live young. Female snakes do not protect their eggs or their young. Newly hatched or newborn snakes are able to feed and fend for themselves; in some species, the young are equipped with an energy supply in the form of a yolk, which sustains them until they can procure a meal. Instinctively, young snakes hide from the birds, mammals, reptiles, and fish that would prey on them. Some snakes double in size during their first year, and some reach full size after two or three years. As a snake grows, it must periodically shed its skin. The husks of outgrown skins, complete with the scale pattern and dark pigments, can be found in areas frequented by snakes. Snakes grow throughout their lives, although their growth rate slows as they age.

Humans kill a great many snakes, both directly, with automobiles and by attacking them, and by destroying their habitats. The number of snakes has dwindled in the last century as asphalt and houses have proliferated across the landscape. Except under unusual circumstances (such as a venomous snake in a yard where children or pets frequently venture), snakes should not be killed. They are a key component of natural ecosystems. Most snakes, including nonvenomous ones, will bite if handled; others will defecate or emit an offensive-smelling musk from glands at the base of the tail; and others will be docile. It is not a good idea to try to catch any snake that you haven't positively indentified.

Eastern Worm Snake *(Carphophis amoenus amoenus).* Found in the southeast corner of Pennsylvania, and in much of the South and Midwest, this small, trim snake is 7 to 12 inches long. It looks like an earthworm: its back is an unpatterned brown, and its undersurfaces are pinkish. Worm snakes are found mainly in the ground, often under rocks and in moist soil of rocky woodlands. They disappear from cleared areas, where increased solar radiation dries out the soil. Worm snakes are sometimes uncovered by people spading their gardens. *Carphophis amoenus* is adept at burrowing, a practice abetted by a narrow, rounded head, a cylindrical body, tiny eyes, smooth scales, and a short tail. During the night, worm snakes sometimes move out onto the surface of the ground. They eat earthworms, soft-bodied insects, slugs, and snails. Larger snakes, opossums, short-tailed shrews, moles, and house cats prey on worm snakes.

Worm snakes emerge from hibernation in March and April. They have two main activity periods, in the spring and in early fall; during summer, if the heat is excessive, worm snakes burrow into the soil and become dormant. In June and July, females lay two to six eggs in mounds of humus and sawdust and in rotting logs and stumps; at times, several females lay their eggs in one communal nest. The eggs hatch in August and September. Worm snakes return to hibernation in October and November.

A study in Kentucky found an average individual home range size of around 300 square yards. During one hour on an April afternoon, herpetologist Carl Ernst and his students collected 108 worm snakes beneath rocks on a little over 100 yards of hillside overlooking the Kentucky River. Worm snakes do not bite when handled but try to "burrow" between human fingers.

Kirtland's Snake *(Clonophis kirtlandii).* This species' limited range centers on Ohio, Indiana, and Illinois, and extends into parts of adjoining states. Listed as an endangered species in Pennsylvania, Kirtland's snake (also known as Kirtland's water snake) is extremely rare here. Historically, it was most common in Allegheny County, where development and water pollution have destroyed much wildlife habitat; it has also been found along the Clarion River. However, no new sightings have been reported in more than twenty years. According to Carl Ernst and Roger Barbour in *Snakes of Eastern North America,* this relict of the tallgrass prairie ecosystem is "in danger of extirpation over much of its range." The species was named after Jared Kirtland, an Ohio naturalist.

Fourteen to 18 inches long, Kirtland's snake is reddish brown, with a pattern of dark blotches on the back and a reddish belly. When frightened, it flattens its body against the ground and becomes rigid. It lives in open, damp habitats such as marsh edges, wet fields and pastures, and the margins of creeks, ponds, canals, and ditches. It feeds mainly on earthworms and slugs. In late summer, females give birth to live young in litters of four to fifteen. Kirtland's snakes use crayfish burrows in wet soil for hibernating and estivating. They are mainly nocturnal and spend the day resting in crayfish burrows and under logs, piles of leaves, and rocks.

Northern Ringneck Snake *(Diadophis punctatus edwardsii).* This 10- to 24-inch snake is slate gray or bluish black with a white, yellow, or orange belly and a neck band of the same color. In the East, its range extends from Nova Scotia to Wisconsin and Georgia. Statewide in Pennsylvania, the ringneck snake lives in many diverse settings, including rocky slopes, gravel pits, dumps, old stone walls, damp coves in dense hemlock woods, gardens, meadows, and moist hardwood forest. The ringneck snake shelters under rocks and debris by day and becomes active at night, particularly during warm, rainy weather. It preys on salamanders, earthworms, slugs, lizards, juvenile snakes, and insects. After seizing a salamander (the redback salamander is a common prey species) or a small snake, a ringneck snake will clamp its prey in its jaws until the victim ceases struggling, and then swallow it. Some researchers believe that the ringneck snake possesses a mildly venomous saliva that immobilizes its prey. Ringneck snakes are themselves eaten by owls, hawks, snakes, bullfrogs, and mammals.

Ringneck snakes are active from March into October. A study of a related subspecies in Kansas found that normal movements within an individual's home range were 230 feet or less. In June or early July, a female ringneck snake will lay two to ten eggs (usually three or four) under rocks and in moist, rotting wood. Several females may use the same site, and communal nests with up to fifty-five eggs have been found. The eggs hatch in late August or September. Ringneck snakes hibernate in mammal burrows, stone walls, brush piles, rotting logs and stumps, and house cellars. A neighbor once showed me a ringneck snake curled up around a saucer beneath a flowerpot on his windowsill; there the snake remained all winter, and when my neighbor checked on it the following spring, it had roused itself and presumably gone outdoors again through some crack in the foundation or siding. Ringneck snakes are fairly docile when handled; they don't usually bite but may release a strong-smelling musk.

Eastern Kingsnake *(Lampropeltis getulus getulus)*. The range of this species may just extend into southeastern Pennsylvania. It inhabits southern New Jersey, Delaware, Maryland, and Virginia, thence south to Florida. Eastern kingsnakes are black with chainlike white or yellow markings. Adults are stout and average 3 to 4 feet in length. They inhabit hardwood, mixed, and pine forests; old fields; upland swamps; and areas near creeks and streams. They often hide beneath boards, bark, logs, and junk. Kingsnakes are so named because they dominate other snakes: in fact, they constrict, kill, and eat snakes of many species, including garter snakes, water snakes, ringneck snakes, hognose snakes, black rat snakes, black racers, and copperheads (the kingsnake is thought to be immune to pitviper venom). They also prey on mice, voles, lizards, skinks, and salamanders, and eat the eggs of turtles and birds.

Female kingsnakes lay nine to seventeen eggs in June or July. The eggs hatch after eight to eleven weeks. When disturbed, a kingsnake vibrates its tail; done in the leaves, this behavior yields a sound much like a rattlesnake's rattle. If handled, a kingsnake may release a pungent musk. It may constrict a person's arm and will tend to chew rather than bite in self-defense.

Eastern Milk Snake *(Lampropeltis triangulum triangulum)*. The name arises from a folk belief, which has no basis in fact, that this snake sucks milk from cows' udders. Milk snakes are 2 to 3 feet long; the largest one ever recorded was 52 inches. The body is marked with red, brown, or gray blotches, and the belly has a black and white checkered pattern. The eastern milk snake inhabits woodlands, old fields, fencerows, rock outcrops, vacant lots in cities, and farmland, where it may venture into barns, basements, and old buildings when hunting for mice. Individuals rest beneath rocks, logs, boards, and trash. The eastern subspecies ranges from southern Canada to northern Georgia and is statewide in Pennsylvania.

Milk snakes are powerful constrictors. They prey on small rodents, shrews, birds and their eggs and nestlings, snakes, frogs, insects, and earthworms. In June and July, females lay six to twenty-five eggs in rotting stumps and logs, beneath rocks, and in sawdust piles; several females may lay their clutches in the same site. The eggs hatch in August and September. The young milk snakes are 5 to 11 inches long and have bright colors and sharp markings.

Should it feel threatened, a milk snake will coil, vibrate its tail, and strike. Many milk snakes are needlessly killed by people who mistake them for copperheads. (The round pupils and a black and white checkerboard-patterned belly distinguish the milk snake from the copperhead.)

Rough Green Snake *(Opheodrys aestivus)*. The rough green snake occurs throughout the South to Florida, Texas, and Mexico. In Pennsylvania, it has been found in the southeastern and southwestern counties; here on the northern fringe of its range, *Opheodrys aestivus* is considered a threatened species.

The rough green snake is a uniform green on the back, with a pale yellow belly. This color scheme blends it in with its habitat: the branches of shrubs and small trees, in fencerows, brushy woods, and vine-tangled thickets edging ponds and streams. Adults are 2 to almost 4 feet long and exceedingly thin; a folk name is "vine snake," referring either to the serpent's slender shape or to its arboreal habitat. The scales on the rough green snake have small keels or ridges, making its skin feel slightly rougher and giving it a drier appearance than the skin of its close relative, the smooth green snake.

Rough green snakes emerge from hibernation in May. They possess an excellent sense of balance and can climb capably. Among the branches, they capture insects and spiders. They eat many caterpillars, grasshoppers, and crickets. Rough green snakes are preyed on by shrikes, hawks, domestic cats, and other predators; the snakes try to escape from their enemies by slithering in among dense leaves and twigs. They are active during the day and sleep coiled up on branches at night. In June and July, females lay their eggs in the soil under rocks and boards, in decaying logs, and in tree cavities. Clutches have three to twelve eggs. Several females may deposit their eggs in the same place. Populations in Pennsylvania are not large, but a study conducted around an Arkansas lake found more than 170 rough green snakes per acre.

Opheodrys aestivus is a docile snake that does not bite when handled, although individuals may give a threat defense by opening the mouth wide to expose its purple-black lining.

Eastern Smooth Green Snake *(Opheodrys vernalis vernalis).* Southern
Canada, New England, Pennsylvania (except for the southeast corner), and
the Great Lakes states are home to the eastern smooth green snake. This
sleek serpent looks much like its relative, the rough green snake, but it is
slightly shorter at 14 to 22 inches; has smooth rather than keeled scales; and
has shiny, wet-looking skin. Smooth green snakes live in meadows, lawns,
weedy fields, thickets, woods edges, marsh borders, and woodland glades.
Sometimes they climb into low bushes, but they spend much less time in
shrubs and trees than do rough green snakes. Many people know this reptile as the "green grass snake."

Smooth green snakes eat mainly insects, along with a few slugs, spiders,
and an occasional small salamander. They mate in May. Females lay three to
thirteen eggs (usually four to six) in June or July, in loose soil under rocks
or in rotting wood or vegetation. The eggs hatch in August after a relatively
short incubation period. (Females may actually incubate their eggs internally, by basking extensively before laying.) In October, smooth green
snakes go into hibernation. They have been found burrowed into gravel
banks and ant mounds. The eastern smooth green snake is our most docile
serpent; when handled, it usually does not bite and rarely discharges musk.
The use of insecticides may harm local populations.

Queen Snake *(Regina septemvittata).* This tan to dark brown water snake
has a yellow stripe along the lower body on each side and a yellow belly.
Adults are slender, 15 to 24 inches long, rarely up to 3 feet. Queen snakes
inhabit shallow, clean streams and small rivers that support crayfish; such
waterways may run through forests, farmland, and even suburban and urban
areas. Queen snakes also live in ponds and lakes. Active by day, they prefer
open stretches where trees do not block out sunlight. They rest under flat
rocks, and they bask on rocks and in the branches of small trees or shrubs
hanging over the water. When startled, they drop down and slither into the
underbrush or swim away through the water. Queen snakes inhabit the
western third of Pennsylvania and the southeastern counties. The species
ranges from Michigan south to the Florida panhandle.

Queen snakes hunt for crayfish among and under streambed rocks; they
concentrate on ones that have recently shed their hard exoskeletons. Their
strong preference for newly molted crayfish makes it hard to keep queen
snakes alive in captivity. Other prey, taken much less frequently, includes
dragonfly nymphs, small catfish, toads, tadpoles, and snails. Queen snakes
mate in April and May, and perhaps also in the fall. From late July to Sep-

tember, females give birth to live young in litters of five to twenty, although ten to twelve are more usual. Raccoons, otters, hellbenders, and great blue herons prey on queen snakes, particularly on juveniles. Queen snakes hibernate in earthen and stone dams and in muskrat burrows; they also use crayfish burrows.

Queen snakes are threatened by water pollution and siltation and the damming and channelizing of streams, all of which destroy crayfish habitat. When captured, a queen snake may thrash about, eject musk, and bite.

Northern Brown Snake *(Storeria dekayi dekayi)*. This small brown snake has two parallel rows of blackish spots down its back. Adults are 9 to 15 inches. The species is found in damp soil and humus in bogs, swamps, freshwater marshes, woods, grasslands, and old fields, and in suburbs and cities, particularly in gardens, vacant lots, cemeteries, and parks. Among our native reptiles, the brown snake is perhaps the most tolerant of what ecologists term "disturbed sites," where humans have drastically changed the environment. Shy and retiring, brown snakes take refuge under logs, flat rocks, bark slabs, and trash. Taxonomists recognize eight subspecies, which live in the eastern half of the continent and as far south as Central America. The northern brown snake is abundant in much of Pennsylvania; it may be absent from the north-central and northeastern counties.

Brown snakes hibernate in ant mounds, abandoned rodent burrows, stone walls, rock crevices, under logs, and in decrepit buildings. They emerge in late March or April. At first, they are active by day, but summer's heat prompts them to take shelter during the day and move about at night. They eat earthworms, slugs, snails, and insects. Adults and young fall prey to domestic cats, opossums, raccoons, skunks, weasels, snakes, and birds, especially hawks. Brown snakes mate in late March and early April. After about four months' gestation, the females give birth to live young in litters of nine to twenty. Newborns are dark brown and have a yellow, collarlike marking on the neck. They are 3 to 4 inches long and a slender sixteenth of an inch in diameter. By the time cold weather drives brown snakes back into hibernation in October and November, the juveniles are 6 to 8 inches long.

Although abundant at one time in cities, brown snakes have suffered population declines, possibly caused by pesticides that kill their invertebrate prey. In 1960, a herpetologist started lifting tar paper and cardboard in a "shanty town" outside Lancaster and found 603 brown snakes colonizing about 5 acres. I doubt that the same feat could be accomplished today in swiftly developing southeastern Pennsylvania.

Northern Redbelly Snake *(Storeria occipitomaculata occipitomaculata)*. Most redbelly snakes have a brown back, pale spotting just behind the head, and a bright red or orange belly. But much color variation exists in this species; some individuals even have black bellies. Adults are normally 8 to 10 inches, to a maximum length of 16 inches. Redbelly snakes inhabit eastern North America; they are statewide in Pennsylvania and are most common in the northern tier and in mountainous terrain. They live in moist forests, woods edges, old fields, wet meadows, borders of swamps, and in and near sphagnum bogs.

Redbelly snakes rest and forage under downed trees, bark piles, rotting wood, and debris, and feed on slugs, earthworms, insects, and snails. Sometimes they climb into low shrubs. In late summer, females give birth to four to nine young. Young and adults are preyed on by larger snakes, mammals, and avian predators. When frightened, a redbelly snake will flatten its head and body against the ground; if picked up, it will curl the upper lip, displaying a startling sneer or grin. Individuals may spray musk but will rarely bite.

Earth Snake *(Virginia valeriae)*. Earth snakes are secretive animals. Although they are found in many parts of eastern North America, little is known about their distribution in Pennsylvania. This snake has a gray, brown, or reddish back with dark flecks; the belly is white or yellowish. Adults are 7 to 10 inches long, making the earth snake one of the smallest of Pennsylvania's serpents. The species inhabits hardwood and mixed pine-and-hardwood forests, abandoned fields, pastures, woods edges, moist lowlands, and wooded areas in cities and towns. Like worm snakes, earth snakes live mainly in the leaf mold and under rocks, logs, boards, and trash, feeding on earthworms plus a few slugs and other small invertebrates.

In August or September, females give birth to two to fourteen live young, with the average litter having five to seven. Several inches of rain will sometimes drive individuals onto the surface of the ground. Earth snakes hibernate from around mid-November until April. They do not bite when captured, although they may release musk and feces. A herpetologist reported collecting a female earth snake in Pennsylvania that played dead by rolling onto its back, but with its head held so that it could watch its captor; placed on its stomach, it immediately rolled over onto its back again.

GARTER SNAKES

Pennsylvania has three species of garter snakes, all belonging to the genus *Thamnophis*. The genus occurs from coast to coast and from southern Canada south to Central America. Of the twenty-two garter snake species currently recognized, fourteen live in North America. One, the eastern garter snake, is our continent's most widely distributed reptile.

Garter snakes are so named because the lines or stripes running lengthwise down their bodies resemble the colorful garters people once wore to hold their stockings up. The average adult garter snake is a little over 2 feet in length. When it encounters an enemy, a garter snake may try to make itself look more imposing by flattening its body to prominently display the markings on its back. At the same time, the snake may try to hide its head. I remember as a boy chasing down and catching a garter snake. I wasn't bothered by the way it thrashed about (I'd gotten a good grip behind its head, so it couldn't bite me), but when the serpent vented a stream of stinking brown goo onto my hands, I dropped it like a hot potato.

Garter snakes prey heavily on frogs and salamanders. They also eat earthworms, insects, and other invertebrates. They are closely related to the water snakes of genus *Nerodia* but are more terrestrial in their habits. Garter snakes hibernate during the winter, mate in the spring, and give birth to live young late in the summer.

Eastern Garter Snake *(Thamnophis sirtalis sirtalis).* This rather stout, medium-size snake is 18 to 26 inches in length, although large, mature specimens have been caught that measured 4 feet. Garter snakes are greenish brown to almost black, with a distinct stripe from head to tail down the center of the back and another less prominent stripe on each side; between the stripes are dark patches or spots. These markings can vary a great deal among individuals. The garter snake has a greenish or yellowish belly.

Eastern garter snakes occur in all sixty-seven counties in Pennsylvania, in lowlands and uplands, in cities, farmland, and wilderness. They do not require open water, but they do need a moist habitat, as may be found in wet meadows, marshes, damp forests, rocky hillsides, old fields, the margins of rivers, creeks, and ponds, farm woodlots, quarries, suburbs, trash dumps, gardens (the species is sometimes called the garden snake), and vacant lots and parks in cities. Garter snakes hide under stones, debris, or other surface cover. Sometimes they climb into low trees or shrubs.

Eastern garter snakes can be found in a variety
of habitats throughout the Northeast.

Amphibians and reptiles—salamanders, frogs, toads, and small snakes—may make up 90 percent of a garter snake's diet. Individuals feed mainly on land and sometimes while swimming in the water. They kill and eat what is readily available, shifting from one type of prey to another. Garter snakes eat small fish, bird nestlings, mice, young chipmunks, shrews, earthworms (often a major part of the diet), slugs, snails, crayfish, millipedes, spiders, assorted insects, and carrion. For large creatures, such as a salamander, the snake may locate its victim by following a scent trail; when the snake closes in, it uses its sight to direct the attack. The garter snake strikes quickly, seizes its prey, disables it, and swallows it as soon as possible. Some herpetologists believe that *Thamnophis sirtalis* has venomous saliva that can immobilize other animals: garter snakes often chew on more active prey, such as frogs and mice, before trying to swallow them.

Garter snakes can tolerate cool temperatures, and they are among the first snakes to leave hibernation in the spring, emerging in late March or early April. In spring, garter snakes move about mainly during the day, although they may hunt at night, taking salamanders, frogs, and toads that congregate to breed and lay eggs in vernal ponds. As the days grow warmer, garter snakes are active in the morning, then again in the early evening and at night from around six until eleven. In the fall, they become active again by day. They can function when the temperature is in the fifties and even the forties, after most other snakes have become torpid. Garter snakes hibernate in moist places where they can keep from becoming dehydrated while holding a body temperature slightly above freezing. Sites include old stone walls, ant mounds, crevices and crannies among rocks, crayfish and mammal burrows, and rotting logs and stumps. Sometimes garter snakes share these places with ribbon snakes, brown snakes, redbelly snakes, green snakes, and ringneck snakes. On warm days from December through February, garter snakes may briefly become active.

Females mature sexually when they are about 22 inches long; males, when they are somewhat shorter. Individuals reach maturity during their second year or in their third spring. Males follow scent trails laid down by the females, whose bodies release chemicals, called pheromones, as they slither across the ground. Different species of snakes secrete different pheromones so that males can more efficiently recognize and locate potential mates. Sometimes people see a number of garter snakes all tangled together, looking like an animated ball of twine: these conglomerations are of multiple males all trying to mate with one female. After mating, the female's reproductive system stores the male's sperm; her ovaries begin to develop eggs, which the sperm later fertilizes. Sperm can stay viable inside the female for as long as one year.

Females give birth three to four months after mating. Most litters arrive from late June into September. Females usually bear eleven to twenty-six offspring, although litters can be as small as three and as numerous as eighty. Larger females tend to have more babies. Newborn garter snakes are 5 to 9 inches long, with sharp markings and bright colors. Immature garter snakes eat mainly earthworms during their first year. The young become food to many predators, including minks, skunks, weasels, raccoons, foxes, cats, dogs, shrews, hawks, owls, herons, pheasants, turkeys, crows, assorted snakes, box turtles, large amphibians such as bullfrogs and spotted salamanders, and fish—even spiders and crayfish have been seen eating small garter snakes. Both young and adult garter snakes die in great numbers on highways, especially in the spring and fall.

Herpetologists studying a population in Kansas discovered that 36 percent of juveniles made it through their first year, and half of all adults in the population survived each year. They estimated a maximum longevity of eight years in the wild. Radio-equipped snakes moved approximately 30 feet per day; the typical pattern was for an adult to shift about for two days, then stay in one place for two and a half days. Home ranges were around 35 acres for males, 23 acres for females. In areas where garter snakes find adequate prey close to good hibernation sites, the snakes probably do not travel very far. Garter snakes have perhaps the highest reproductive potential of all North American snakes, and they can thrive even in areas where humans have radically changed the environment.

Eastern garter snakes defend themselves aggressively. They emit both musk and feces. Young snakes in particular try to bite, striking so hard that they sometimes leave the ground. Handle garter snakes with care: a bite from a large adult can bleed copiously.

Shorthead Garter Snake *(Thamnophis brachystoma).* Shorthead garter snakes are dark brown to blackish, with one tan stripe on the back and one on each side and with no black spots between the stripes. They have short bodies and blunt heads no wider than their necks. They live in meadows, old fields, and along the edges of marshy areas among weedy growth. They eat earthworms almost exclusively. The shorthead garter snake is less fecund than the eastern garter snake: litters contain five to fourteen young. The species has a small range: northwestern Pennsylvania and southwestern New York. Herpetologists speculate that *Thamnophis brachystoma* is a remnant population that, thousands of years in the past, was cut off by glaciers from the main population to the west, which evolved into a separate species, Butler's garter snake *(Thamnophis butleri),* of Ohio, Michigan, Indiana, and Wisconsin.

Ribbon Snake *(Thamnophis sauritus).* The ribbon snake is a type of garter snake. Ribbon snakes inhabit much of Pennsylvania, with the exception of the mountains and plateaus of the Alleghenies. Adults are 18 to 26 inches, with a few individuals growing as long as 3 feet. Three yellow stripes—one on the back and one on each side—stand out against a reddish brown to an almost black body. Ribbon snakes are hard to identify unless brought to hand—which is no mean feat, since they're skittish, quick to flee through grass and weeds, and if apprehended, disposed to thrash violently. In nature, ribbon snakes are often mistaken for eastern garter snakes. But ribbon snakes are much slimmer than garter snakes, have proportionately longer tails (up to one-third of the total body length), and are swifter and more agile. Also, ribbon snakes favor damper habitats. Semiaquatic is how most experts classify this species. Look for *Thamnophis sauritus* in and near shallow water on the fringes of swamps, bogs, streams, lakes, and ponds.

Ribbon snakes swim capably, keeping their heads above the water rather than diving and swimming underwater like true water snakes. They forage mainly in the early morning and the evening, hunting down amphibians, including tadpoles, toads, small frogs, and salamanders. They also take small fish, spiders, and insects; only rarely do they eat earthworms. By day, they may bask in the branches of trees and shrubs hanging over the water. Ribbon snakes mate in April and May, and females give birth to live young in July and August. Litters range from three to twenty-five young, with ten to twelve the average. A study in Michigan found a population density of thirteen ribbon snakes per acre of habitat. Many predators take ribbon snakes, including raccoons, otters, minks, herons, kingsnakes, snapping turtles, bullfrogs, bass, and pickerel.

HOGNOSE SNAKE

The eastern hognose snake, *Heterodon platyrhinos,* ranges from southern New England south to Florida and west to Minnesota and Texas. The species is found in much of Pennsylvania: the northeastern, southeastern, and south-central counties, and in the Lake Erie drainage basin. This medium-size snake exhibits several adaptations for bufophagy—a fancy word for the eating of toads, a favorite prey.

Biology. Hognose snakes can grow to 45 inches, although most adults are a stout-bodied 20 to 33 inches in length. The head and neck are broad. Most hognose snakes are colored yellow, orange, reddish, or tan, overlaid with dark brown or black blotches. A fair number of adults become melanistic—black in color—as they age; some are plain gray. The undersurfaces are usually yellow, light gray, or pinkish, with gray or greenish mottling. The tail, extending back from the vent, is paler than the rest of the belly. The best field mark is this snake's namesake snout, which turns up abruptly at the end, even in juveniles. A pronounced ridge or keel runs along the top of the snout, whose shape makes it a useful digging tool.

Another way to identify a hognose snake is by its behavior. If it feels it is in danger, a hognose snake will suck in air, spread its neck into a fearsome-looking hood, hiss loudly, and lash out with its head in the direction of its foe—but without biting, for it keeps its mouth shut. Some people know the hognose snake by the names puff adder and blowing viper, and, after watching it inflate itself and strike, feel motivated to kill it, on the unfortunate assumption that any snake that looks so ferocious must surely be venomous. Should its bluster fail to scare off a would-be predator, a hognose snake will writhe as if mortally wounded, then flop onto its back and play dead. Its mouth hangs open, and its tongue sticks out. Sometimes the snake regurgitates. Turn one of these bluffers onto its belly, and it will usually twist over onto its back again.

Snakes' bodies have become especially modified to let them swallow prey that is larger than themselves, and *Heterodon platyrhinos* illustrates this adaptation well. A hognose snake can eat a toad whose body diameter is several times that of the snake's. The snake's sharp, rearward-curving teeth hold the toad fast. A structure called the quadrate bone connects the snake's lower jaw to the skull on each side. The quadrate bone acts as an extender, letting the snake drop its lower jaw down to radically increase the gape of its mouth. The sides of the lower jaw are linked at the chin by a flexible muscle, so that each side can be splayed sideways and moved independently. The snake edges one side of its jaw forward, then the other, gradually engulfing its vicim. The skin on a snake's neck is highly elastic and stretches to accommodate its meal.

At the back of its jaws, the hognose snake has a pair of large teeth to puncture toads and deflate them. The three hognose snakes that I have found latched onto toads all had caught their prey from behind. I backed off and watched as each consumed its still-living captive, which now and then would struggle mightily or push back with its forefeet against the snake's snout. After about half an hour, the toad was a fist-size lump in the snake's neck. The skin of toads secretes poisonous compounds, but hognose snakes produce enzymes in their digestive tracts to neutralize the toxins. The saliva of the hognose snake may be mildly venomous and may help kill or immobilize its prey. People who have been scratched by hognose snakes' teeth have reported feeling a burning pain at the wound site, much like a bee sting, along with discoloration and swelling.

In Virginia, a herpetologist discovered that toads made up 40 percent of hognose snakes' diets; frogs were 30 percent; small mammals accounted for 19 percent; and salamanders constituted 11 percent. Observers have reported *Heterodon platyrhinos* preying on small snakes, lizards and lizard eggs, hatchling turtles, insects, fish, mice, chipmunks, earthworms, centipedes, spiders, and snails.

Hognose snakes are active only during the day and mainly in the morning. They bask on rocks, gravel banks, and stone walls, and in grassy areas. To stay cool during hot weather, they dig down into the soil, often enlarging the tunnel or runway of a small mammal. The hognose snake is a capable excavator: an individual will set its body and push its pointed snout down into the earth, shoving the dirt aside or lifting it out on top of the broad, flattened skull. Hognose snakes also burrow to take shelter at night and to find toads that have hidden themselves in damp soil or humus.

Hognose snakes hibernate in burrows they dig themselves. They also may spend the winter under rocks, in rotting stumps or logs, and beneath piles of trash. They emerge in April and mate in the spring. In June and early July, the females lay eggs, usually fifteen to twenty-five of them, in loose soil or beneath rocks. Clutches vary in size from four to sixty eggs. The eggs hatch in August or early September; the hatchlings are about 7 inches long. Many animals kill and eat hognose snakes, especially the juveniles. Kingsnakes, black racers, black rat snakes, red-tailed hawks, and barred owls have been documented as predators, and no doubt many other creatures also kill hognose snakes. Hognose snakes become sexually mature when they are between one and a half and two years old.

Although they almost never bite, hognose snakes aren't a lot of fun to handle; they thrash and squirm and eject a stinking musk before feigning death. If they get used to captivity, individuals no longer bother playing dead.

Habitat. Hognose snakes prefer dry, open sites with well-drained sandy or loamy soils in which they can burrow. I have also encountered them in open woods near my home, where the soils are clayey and sand is not to be found. Hognose snakes live on woods edges and forested hillsides, and near openings such as woods roads. They also inhabit cultivated and abandoned fields and overgrown areas in cities and suburbs. They avoid wetlands.

Population. Populations of *Heterodon platyrhinos* seem to be spotty throughout the species' range, and hognose snakes are not considered to be abundant in Pennsylvania. To date, herpetologists have not studied the movements of hognose snakes in much detail. In a Kansas study, recaptured individuals had moved up to 950 feet, with even greater shifts during the breeding season. Since hognose snakes feed heavily on toads, a drop in the population of these amphibians—caused by pesticides or by ever-increasing automobile traffic on an ever-expanding road network—could seriously harm the hognose snake population.

BLACK RACER AND BLACK RAT SNAKE

Black racers and black rat snakes are common throughout Pennsylvania. Both are large predators that kill and eat a variety of animals, including other snakes. People often mistake one for the other, despite significant differences. The black racer is dark on both its top and bottom, save for some white markings on the chin; the black rat snake has a black back and a pale belly. The racer has a head about the same diameter as its body; the rat snake has a squarish head that is broader than its body. The racer has a cylindrical body, as do most other snakes; the rat snake's frame is flat-bottomed, shaped like a bread loaf when viewed in cross section. The racer is swift and alert, and the rat snake is slower and more phlegmatic in temperament.

Northern Black Racer (*Coluber constrictor constrictor*). Adults are 36 to 60 inches in length, occasionally up to 6 feet. The body is long and slender, the head small, and the eyes large. The scales are smooth and satiny. The chin and throat usually show some areas of white, and the belly is dark gray or bluish black. Eleven subspecies of racers live from Canada south through the forty-eight contiguous United States, except for parts of the desert Southwest, and into Mexico. The northern black racer ranges from Maine to Alabama.

Racers are active by day, particularly in the morning. At night and during rainy weather, they shelter beneath stones, boards, bark slabs, downed timber, and debris. They remain active during hot weather that drives most other snakes into cool hideouts. Black racers prefer a dry habitat, although they may turn up almost anywhere, including marshes and swamps. They are found in rocky outcrops, hillsides, powerline rights-of-way, grassy areas, open forest near grassy tracts, and the brushy edges between grassy and wooded habitats. Racers live in farming areas, where they hunt around barns and old buildings, and in cities and suburbs.

How terrifying to be a rodent and to view, proceeding toward you above the grass, the black head held erect and level, the unblinking reptilian eye alert to any movement. Racers go coursing through the cover, checking out every nook and cranny in search of prey. They resemble weasels or minks in their relentless hunting. Black racers climb into shrubs and low trees, they swim, and they change directions quickly; they are probably the fastest of our snakes (a Florida subspecies was clocked at 3.6 miles per hour). They are also among the most aggressive. Carl Ernst and

Roger Barbour, writing in *Snakes of Eastern North America,* abandon controlled, scientific prose when describing the black racer: "Generally, this snake has a rotten disposition. When handled, if it cannot bite, it will defecate and spray musk all over you." Startled, a black racer usually flees into tall grass or down a woodchuck hole. Or it may coil and vibrate its tail tip in the leaves, making a sound like a rattlesnake's warning buzz. Cornered, it may strike wildly and even advance toward its adversary. The racer's teeth are sharp but small; on humans, they generally cause wounds no more serious than brier scratches.

Black racers prey on insects, salamanders, frogs (especially treefrogs), toads, small turtles, lizards, skinks, snakes (including water, worm, ringneck, green, and garter snakes; also, small copperheads and rattlesnakes have been found in the bellies of large racers), birds and their eggs and nestlings (ground-nesting species are frequently taken), moles, shrews, voles, mice, chipmunks, rats, rabbits, and flying squirrels. Juvenile racers take more insect food, and adults eat more rodents and reptiles. A Virginia study found that snakes made up 26 percent, by volume, of prey taken. Other researchers have shown higher percentages of insect, bird, and mammal prey. No doubt the black racer takes whatever creatures are abundant and catchable at any given time.

Despite the species' Latin name, this snake is not a constrictor. The racer grabs its prey in its mouth. A small creature will be eaten alive, immediately. After tackling a larger animal, the snake will use one or more coils of its body to press its victim to the ground, while biting and chewing on it. In some cases the victim may be suffocated by pressure from the snake's body; in other instances it may succumb to wounds or to stress.

Black racers hibernate starting in September or October, secreting themselves in abandoned woodchuck burrows, rotting logs and stumps, gravel banks, and rock crevices. At some sites, many black racers cluster together, along with garter snakes, black rat snakes, copperheads, and rattlesnakes. As the weather warms in spring, racers leave their hibernation and bask on rocks near the den. Later they spread out into surrounding areas. Black racers mate in April and May, and females lay eggs from mid-June into August. Four to twenty-five eggs (usually around ten or eleven) are deposited in rotting wood, in mammal burrows, beneath rocks or boards, or buried in a shallow hole grubbed into sandy ground. The eggs are leathery and white, 1 inch or slightly longer, and have a distinctive granular texture: their many small bumps make them look as if they've been sprinkled with salt.

After six to nine weeks of solar incubation, the eggs hatch. The juveniles, 8 to 13 inches long, don't look anything like the adults. Young racers are gray, with dark gray to reddish brown blotches. These markings fade as the snakes grow larger. The uniform black coloration is achieved after about three years, by which time the snakes are sexually mature.

Except during hibernation, black racers are solitary. A Kansas study found a population density of 1.2 to 2.8 adults per acre and an adult survival rate of 62 percent per year; the researchers predicted a natural life expectancy of ten years. Many racers are killed by automobiles and the mowing of hayfields.

Black Rat Snake *(Elapha obsoleta obsoleta).* Adults are 42 to 72 inches, with the record a whopping 101 inches, or almost 8.5 feet. The back is a shiny black or dark brown, and the belly is grayish. *Elaphe obsoleta* occurs in five subspecies, or races, east of the Mississippi River. Folk names for this well-known reptile include mountain blacksnake and pilot blacksnake, the latter from the belief that rat snakes guide copperheads and rattlesnakes to their dens.

Black rat snakes use many different habitats, including hardwood forests, brushy fields, farming areas, forested wetlands, and patches of woods in suburbs and cities. They often live in barns and outbuildings. When we replaced the metal roof on our house, we found several skins shed by black rat snakes; I have also met with these reptiles in our shed, where, I am certain, they nab the mice and flying squirrels that also reside there. Black rat snakes are active by day and by night; in summer, they often move about just after sunset.

Black rat snakes hunt by smell and by sight. They take almost every type of small or immature rodent, including chipmunks, gray squirrels, voles, mice, and rats. *Elaphe obsoleta* is an accomplished climber. Think of a loaf of bread: flat on the bottom, with steeply angled sides and a rounded top. That is the shape of the black rat snake's body. The flat belly and the near-vertical sides let the snake wedge its body into bark furrows and cracks in rock faces, helping the animal to climb. I once found a black rat snake sunning itself 4 feet up on the stone wall of our house; its body was wedged into the joints, so that it looked like black mortar between the stones. Black rat snakes scale trees to take the young of cavity-nesting birds; they climb into shrubs and brier tangles to catch birds and eat their eggs and nestlings. They also catch frogs, lizards, snakes (including smaller black rat snakes), cottontails, snails, and insects. Black rat snakes bolt small creatures immedi-

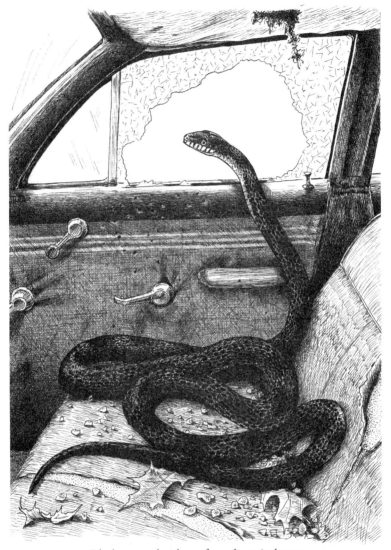

Black rat snakes hunt for rodents in barns,
old foundations, trash heaps, and junked cars.

ately. They constrict large ones until they are dead or are disabled sufficiently to be swallowed.

During winter, black rat snakes hibernate in caves, rock crevices, hollow trees and logs, mammal burrows, piles of rotting timber, old buildings,
and stone walls. In rocky areas and on talus slopes, they may share space with
black racers, copperheads, and rattlesnakes. Black rat snakes mate in the

spring; I have found entwined pairs in the branches of small saplings a few feet above the ground. Five to seven weeks after mating, a female lays eight to fourteen eggs in piles of rotting vegetation, manure, and sawdust; beneath rocks; and in hollow logs or stumps. The eggs are elongated and white, and the leathery shells have a granular surface. The eggs hatch from late July until mid-September. Young black rat snakes are 10 to 16 inches long; they are pale gray with dark gray blotches on the back. These markings dim as the snakes grow and are usually gone by adulthood, when individuals have become about 4 feet long.

Most individuals mature sexually in their fourth year. Adults have home ranges of 23 to 29 acres and may range up to 650 yards. During spring, summer, and fall, a snake may come back to the same shelter spot when not hunting. Young black rat snakes are preyed on by other snakes, large birds, bobcats, foxes, otters, and other mammals. Great horned owls and red-tailed hawks take some adults. Many others die when they try to cross roads and are run over by vehicles.

Some black rat snakes may be quite calm and may even let people pick them up without struggling. Others are more defensive. A truly agitated rat snake will vibrate its tail in the leaves, raise its body and head up to an impressive height, bite, and spray a foul musk. In the past, many people feared black snakes and killed them on sight. Attitudes are changing, and people are growing more aware of how necessary and beneficial snakes are. When Ernst and Barbour, authors of *Snakes of Eastern North America,* raised black rat snakes at the University of Kentucky, local farmers would show up at the end of the academic year, take home any excess snakes, and turn them loose in barns and corncribs for rodent control.

WATER SNAKE

The northern water snake, *Nerodia sipedon sipedon,* lives statewide in Pennsylvania's streams and lakes and is the largest of our three native semiaquatic snakes (the others are the queen snake and Kirtland's snake, smaller water-loving species). Northern water snakes range from Maine and southern Canada south to Mississippi, and west through the Great Lake states and Kansas.

Biology. Northern water snakes vary remarkably in color and pattern. The best way to identify one is by its presence in a watery habitat and its large size: 24 to over 40 inches, with some individuals surpassing 50 inches. Most water snakes have thick, heavy bodies, and females are larger and heavier than males. The background color of the body can range from pallid gray to deep, dark brown, upon which are superimposed darker bands and blotches that may be any color from rusty brown to black; these dark markings are larger than the paler intervals separating them. The pale belly also may show a pattern. Some adults become so dark with age that the markings on their backs no longer show up.

Water snakes emerge from hibernation in April. They bask on logs, rocks, and debris at the edges of streams or stretched out on tree limbs; alert and very shy, they will slip into the water or drop into it from above at the sight or sound of a potential predator. Excellent swimmers and divers, water snakes usually skim along with the body submerged and the head held above the water's surface. In spring and fall, water snakes are active mainly during the warmth of the day. In summer, they do much of their foraging between twilight and around midnight, except in cold, swift streams, where they may be active only by day. Under laboratory conditions, water snakes can stay beneath the water for over sixty-five minutes if undisturbed, with the heart rate falling to 9 percent of the normal resting rate.

Water snakes hunt by prowling in the shallows, using both sight and scent to find prey in the water and along waterway margins. They eat fish (over thirty species have been documented) and amphibians, including frogs, toads, tadpoles, and salamanders. Less frequently, they take crayfish, insects, and small mammals. They usually swallow their prey alive. Water snakes sometimes eat carrion. Various studies have shown that fish make up more than half of the typical diet. Herpetologists believe that under natural conditions, water snakes kill mainly sick and less vigorous fish; in streams and ponds, they thin out dense populations that otherwise would produce only stunted fish.

Females take two to three years to become sexually mature, and males mature after twenty-one to twenty-four months. Water snakes mate from April into June. Most matings happen on snags, logs, vegetation, and dry ground near water. After about two months' gestation, females give birth to live young in late summer or early fall. A typical litter has around twenty-five young. There is only one litter per year. Juvenile water snakes are 6 to 10 inches long and have sharp black blotches against a pale gray or brown background. Large game fish eat young water snakes, and so do many other predators. The adults are preyed on by kingsnakes, raccoons, snapping turtles, and others.

*Northern water snakes are long, sturdily built, shy,
and, when threatened, ferocious in their defense.*

In good habitats, water snakes may not move about much: an individual may stay in a favored pond or stretch of creek for days on end, traveling not at all or from 15 to 50 feet. Water snakes may move farther when seeking places in which to overwinter. They hibernate in stone causeways, ant mounds, crayfish burrows, the bank burrows and lodges of muskrats and beavers, meadow vole tunnels in low-lying areas, and rock crevices or hollow logs on slopes above the water.

A water snake will try to escape if confronted by a human. Cornered, the snake will flatten its body, discharge musk, and strike hard and repeatedly. The long teeth of this serpent—which help it snag and hold struggling fish—can inflict a nasty wound. The water snake's aggressiveness and vile temper have led many people to believe that it is a dangerous serpent, perhaps even a water moccasin; in fact, the water moccasin or cottonmouth, *Agkistrodon piscivorus,* is found only as far north as southern Virginia and is absent from Pennsylvania.

Habitat. Water snakes live in swamps, marshes, beaver ponds, and the borders of lakes, rivers, streams, and brooks; they thrive in wilderness situations and in waters surrounded by city parks and golf courses. They prefer slow waters but will also use swift-flowing streams.

Population. The water snake is an abundant reptile that frequently goes unnoticed, because it is shy and keeps itself hidden in the plants that grow thickly along the borders of most bodies of water. This serpent depends on a steady supply of prey, which, in turn, requires fresh, unpolluted water: water snake populations have fallen in urban areas where wetland habitats have been polluted or drained.

COPPERHEAD

The northern copperhead, *Agkistrodon contortrix mokasen,* is one of three venomous pitvipers found in Pennsylvania (the other two are the timber rattlesnake and eastern massasauga). A pitviper possesses specialized sensory equipment: small pits or holes on the face, one between the nostril and the eye on each side of the head, that pick up heat given off by objects in the environment, including warm-blooded prey. This heat sense helps a snake find prey and strike it accurately. Pitvipers also produce venom, delivered through their fangs when they bite.

Copperheads range from southern New England west to eastern Kansas and south to the Florida panhandle and Texas. *Agkistrodon contortrix* lives in suitable habitats in about three-quarters of Pennsylvania, except for the northern edge.

Biology. The body of an adult copperhead is banded in earth tones, a camouflaging pattern that helps the snake ambush its prey and conceals it from predators. The hourglass-shaped bands are dark brown to chestnut brown, set against a paler background. The head—with the classic broad, flat viper shape, triangular when viewed from above—is an unpatterned coppery or bronze color. In bright light, the elliptical pupils of the eyes can easily be discerned, but in dim light the pupils open fully and appear to be almost round. The body is sturdy. Most adults are a bit less than 3 feet in length, but some reach 4 feet or longer, and the record is 53 inches. Often people mistake banded nonvenomous species—particularly milk snakes, water snakes, and hognose snakes—for copperheads, frequently with fatal results for the harmless serpents.

Like our other venomous snakes, copperheads have hollow, rearward-curving fangs on the upper jaw. At rest, a snake keeps these fangs folded back against its gums, but when the mouth opens as the snake strikes, the fangs stand up and project forward. If the fangs are damaged or become worn, they drop out and new ones grow in. The fangs connect to glands that produce and store a protein-based venom that acts as a hemotoxin, attacking red blood cells. The copperhead's venom is not as potent as that of the timber rattlesnake. Larger copperheads can deliver more venom than smaller ones can.

Copperheads emerge from hibernation in April. They bask on rocks near their dens. As the weather warms, they move into nearby cover. Adults eat mainly small mammals such as voles, mice, chipmunks, shrews, moles, and young rabbits. They also take birds, salamanders, frogs, lizards, small snakes, and insects, including large caterpillars. Young copperheads eat more insects and other invertebrates than adults do. During outbreaks of period-ical cicadas, copperheads feast on these plentiful insects. Copperheads hunt mainly by lying in wait, and they also move stealthily through rocks and ground cover. A copperhead will grab a small animal and hold on until it dies; with larger prey, the snake strikes and delivers venom, lets the animal run off and die, and then follows its scent trail.

Although copperheads occasionally climb into low shrubs or trees, they spend most of their time on the ground. They can swim. They hide under

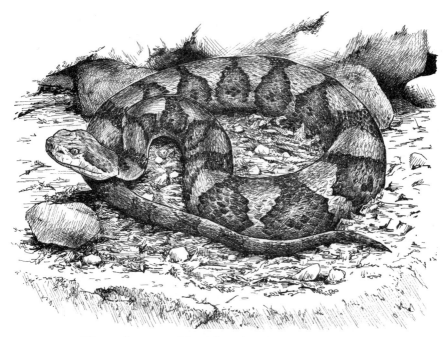

The earth tones and hourglass-shaped bands on its body
help blend the copperhead into ground cover.

stones, leaves, fallen bark, tin scraps, and boards, often around ramshackle rural buildings; they also crawl into sawdust piles.

Copperheads mate in late July and August. During the mating season, males engage in a ritualized behavior known as a "combat dance," in which two rivals coil about one another, stare fixedly, and seek in some manner to establish dominance. Females also adopt fighting postures similar to those used by the males, and male copperheads that flee from displaying females do not get a chance to mate. In this way, the strongest, the most aggressive, and presumably the fittest males do most of the breeding. Several males may mate with the same female; females store the males' sperm inside their bodies, with the sperm remaining viable until ovulation takes place in the spring. Female copperheads mature sexually at about three years, males at two years. At the most, females bear young every other year.

After 105 to 110 days' gestation, one to twenty young (most commonly, four to eight) are born alive, each enclosed in an oblong, semitransparent sac. The young snake fights its way out of this wrapper and stretches out to a length of about 9 inches. It is patterned like an adult, although with a paler background color, and possesses a bright yellow tail tip. A juvenile copperhead may hold its tail upright and wriggle it to attract a frog or a salamander, which then becomes a meal. Newborns have fangs (functional ones as well as replacements buried in the gums behind the exposed fangs) and can deliver venom sufficient to paralyze a mouse. Usually, however, young copperheads prey on insects.

In September and October, copperheads slither back to their dens. The same hibernation sites may be used year after year: caves, crevices among rocks, gravel banks, old stone walls, building foundations, hollow stumps and logs, sawdust piles, and mammal burrows. Several copperheads may share a den, and rattlesnakes, black racers, and black rat snakes also may use the same hibernaculum.

When it hears a human or other large animal, a copperhead will usually freeze in place, the body coiled, the head held up at a 45-degree angle. If approached closely, the snake may vibrate its tail in the leaves, but it is not likely to strike. Most people who are bitten have either stepped on a copperhead accidentally, placed their hands too close to a hidden snake, or tried to capture the serpent. Fortunately, copperhead bites are almost never fatal, but a bitten person should still seek medical attention right away.

To lessen your chances of being bitten, remain aware of when and where copperheads are most active. Move cautiously on warm, humid nights. Around rural homes and in children's play areas, clean up bark, old sheets of roofing, and plywood scraps, beneath which copperheads may shelter. Keep vegetation trimmed, making it less attractive to the mice and voles that copperheads hunt. When hiking or climbing, watch where you put your feet and hands. And be sure to stay out of striking distance of any snake when trying to identify it.

Habitat. Copperheads favor forested hillsides with talus slides and rock outcroppings. They are also found near sawdust heaps, stone walls, and rock piles. They live in old fields, hedgerows, forest edges, blueberry thickets, high ground in swamps and marshes, suburban woodlots, auto junkyards, and ravines cutting through farmland and urban areas. In Pennsylvania, herpetologist Howard Reinert found that in areas where timber rattlesnakes and copperheads coincide, copperheads use more-open areas with higher densities of rocks and less surface vegetation.

Quarrying of rocky ridges destroys copperhead habitat. Copperheads do not need remote wilderness to survive, but roads and other development can cut off den sites from summer foraging areas, creating zones of danger through which the snakes must pass several times a year.

Population. The population status of *Agkistrodon contortrix* is largely unknown in Pennsylvania and much of the Northeast. Copperheads seem fairly adaptable to different habitats and human activity. Thanks to their cryptic colors and docile temperaments, they have survived in areas where rattlesnakes have been wiped out.

A copperhead may move up to 2 miles from the den site in search of food and a mate. In a long-term study in Kansas, researchers found a density of 2.4 to 3.6 copperheads per acre. The scientists estimated that most copperheads lived for thirteen years in the wild, and 71 percent of the adult population survived each year. Kingsnakes, black racers, milk snakes, and hawks prey on copperheads; these and other predators, including shrews, moles, and opossums, also take juveniles.

TIMBER RATTLESNAKE
AND MASSASAUGA

The timber rattlesnake and the eastern massasauga are Pennsylvania's only rattlesnakes. These reptiles have, on the end of the tail, a structure unique in nature: a rattle, a set of loose, interlocking segments composed of keratin, a protein that forms horns, hair, and nails in other animals. Vibrated rapidly, the rattle makes a harsh, buzzing sound. I remember vividly, in a wooded mountain gap in Centre County, lifting my foot over a log and hearing a sound of pure danger—like a live electric wire or a chunk of fat thrown into a red-hot skillet. Before my mind even registered *"Snake!"* my body had thrown itself backward. I peered over the log at a rattlesnake coiled where I had almost set foot.

Rattlesnakes belong to the group known as the pitvipers, named for pit-shaped, heat-sensing organs between the eyes and nostrils. Pitvipers bite, both to kill their prey and to defend themselves, delivering venom through hollow upper fangs about half an inch long. Rattlesnakes can indeed be dangerous, but they rarely bite if they're not molested. They have been killed on sight by too many people for too many years. The massasauga was never abundant in Pennsylvania; today it is an endangered species, and people are forbidden to kill it. The timber rattlesnake has not yet been given full protection, although the snake's numbers have fallen drastically over the last century as humans have persecuted it while destroying and fragmenting its habitat.

Timber Rattlesnake *(Crotalus horridus horridus).* Over thirty species of rattlesnakes are spread from southern Canada into South America. The timber rattlesnake occurs from southern New Hampshire to Georgia, and west to Minnesota, Kansas, and Texas. Timber rattlesnakes live across Pennsylvania from the northeast to the southwest, mainly in wooded uplands. It's possible that the largest remaining contiguous population of rattlesnakes in the Northeast is centered on the mountains of north-central Pennsylvania.

Adult timber rattlesnakes are heavy-bodied, generally 3 to 4 feet in length and rarely as long as 5 feet. Males are about 15 percent larger than females. Timber rattlesnakes come in yellow and black color phases. In yellow-phase individuals, the head is yellow, and on the body, black or dark brown crossbands stand out against a yellow or tan background. In black-phase snakes, the head is black, and dark stippling dulls or may even blot out the body's crossband pattern. In both color phases, the tail is a velvety-looking black. Color is not linked to gender. In some parts of Pennsylvania, yellow is the more usual color, but overall, black-phase timber rattlesnakes probably outnumber yellow-phase individuals.

Nonvenomous snakes have two rows of scales on the underside of the tail, extending rearward from the vent, or excretory opening; pit vipers, including the rattlesnakes, have just one row of scales. Nonvenomous snakes have round pupils, and venomous snakes have elliptical pupils like those of cats.

Venom-producing glands, one on each side of the skull, give the rattlesnake its broad, lance-shaped head; the head is much wider than the neck. At the body's other end is the useful rattle. The number of rattle segments cannot be used to accurately gauge a snake's age. Each time a snake sheds its skin, it adds another segment to the base of the rattle. Young snakes may shed as often as three or four times a year; most adults average one to two sheddings per year, but some shed as many as four times in a season. As a snake moves about in its habitat, its rattle keeps getting broken or scraped off.

The lives of timber rattlesnakes revolve around their denning areas: rocky ledges, talus slides, and boulder fields, which generally face southward, yielding the maximum exposure to sunlight in the spring and fall. Rattlesnakes hibernate during the winter, crawling below the frost line in cracks, crevices, and small caves in the rocks, where they maintain a body temperature of about 50 degrees Fahrenheit. When the air temperature approaches 60 degrees, usually in late April or early May, rattlesnakes emerge and bask on exposed rocks. At night, or if the weather turns cold, they may retreat back into their dens. As the days grow warmer, the snakes expand their movements and disperse into the surrounding forest.

Timber rattlesnakes inhabit upland hardwood and mixed hardwood-and-pine forests. In summer, they use open woods, blueberry and huckleberry patches, mountain laurel thickets, grassy fields, and brushy cutover areas. Studies in Pennsylvania found that home ranges of adult males covered around 250 forested acres; females' ranges were half as large. An individual snake moves in a circuit through its range, stopping periodically to hunt or to search for a mate; by autumn, it ends up back at the den, which it enters in October or November. Rattlesnakes may hibernate in groups of fifty or more in remote or protected areas where populations have not been devastated by humans. Sometimes they share their dens with copperheads, black racers, and black rat snakes.

The broad, lance-shaped head of the timber rattlesnake accommodates
a pair of venom glands. Timber rattlesnakes often lie on rocks,
soaking up the warmth of the sun.

Rattlesnakes prey mainly on small mammals, including white-footed mice, deer mice, jumping mice, voles, eastern woodrats, chipmunks, squirrels, cottontails, and shrews. Occasionally they eat birds. A snake will lie coiled beside a log, with its chin resting against the surface; when an animal runs along on top of the log, the snake feels the vibrations. It uses its vision or its heat-sensing capabilities to direct a strike. The fangs inject venom into the prey, which runs off and dies; the snake sniffs out its trail and follows. Studies have shown that rattlesnakes do most of their ambush feeding between 9:00 P.M. and 8:00 A.M. Rattlesnakes may climb into shrubs or small trees, either following the trails of prey or looking for basking sites. Rattlesnakes swim capably. They are occasionally killed by deer, dogs, hawks, and other snakes, but humans are the major predators.

Herpetologists long believed that timber rattlesnakes bred in the spring like many other snakes. However, recent studies have shown that they mate in July, August, and September. Competing males fight by linking their bodies, with each snake raising its head and trying to push its opponent off balance and pin it to the ground. After mating, the female stores the male's sperm in her body. In late May or June of the following year, she ovulates.

A pregnant, or gravid, female requires a habitat that is different from those used by males and by females that are not gravid. A gravid female typically remains at or within 500 yards of the overwintering den, in an area that herpetologists term a "birthing rookery." She basks on rocks, regulating her temperature to foster the development of the embryos inside her body. Females are thought not to eat during gestation. They give birth to live young in August, September, and October. Litters have five to seventeen young, with six to ten the norm.

Newborn rattlesnakes are 10 to 14 inches long and have fangs that can inject venom. The rattle consists of a single button. The young are banded and look much like yellow-phase adults, except their colors are grayish and duller. Juveniles may stay with their mother for up to two weeks. In autumn, they follow the adults' scent trails back to the hibernating dens. They are preyed on by birds, mammals, and snakes. Some observers believe that wild turkeys eat many young rattlesnakes.

The timber rattlesnake has a fairly low reproductive potential. Females and males do not reach sexual maturity until they are four or five years old. Females may not bear litters until they are five to eight years of age. Litters arrive at two-, three-, or four-year intervals, depending on the quality of the habitat and the amount of food that a female can obtain: she must build up adequate fat reserves before channeling energy to a new brood. A female may reproduce three to ten or more times in her life, based on a life expectancy of sixteen to thirty years; however, many females are killed before they live out their full span. After local rattlesnake populations are depleted, they recover slowly, if at all.

Two major factors have caused the rattlesnake to decline in Pennsylvania. One factor is habitat loss. People are building homes and cabins in primary summer foraging areas; when snakes disperse from their dens, some are killed when crossing roads, and others are dispatched by homeowners. Quarrying operations destroy rocky denning areas. Natural forces are also at work: as trees mature, they block out sunlight, making basking sites and birthing rookeries less attractive. Herpetologists have suggested that these critical habitats on public lands be inventoried and managed by cutting back brush or trees shading the sites. Easements purchased from private landowners could protect dens, migration corridors, and foraging habitats.

The second factor endangering rattlesnakes is disturbance and persecution by humans. Many people know the locations of ancestral snake dens and go there in spring and autumn to kill or capture snakes. The Pennsylvania Fish and Boat Commission, which has jurisdiction over reptiles and amphibians in the state, permits limited hunting of *Crotalus horridus* (six weeks in June and July; each hunter is allowed one snake per year) and allows eight to ten organized snake hunts annually. These practices kill and injure snakes and further disrupt a beleaguered population.

Most northeastern states protect the timber rattlesnake as an endangered or threatened species. Many conservationists believe that Pennsylvania should follow suit. They urge the state to identify and protect the snakes' habitat: around four hundred denning sites have been recorded, but it is not certain how many of those sites are active today. Nor do biologists know how many timber rattlesnakes remain in Pennsylvania.

If possible, people should refrain from killing rattlesnakes. Pitvipers are fascinating, beautiful creatures uniquely adapted to the eastern woods, and they play an important role in keeping the ecosystem in balance. Clark Shiffer, former herpetology and endangered species coordinator for the Pennsylvania Fish and Boat Commission, has described the rattlesnake as "an animal that will need more of our help in the future, so that it and the wildness it represents are not lost to us and to those that follow."

A word about snakebite: a rattlesnake may not rattle before it strikes. A snake may feign a strike or strike with the mouth closed; some actual bites do not deliver venom. A person is far less apt to die from snakebite than by being hit by lightning, suffering an allergic reaction to a bee sting, or perishing in a car accident. To lower your chances of being bitten, learn about the life history of *Crotalus horridus* so that you can keep away from places where rattlesnakes occur and realize when you should be on the lookout for them. Learn to distinguish venomous and nonvenomous snakes. If bitten by a venomous snake, keep the stricken body part lower than the heart, walk slowly out of the woods, and get immediate medical treatment.

Eastern Massasauga *(Sistrurus catenatus catenatus)*. Also known as the "swamp rattler," the eastern massasauga ranges from western Pennsylvania and southern Ontario west through Ohio, Michigan, and several Midwestern states. This small rattlesnake (20 to 30 inches) is dusky gray to black, with rounded dark blotches on its back and a black, yellow-marked belly. Massasaugas live in marshes and wet meadows bordering old fields—a combination of damp and dry habitats found in relict prairie terrain in parts of Mercer, Butler, and Venango counties.

Massasaugas hibernate in wet soil, often in crayfish burrows, 6 to 24 inches below the ground. In spring, they emerge in April. They remain in lowland habitats for about a month, basking and feeding on crayfish and frogs; as summer progresses, they shift to higher, drier ground and prey on rodents and insects. A study in Ontario found that massasaugas often used open areas, such as woods roads and trails, with individuals moving an average of slightly more than 60 yards in a day; a Pennsylvania survey documented movements of less than 10 yards per day. Excellent swimmers, massasaugas enter the water readily. On land, they are rather sluggish and may not rattle even when people get quite close to them.

Massasaugas mate in mid to late summer. Young are born the following year in August or early September. Females begin to breed at two years of age; they give birth every other year, bearing an average of six or seven young per litter. Massasaugas return to lowland habitats in the fall. Adults enter hibernation around mid-October; juveniles may go into hibernation slightly later than the adults.

Herpetologist Howard Reinert has identified ten historic metapopulations (clusters of locations where massasaugas have been found) in Pennsylvania, of which three or four may still support the snakes. Gravel mining, highway construction, dam building, urbanization, and forest succession harm massasauga habitat. One of the biggest threats to the species in Pennsylvania comes from the mining of coal deposits near glacial wetlands where the snakes live. The eastern massasauga is considered an endangered species in Pennsylvania. It is a candidate for the U.S. threatened and endangered species list, and if so classified, it will receive additional protection.

Brown, W. S. *Biology, Status, and Management of the Timber Rattlesnake (Crotalus horridus): A Guide for Conservation.* Society for the Study of Amphibians and Reptiles, 1993.

Brennan, C. E. *Rattler Tales from Northcentral Pennsylvania.* Pittsburgh: University of Pittsburgh Press, 1995.

Reinert, H. "A Profile and Impact Assessment of Organized Rattlesnake Hunts in Pennsylvania." *Journal of the Pennsylvania Academy of Science* 64 (1990): 136-44.

INDEX OF
SCIENTIFIC NAMES

INDEX OF
COMMON NAMES